GW00838827

PHILIP'S

DESK
REFERENCE
ATLAS

PHILIP'S

DESK
REFERENCE
ATLAS

Contents

Cartography by Philip's

Text
Keith Lye

Executive Editor
Caroline Rayner

Executive Art Editor
Alison Myer

Editor
Kara Turner

Production
Claudette Morris

Picture Research
Claire Gouldstone

Picture Acknowledgements
Zefa Picture Library /Tom V. Sant
/Geosphere Project Front cover and spine,
main title page Robert Harding Picture
Library /Photri 1 Image Bank /Lionel Brown
10 Rex Features /Sipa 6, 24 Still Pictures 26,
/Anne Piantanida 8, /Chris Caldicott 16, /Mark
Edwards 18, 20, /Hartmut Schwarzbach 14, 22,
/Luke White 4 Tony Stone Images /Kevin
Kelley 2, /Art Wolfe 12

First published in Great Britain in 1996
by George Philip Limited,
an imprint of Reed Books, Michelin House,
81 Fulham Road, London SW3 6RB, and
Auckland, Melbourne, Singapore and Toronto

© 1996 Reed International Books Limited

A CIP catalogue record for this book is
available from the British Library.

ISBN 0–540–06382–7

Printed and bound in China

World Statistics

vi–ix	Countries
x–xi	Cities
xii–xv	Physical
xvi	Climate

The Earth in Focus

2–3	The Universe & Solar System
4–5	The Changing Earth
6–7	Earthquakes & Volcanoes
8–9	Water & Ice
10–11	Weather & Climate
12–13	Landscape & Vegetation
14–15	Population
16–17	Languages & Religions
18–19	Agriculture & Industry
20–21	Trade & Commerce
22–23	Transport & Travel
24–27	International Organizations
28–29	Regions in the News
30–32	World Flags

World Maps

1	General Reference	52–53	North-east Africa
2–3	Northern Hemisphere	54–55	North-west Africa
4–5	Southern Hemisphere	56–57	Central Africa
6–7	Europe	58–59	Southern Africa
8–9	Scandinavia	60–61	Australia
10–11	British Isles	62–63	South-east Australia
12–13	France	64–65	New Zealand, Central &
14–15	Germany & Benelux		South-west Pacific
16–17	Central Europe	66–67	North America
18–19	Spain & Portugal	68–69	Eastern Canada
20–21	Italy & the Adriatic	70–71	Western Canada
22–23	Greece & the Balkans	72–73	North-east USA
24–25	Eastern Europe	74–75	South-east USA
26–27	Asia	76–77	North-central USA
28–29	Western Russia & Central Asia	78–79	Southern USA
30–31	Eastern Siberia	80–81	North-west USA
32–33	Japan	82–83	South-west USA
34–35	China	84–85	Mexico
36–37	South-east Asia	86–87	Caribbean & Central
38–39	Philippines & Eastern Indonesia		America
40–41	Eastern India, Bangladesh &	88–89	South America
	Burma	90–91	South America – North-west
42–43	Western India & Pakistan	92–93	South America – North-east
44–45	Iran, the Gulf & Afghanistan	94–95	South America – South
46–47	The Middle East & Turkey	96	Antarctica
48–49	Arabia & the Horn of Africa		
50–51	Africa	97–160	**Index to World Maps**

World Statistics – Countries

Listed below are all the countries of the world; the more important territories are also included. If a territory is not completely independent, then the country it is associated with is named. The area figures give the total area of land, inland water and ice. Annual income is the GNP per capita. The figures are the latest available, usually 1995.

Country / Territory	Area (1,000 sq km)	Area (1,000 sq mls)	Population (1,000s)	Capital City	Annual Income US$
Afghanistan	652	252	19,509	Kabul	220
Albania	28.8	11.1	3,458	Tirana	340
Algeria	2,382	920	25,012	Algiers	1,650
Andorra	0.45	0.17	65	Andorra la Vella	14,000
Angola	1,247	481	10,020	Luanda	600
Argentina	2,767	1,068	34,663	Buenos Aires	7,290
Armenia	29.8	11.5	3,603	Yerevan	660
Australia	7,687	2,968	18,107	Canberra	17,510
Austria	83.9	32.4	8,004	Vienna	23,120
Azerbaijan	86.6	33.4	7,559	Baku	730
Azores (Port.)	2.2	0.87	238	Ponta Delgada	na
Bahamas	13.9	5.4	277	Nassau	11,500
Bahrain	0.68	0.26	558	Manama	7,870
Bangladesh	144	56	118,342	Dhaka	220
Barbados	0.43	0.17	263	Bridgetown	6,240
Belarus	207.6	80.1	10,500	Minsk	2,930
Belgium	30.5	11.8	10,140	Brussels	21,210
Belize	23	8.9	216	Belmopan	2,440
Benin	113	43	5,381	Porto-Novo	420
Bhutan	47	18.1	1,639	Thimphu	170
Bolivia	1,099	424	7,900	La Paz/Sucre	770
Bosnia-Herzegovina	51	20	3,800	Sarajevo	2,500
Botswana	582	225	1,481	Gaborone	2,590
Brazil	8,512	3,286	161,416	Brasília	3,020
Brunei	5.8	2.2	284	Bandar Seri Begawan	9,000
Bulgaria	111	43	8,771	Sofia	1,160
Burkina Faso	274	106	10,326	Ouagadougou	300
Burma (= Myanmar)	677	261	46,580	Rangoon	950
Burundi	27.8	10.7	6,412	Bujumbura	180
Cambodia	181	70	10,452	Phnom Penh	600
Cameroon	475	184	13,232	Yaoundé	770
Canada	9,976	3,852	29,972	Ottawa	20,670
Canary Is. (Spain)	7.3	2.8	1,494	Las Palmas/Santa Cruz	na
Cape Verde Is.	4	1.6	386	Praia	870
Central African Republic	623	241	3,294	Bangui	390
Chad	1,284	496	6,314	Ndjaména	200
Chile	757	292	14,271	Santiago	3,070
China	9,597	3,705	1,226,944	Beijing	490
Colombia	1,139	440	34,948	Bogotá	1,400
Comoros	2.2	0.86	654	Moroni	520
Congo	342	132	2,593	Brazzaville	920
Costa Rica	51.1	19.7	3,436	San José	2,160
Croatia	56.5	21.8	4,900	Zagreb	4,500
Cuba	111	43	11,050	Havana	1,250
Cyprus	9.3	3.6	742	Nicosia	10,380

Country / Territory	Area (1,000 sq km)	Area (1,000 sq mls)	Population (1,000s)	Capital City	Annual Income US$
Czech Republic	78.9	30.4	10,500	Prague	2,730
Denmark	43.1	16.6	5,229	Copenhagen	26,510
Djibouti	23.2	9	603	Djibouti	780
Dominica	0.75	0.29	89	Roseau	2,680
Dominican Republic	48.7	18.8	7,818	Santo Domingo	1,080
Ecuador	284	109	11,384	Quito	1,170
Egypt	1,001	387	64,100	Cairo	660
El Salvador	21	8.1	5,743	San Salvador	1,320
Equatorial Guinea	28.1	10.8	400	Malabo	360
Eritrea	94	36	3,850	Asmara	500
Estonia	44.7	17.3	1,531	Tallinn	3,040
Ethiopia	1,128	436	51,600	Addis Ababa	100
Fiji	18.3	7.1	773	Suva	2,140
Finland	338	131	5,125	Helsinki	18,970
France	552	213	58,286	Paris	22,360
French Guiana (Fr.)	90	34.7	154	Cayenne	5,000
French Polynesia (Fr.)	4	1.5	217	Papeete	7,000
Gabon	268	103	1,316	Libreville	4,050
Gambia, The	11.3	4.4	1,144	Banjul	360
Georgia	69.7	26.9	5,448	Tbilisi	560
Germany	357	138	82,000	Berlin/Bonn	23,560
Ghana	239	92	17,462	Accra	430
Greece	132	51	10,510	Athens	7,390
Grenada	0.34	0.13	94	St George's	2,410
Guadeloupe (Fr.)	1.7	0.66	443	Basse-Terre	9,000
Guatemala	109	42	10,624	Guatemala City	1,110
Guinea	246	95	6,702	Conakry	510
Guinea-Bissau	36.1	13.9	1,073	Bissau	220
Guyana	215	83	832	Georgetown	350
Haiti	27.8	10.7	7,180	Port-au-Prince	800
Honduras	112	43	5,940	Tegucigalpa	580
Hong Kong (UK)	1.1	0.40	6,000	–	17,860
Hungary	93	35.9	10,500	Budapest	3,330
Iceland	103	40	269	Reykjavik	23,620
India	3,288	1,269	942,989	New Delhi	290
Indonesia	1,905	735	198,644	Jakarta	730
Iran	1,648	636	68,884	Tehran	4,750
Iraq	438	169	20,184	Baghdad	2,000
Ireland	70.3	27.1	3,589	Dublin	12,580
Israel	27	10.3	5,696	Jerusalem	13,760
Italy	301	116	57,181	Rome	19,620
Ivory Coast	322	125	14,271	Yamoussoukro	630
Jamaica	11	4.2	2,700	Kingston	1,390
Japan	378	146	125,156	Tokyo	31,450
Jordan	89.2	34.4	5,547	Amman	1,190
Kazakstan	2,717	1,049	17,099	Alma-Ata	1,540
Kenya	580	224	28,240	Nairobi	270
Korea, North	121	47	23,931	Pyongyang	1,100
Korea, South	99	38.2	45,088	Seoul	7,670
Kuwait	17.8	6.9	1,668	Kuwait City	23,350

World Statistics – Countries

Country / Territory	Area (1,000 sq km)	Area (1,000 sq mls)	Population (1,000s)	Capital City	Annual Income US$
Kyrgyzstan	198.5	76.6	4,738	Bishkek	830
Laos	237	91	4,906	Vientiane	290
Latvia	65	25	2,558	Riga	2,030
Lebanon	10.4	4	2,971	Beirut	1,750
Lesotho	30.4	11.7	2,064	Maseru	660
Liberia	111	43	3,092	Monrovia	800
Libya	1,760	679	5,410	Tripoli	6,500
Lithuania	65.2	25.2	3,735	Vilnius	1,310
Luxembourg	2.6	1	408	Luxembourg	35,850
Macau (Port.)	0.02	0.006	490	Macau	7,500
Macedonia	25.7	9.9	2,173	Skopje	730
Madagascar	587	227	15,206	Antananarivo	240
Madeira (Port.)	0.81	0.31	253	Funchal	na
Malawi	118	46	9,800	Lilongwe	220
Malaysia	330	127	20,174	Kuala Lumpur	3,160
Maldives	0.30	0.12	254	Malé	820
Mali	1,240	479	10,700	Bamako	300
Malta	0.32	0.12	367	Valletta	6,800
Martinique (Fr.)	1.1	0.42	384	Fort-de-France	3,500
Mauritania	1,025	396	2,268	Nouakchott	510
Mauritius	2.0	0.72	1,112	Port Louis	2,980
Mexico	1,958	756	93,342	Mexico City	3,750
Micronesia, Fed. States of	0.70	0.27	125	Palikir	1,560
Moldova	33.7	13	4,434	Kishinev	1,180
Mongolia	1,567	605	2,408	Ulan Bator	400
Morocco	447	172	26,857	Rabat	1,030
Mozambique	802	309	17,800	Maputo	80
Namibia	825	318	1,610	Windhoek	1,660
Nepal	141	54	21,953	Katmandu	160
Netherlands	41.5	16	15,495	Amsterdam/The Hague	20,710
Netherlands Antilles (Neths)	0.99	0.38	199	Willemstad	9,700
New Caledonia (Fr.)	19	7.3	181	Nouméa	6,000
New Zealand	269	104	3,567	Wellington	12,900
Nicaragua	130	50	4,544	Managua	360
Niger	1,267	489	9,149	Niamey	270
Nigeria	924	357	88,515	Abuja	310
Norway	324	125	4,361	Oslo	26,340
Oman	212	82	2,252	Muscat	5,600
Pakistan	796	307	143,595	Islamabad	430
Panama	77.1	29.8	2,629	Panama City	2,580
Papua New Guinea	463	179	4,292	Port Moresby	1,120
Paraguay	407	157	4,979	Asunción	1,500
Peru	1,285	496	23,588	Lima	1,490
Philippines	300	116	67,167	Manila	830
Poland	313	121	38,587	Warsaw	2,270
Portugal	92.4	35.7	10,600	Lisbon	7,890
Puerto Rico (US)	9	3.5	3,689	San Juan	7,020
Qatar	11	4.2	594	Doha	15,140
Réunion (Fr.)	2.5	0.97	655	Saint-Denis	3,900
Romania	238	92	22,863	Bucharest	1,120

Country / Territory	Area (1,000 sq km)	Area (1,000 sq mls)	Population (1,000s)	Capital City	Annual Income US$
Russia	17,075	6,592	148,385	Moscow	2,350
Rwanda	26.3	10.2	7,899	Kigali	200
St Lucia	0.62	0.24	147	Castries	3,040
St Vincent & Grenadines	0.39	0.15	111	Kingstown	1,730
São Tomé & Príncipe	0.96	0.37	133	São Tomé	330
Saudi Arabia	2,150	830	18,395	Riyadh	8,000
Senegal	197	76	8,308	Dakar	730
Sierra Leone	71.7	27.7	4,467	Freetown	140
Singapore	0.62	0.24	2,990	Singapore	19,310
Slovak Republic	49	18.9	5,400	Bratislava	1,900
Slovenia	20.3	7.8	2,000	Ljubljana	6,310
Solomon Is.	28.9	11.2	378	Honiara	750
Somalia	638	246	9,180	Mogadishu	500
South Africa	1,220	471	44,000	Pretoria/Cape Town/ Bloemfontein	2,900
Spain	505	195	39,664	Madrid	13,650
Sri Lanka	65.6	25.3	18,359	Colombo	600
Sudan	2,506	967	29,980	Khartoum	750
Surinam	163	63	421	Paramaribo	1,210
Swaziland	17.4	6.7	849	Mbabane	1,050
Sweden	450	174	8,893	Stockholm	24,830
Switzerland	41.3	15.9	7,2681	Bern	36,410
Syria	185	71	14,614	Damascus	5,700
Taiwan	36	13.9	21,100	Taipei	11,000
Tajikistan	143.1	55.2	6,102	Dushanbe	470
Tanzania	945	365	29,710	Dodoma	100
Thailand	513	198	58,432	Bangkok	2,040
Togo	56.8	21.9	4,140	Lomé	330
Trinidad & Tobago	5.1	2	1,295	Port of Spain	3,730
Tunisia	164	63	8,906	Tunis	1,780
Turkey	779	301	61,303	Ankara	2,120
Turkmenistan	488.1	188.5	4,100	Ashkhabad	1,400
Uganda	236	91	20,466	Kampala	190
Ukraine	603.7	233.1	52,027	Kiev	1,910
United Arab Emirates	83.6	32.3	2,800	Abu Dhabi	22,470
United Kingdom	243.3	94	58,306	London	17,970
United States of America	9,373	3,619	263,563	Washington, DC	24,750
Uruguay	177	68	3,186	Montevideo	3,910
Uzbekistan	447.4	172.7	22,833	Tashkent	960
Vanuatu	12.2	4.7	167	Port-Vila	1,230
Venezuela	912	352	21,800	Caracas	2,840
Vietnam	332	127	74,580	Hanoi	170
Virgin Is. (US)	0.34	0.13	105	Charlotte Amalie	12,000
Western Sahara	266	103	220	El Aaiún	300
Western Samoa	2.8	1.1	169	Apia	980
Yemen	528	204	14,609	Sana	800
Yugoslavia	102.3	39.5	10,881	Belgrade	1,000
Zaïre	2,345	905	44,504	Kinshasa	500
Zambia	753	291	9,500	Lusaka	370
Zimbabwe	391	151	11,453	Harare	540

World Statistics – Cities

Listed below are all the cities with more than 600,000 inhabitants (only cities with more than 1 million inhabitants are included for China, Brazil and India). The figures are taken from the most recent census, and as far as possible are for the metropolitan area, e.g. greater New York, Mexico or London. The figures are in thousands.

	Population (1,000s)		Population (1,000s)		Population (1,000s)		Population (1,000s)
Afghanistan		Edmonton	840	**Dominican Republic**		Nagpur	1,661
Kābul	1,424	Hamilton	600	Santo Domingo	2,200	Patna	1,099
Algeria		Montréal	3,127	**Ecuador**		Pune	2,485
Algiers	1,722	Ottawa–Hull	921	Guayaquil	1,508	Surat	1,517
Oran	664	Québec	646	Quito	1,101	Vadodara	1,115
Angola		Toronto	3,893	**Egypt**		Varanasi	1,026
Luanda	1,544	Vancouver	1,603	Alexandria	3,380	Vishakhapatnam	1,052
Argentina		Winnipeg	652	Cairo	6,800	**Indonesia**	
Buenos Aires	11,256	**Chile**		El Gîza	2,144	Bandung	2,027
Córdoba	1,198	Santiago	5,343	Shubra el Kheima	834	Jakarta	8,259
La Plata	640	**China**		**El Salvador**		Malang	650
Mendoza	775	Anshan	1,204	San Salvador	1,522	Medan	1,686
Rosario	1,096	Beijing	6,690	**Ethiopia**		Palembang	1,084
San Miguel de Tucumán	622	Changchun	2,470	Addis Ababa	2,213	Semarang	1,005
Armenia		Changsha	1,510	**France**		Surabaya	2,421
Yerevan	1,254	Chengdu	2,760	Bordeaux	696	Ujung Pandang	913
Australia		Chongqing	3,870	Lille	959	**Iran**	
Adelaide	1,070	Dalian	2,400	Lyons	1,262	Ahvaz	725
Brisbane	777	Fushun	1,202	Marseilles	1,087	Bakhtaran	624
Melbourne	3,081	Fuzhou	1,380	Paris	9,319	Esfahan	1,127
Perth	1,193	Guangzhou	3,750	Toulouse	650	Mashhad	1,759
Sydney	3,657	Guiyang	1,080	**Georgia**		Qom	681
Austria		Hangzhou	1,790	Tbilisi	1,279	Shiraz	965
Vienna	1,560	Harbin	3,120	**Germany**		Tabriz	1,089
Azerbaijan		Hefei	1,110	Berlin	3,475	Tehran	6,476
Baku	1,149	Jilin	1,037	Cologne	693	**Iraq**	
Bangladesh		Jinan	2,150	Dortmund	602	Al Mawsil	664
Chittagong	2,041	Kunming	1,500	Essen	622	Arbil	770
Dhaka	6,105	Lanzhou	1,340	Frankfurt	660	As Sulaymaniyah	952
Khulna	877	Linhai	1,012	Hamburg	1,703	Baghdad	3,841
Belarus		Macheng	1,010	Munich	1,256	Diyala	961
Minsk	1,613	Nanchang	1,440	**Ghana**		**Ireland**	
Belgium		Nanjing	2,490	Accra	965	Dublin	1,024
Brussels	952	Ningbo	1,100	**Greece**		**Italy**	
Bolivia		Qingdao	2,300	Athens	3,097	Genoa	668
La Paz	1,126	Qiqihar	1,070	**Guatemala**		Milan	1,359
Santa Cruz	695	Shanghai	8,930	Guatemala	2,000	Naples	1,072
Brazil		Shenyang	4,050	**Guinea**		Palermo	697
Belém	1,246	Shijiazhuang	1,610	Conakry	810	Rome	2,723
Belo Horizonte	2,049	Taiyuan	1,720	**Haiti**		Turin	953
Brasília	1,596	Tangshan	1,044	Port-au-Prince	1,402	**Ivory Coast**	
Curitiba	1,290	Tianjin	5,000	**Honduras**		Abidjan	1,929
Fortaleza	1,758	Ürümqi	1,130	Tegucigalpa	679	**Jamaica**	
Manaus	1,011	Wuhan	3,870	**Hong Kong**		Kingston	644
Nova Iguaçu	1,286	Xi'an	2,410	Hong Kong	6,149	**Japan**	
Pôrto Alegre	1,263	Zhengzhou	1,690	**Hungary**		Chiba	851
Recife	1,290	Zibo	2,400	Budapest	2,009	Fukuoka	1,269
Rio de Janeiro	5,336	**Colombia**		**India**		Hiroshima	1,102
Salvador	2,056	Barranquilla	1,049	Ahmadabad	3,298	Kawasaki	1,200
São Paulo	9,480	Bogotá	5,132	Bangalore	4,087	Kitakyushu	1,020
Bulgaria		Cali	1,687	Bhopal	1,064	Kobe	1,509
Sofia	1,221	Cartagena	726	Bombay (Mumbai)	12,572	Kumamoto	640
Burkina Faso		Medellín	1,608	Calcutta	10,916	Kyoto	1,452
Ouagadougou	634	**Congo**		Coimbatore	1,136	Nagoya	2,159
Burma (Myanmar)		Brazzaville	938	Delhi	7,207	Okayama	605
Rangoon	2,513	**Croatia**		Hyderabad	4,280	Osaka	2,589
Cambodia		Zagreb	931	Indore	1,104	Sakai	806
Phnom Penh	900	**Cuba**		Jaipur	1,514	Sapporo	1,732
Cameroon		Havana	2,119	Kanpur	2,111	Sendai	951
Douala	884	**Czech Republic**		Lucknow	1,642	Tokyo	11,927
Yaoundé	750	Prague	1,216	Ludhiana	1,012	Yokohama	3,288
Canada		**Denmark**		Madras	5,361	**Jordan**	
Calgary	754	Copenhagen	1,337	Madurai	1,094	Amman	1,272

	Population (1,000s)
Az-Zarqa	605
Kazakstan	
Alma-Ata (Almaty)	1,147
Qaraghandy	613
Kenya	
Nairobi	1,429
Korea, North	
Chinnampo	691
Chongjin	754
Hamhung	775
Pyongyang	2,639
Korea, South	
Inchon	1,818
Kwangju	1,145
Puchon	668
Pusan	3,798
Seoul	10,628
Suwon	645
Taegu	2,229
Taejon	1,062
Ulsan	683
Kyrgyzstan	
Bishkek	628
Latvia	
Riga	840
Lebanon	
Beirut	1,500
Libya	
Tripoli	990
Madagascar	
Antananarivo	1,053
Malaysia	
Kuala Lumpur	1,145
Mali	
Bamako	746
Mauritania	
Nouakchott	600
Mexico	
Ciudad Juárez	798
Culiacán Rosales	602
Guadalajara	2,847
León	872
Mexicali	602
Mexico City	15,048
Monterrey	2,522
Puebla	1,055
Tijuana	743
Moldova	
Chişinău (Kishinev)	667
Mongolia	
Ulan Bator	601
Morocco	
Casablanca	3,079
Fès	735
Marrakesh	665
Oujda	661
Rabat–Salé	1,344
Mozambique	
Maputo	1,070
Netherlands	
Amsterdam	1,091
Rotterdam	1,069
The Hague	694
New Zealand	
Auckland	896
Nicaragua	
Managua	974
Nigeria	
Ibadan	1,295

	Population (1,000s)
Kano	700
Lagos	1,347
Ogbomosho	661
Norway	
Oslo	714
Pakistan	
Faisalabad	1,104
Gujranwala	659
Hyderabad	752
Karachi	5,181
Lahore	2,953
Multan	722
Rawalpindi	795
Paraguay	
Asunción	945
Peru	
Arequipa	620
Lima–Callao	6,601
Philippines	
Caloocan	629
Cebu	641
Davao	868
Manila	6,720
Quezon City	1,667
Poland	
Kraków	751
Lodz	847
Warsaw	1,655
Wroclaw	643
Portugal	
Lisbon	2,561
Oporto	1,174
Puerto Rico	
San Juan	1,816
Romania	
Bucharest	2,067
Russia	
Barnaul	665
Chelyabinsk	1,170
Irkutsk	644
Izhevsk	651
Kazan	1,107
Khabarovsk	626
Krasnodar	751
Krasnoyarsk	925
Moscow	8,957
Nizhniy Novgorod	1,451
Novokuznetsk	614
Novosibirsk	1,472
Omsk	1,193
Perm	1,108
Rostov	1,027
St Petersburg	5,004
Samara	1,271
Saratov	916
Simbirsk	638
Togliatti	677
Ufa	1,100
Vladivostok	675
Volgograd	1,031
Voronezh	958
Yaroslavl	637
Yekaterinburg	1,413
Saudi Arabia	
Jedda	1,400
Mecca	618
Riyadh	2,000
Senegal	
Dakar	1,730

	Population (1,000s)
Singapore	
Singapore	2,874
Somalia	
Mogadishu	1,000
South Africa	
Cape Town	1,912
Durban	1,137
East Rand	1,379
Johannesburg	1,196
Port Elizabeth	853
Pretoria	1,080
Vanderbijlpark–Ver.	774
West Rand	870
Spain	
Barcelona	1,631
Madrid	3,041
Sevilla	714
Valencia	764
Zaragoza	607
Sri Lanka	
Colombo	1,863
Sweden	
Göteburg	783
Stockholm	1,539
Switzerland	
Zürich	840
Syria	
Aleppo	1,445
Damascus	1,451
Taiwan	
Kaohsiung	1,405
T'aichung	817
T'ainan	700
T'aipei	2,653
Tajikistan	
Dushanbe	602
Tanzania	
Dar-es-Salaam	1,361
Thailand	
Bangkok	5,876
Tunisia	
Tunis	1,395
Turkey	
Adana	916
Ankara	2,559
Bursa	835
Gaziantep	603
Istanbul	6,620
Izmir	1,757
Uganda	
Kampala	773
Ukraine	
Dnipropetrovsk	1,190
Donetsk	1,121
Kharkiv	1,622
Kiev (Kyyiv)	2,643
Kryvyy Rih	729
Lviv	807
Odesa	1,096
Zaporizhye	898
United Kingdom	
Birmingham	1,400
Glasgow	730
Liverpool	1,060
London	6,378
Manchester	1,669
Newcastle	617
United States	
Atlanta	3,143

	Population (1,000s)
Baltimore	2,434
Boston	5,439
Buffalo	1,194
Charlotte	1,212
Chicago	8,410
Cincinnati	1,865
Cleveland	2,890
Columbus	1,394
Dallas	4,215
Denver	2,089
Detroit	5,246
Hartford	1,156
Houston	3,962
Indianapolis	1,424
Kansas City	1,617
Jacksonville	661
Los Angeles	15,048
Memphis	610
Miami	3,309
Milwaukee	1,629
Minneapolis–St Paul	2,618
New Orleans	1,303
New York	19,670
Norfolk	1,497
Oklahoma	984
Omaha	656
Philadelphia	5,939
Phoenix	2,330
Pittsburgh	2,406
Portland	1,897
St Louis	2,519
Sacramento	1,563
Salt Lake City	1,128
San Antonio	1,379
San Diego	2,601
San Francisco	6,410
San Jose	801
Seattle	3,131
Tampa	2,107
Washington, DC	4,360
Uruguay	
Montevideo	1,384
Uzbekistan	
Tashkent	2,094
Venezuela	
Barquisimento	745
Caracas	2,784
Maracaibo	1,364
Maracay	800
Valencia	1,032
Vietnam	
Haiphong	1,448
Hanoi	3,056
Ho Chi Minh City	3,924
Yugoslavia (Serbia and Montenegro)	
Belgrade	1,137
Zaïre	
Kinshasa	3,804
Lubumbashi	739
Mbuji-Mayi	613
Zambia	
Lusaka	982
Zimbabwe	
Bulawayo	622
Harare	1,189

World Statistics – Physical

Under each subject heading, the statistics are listed by continent. The figures are in size order beginning with the largest, longest or deepest, and are rounded as appropriate. Both metric and imperial measurements are given. The lists are complete down to the > mark; below this mark they are selective.

Land & Water

	km²	miles²	%
The World	509,450,000	196,672,000	–
Land	149,450,000	57,688,000	29.3
Water	360,000,000	138,984,000	70.7
Asia	44,500,000	17,177,000	29.8
Africa	30,302,000	11,697,000	20.3
North America	24,241,000	9,357,000	16.2
South America	17,793,000	6,868,000	11.9
Antarctica	14,100,000	5,443,000	9.4
Europe	9,957,000	3,843,000	6.7
Australia & Oceania	8,557,000	3,303,000	5.7
Pacific Ocean	179,679,000	69,356,000	49.9
Atlantic Ocean	92,373,000	35,657,000	25.7
Indian Ocean	73,917,000	28,532,000	20.5
Arctic Ocean	14,090,000	5,439,000	3.9

Seas

Pacific Ocean

	km²	miles²
South China Sea	2,974,600	1,148,500
Bering Sea	2,268,000	875,000
Sea of Okhotsk	1,528,000	590,000
East China & Yellow	1,249,000	482,000
Sea of Japan	1,008,000	389,000
Gulf of California	162,000	62,500
Bass Strait	75,000	29,000

Atlantic Ocean

	km²	miles²
Caribbean Sea	2,766,000	1,068,000
Mediterranean Sea	2,516,000	971,000
Gulf of Mexico	1,543,000	596,000
Hudson Bay	1,232,000	476,000
North Sea	575,000	223,000
Black Sea	462,000	178,000
Baltic Sea	422,170	163,000
Gulf of St Lawrence	238,000	92,000

Indian Ocean

	km²	miles²
Red Sea	438,000	169,000
The Gulf	239,000	92,000

Mountains

Europe

		m	ft
Mont Blanc	France/Italy	4,807	15,771
Monte Rosa	Italy/Switzerland	4,634	15,203
Dom	Switzerland	4,545	14,911
Liskamm	Switzerland	4,527	14,852
Weisshorn	Switzerland	4,505	14,780
Taschorn	Switzerland	4,490	14,730
Matterhorn/Cervino	Italy/Switzerland	4,478	14,691
Mont Maudit	France/Italy	4,465	14,649
Dent Blanche	Switzerland	4,356	14,291
Nedelhorn	Switzerland	4,327	14,196
> Grandes Jorasses	France/Italy	4,208	13,806
Jungfrau	Switzerland	4,158	13,642
Barre des Ecrins	France	4,103	13,461
Gran Paradiso	Italy	4,061	13,323
Piz Bernina	Italy/Switzerland	4,049	13,284
Eiger	Switzerland	3,970	13,025

Europe (cont.)

		m	ft
Monte Viso	Italy	3,841	12,602
Grossglockner	Austria	3,797	12,457
Wildspitze	Austria	3,772	12,382
Monte Disgrazia	Italy	3,678	12,066
Mulhacén	Spain	3,478	11,411
Pico de Aneto	Spain	3,404	11,168
Marmolada	Italy	3,342	10,964
Etna	Italy	3,340	10,958
Zugspitze	Germany	2,962	9,718
Musala	Bulgaria	2,925	9,596
Olympus	Greece	2,917	9,570
Triglav	Slovenia	2,863	9,393
Monte Cinto	France (Corsica)	2,710	8,891
Gerlachovka	Slovak Republic	2,655	8,711
Torre de Cerrado	Spain	2,648	8,688
Galdhöpiggen	Norway	2,468	8,100
Hvannadalshnúkur	Iceland	2,119	6,952
Kebnekaise	Sweden	2,117	6,946
Ben Nevis	UK	1,343	4,406

Asia

		m	ft
Everest	China/Nepal	8,848	29,029
K2 (Godwin Austen)	China/Kashmir	8,611	28,251
Kanchenjunga	India/Nepal	8,598	28,208
Lhotse	China/Nepal	8,516	27,939
Makalu	China/Nepal	8,481	27,824
Cho Oyu	China/Nepal	8,201	26,906
Dhaulagiri	Nepal	8,172	26,811
Manaslu	Nepal	8,156	26,758
Nanga Parbat	Kashmir	8,126	26,660
Annapurna	Nepal	8,078	26,502
Gasherbrum	China/Kashmir	8,068	26,469
Broad Peak	China/Kashmir	8,051	26,414
Xixabangma	China	8,012	26,286
Kangbachen	India/Nepal	7,902	25,925
Jannu	India/Nepal	7,902	25,925
Gayachung Kang	Nepal	7,897	25,909
Himalchuli	Nepal	7,893	25,896
Disteghil Sar	Kashmir	7,885	25,869
Nuptse	Nepal	7,879	25,849
Khunyang Chhish	Kashmir	7,852	25,761
Masherbrum	Kashmir	7,821	25,659
Nanda Devi	India	7,817	25,646
Rakaposhi	Kashmir	7,788	25,551
Batura	Kashmir	7,785	25,541
Namche Barwa	China	7,756	25,446
Kamet	India	7,756	25,446
Soltoro Kangri	Kashmir	7,742	25,400
Gurla Mandhata	China	7,728	25,354
Trivor	Pakistan	7,720	25,328
> Kongur Shan	China	7,719	25,324
Tirich Mir	Pakistan	7,690	25,229
K'ula Shan	Bhutan/China	7,543	24,747
Pik Kommunizma	Tajikistan	7,495	24,590
Elbrus	Russia	5,642	18,510
Demavend	Iran	5,604	18,386
Ararat	Turkey	5,165	16,945
Gunong Kinabalu	Malaysia (Borneo)	4,101	13,455
Yu Shan	Taiwan	3,997	13,113
Fuji-San	Japan	3,776	12,388

Africa

		m	ft
Kilimanjaro	Tanzania	5,895	19,340
Mt Kenya	Kenya	5,199	17,057
Ruwenzori (Margherita)	Uganda/Zaire	5,109	16,762
Ras Dashan	Ethiopia	4,620	15,157

Africa (cont.)

		m	ft
Meru	Tanzania	4,565	14,977
Karisimbi	Rwanda/Zaire	4,507	14,787
Mt Elgon	Kenya/Uganda	4,321	14,176
Batu	Ethiopia	4,307	14,130
Guna	Ethiopia	4,231	13,882
Toubkal	Morocco	4,165	13,665
Irhil Mgoun	Morocco	4,071	13,356
Mt Cameroon	Cameroon	4,070	13,353
Amba Ferit	Ethiopia	3,875	13,042
Pico del Teide	Spain (Tenerife)	3,718	12,198
Thabana Ntlenyana	Lesotho	3,482	11,424
Emi Koussi	Chad	3,415	11,204
Mt aux Sources	Lesotho/South Africa	3,282	10,768
Mt Piton	Réunion	3,069	10,069

Oceania

		m	ft
Puncak Jaya	Indonesia	5,029	16,499
Puncak Trikora	Indonesia	4,750	15,584
Puncak Mandala	Indonesia	4,702	15,427
Mt Wilhelm	Papua New Guinea	4,508	14,790
Mauna Kea	USA (Hawaii)	4,205	13,796
Mauna Loa	USA (Hawaii)	4,170	13,681
Mt Cook	New Zealand	3,753	12,313
Mt Balbi	Solomon Is.	2,439	8,002
Orohena	Tahiti	2,241	7,352
Mt Kosciusko	Australia	2,237	7,339

North America

		m	ft
Mt McKinley (Denali)	USA (Alaska)	6,194	20,321
Mt Logan	Canada	5,959	19,551
Citlaltepetl	Mexico	5,700	18,701
Mt St Elias	USA/Canada	5,489	18,008
Popocatepetl	Mexico	5,452	17,887
Mt Foraker	USA (Alaska)	5,304	17,401
Ixtaccihuatl	Mexico	5,286	17,342
Lucania	Canada	5,227	17,149
Mt Steele	Canada	5,073	16,644
Mt Bona	USA (Alaska)	5,005	16,420
Mt Blackburn	USA (Alaska)	4,996	16,391
Mt Sanford	USA (Alaska)	4,940	16,207
Mt Wood	Canada	4,848	15,905
Nevado de Toluca	Mexico	4,670	15,321
Mt Fairweather	USA (Alaska)	4,663	15,298
Mt Hunter	USA (Alaska)	4,442	15,573
Mt Whitney	USA	4,418	14,495
Mt Elbert	USA	4,399	14,432
Mt Harvard	USA	4,395	14,419
Mt Rainier	USA	4,392	14,409
Blanca Peak	USA	4,372	14,344
Longs Peak	USA	4,345	14,255
Tajumulco	Guatemala	4,220	13,845
Grand Teton	USA	4,197	13,770
Mt Waddington	Canada	3,994	13,104
Mt Robson	Canada	3,954	12,972
Chirripó Grande	Costa Rica	3,837	12,589
Mt Assiniboine	Canada	3,619	11,873
Pico Duarte	Dominican Rep.	3,175	10,417

South America

		m	ft
Aconcagua	Argentina	6,960	22,834
Bonete	Argentina	6,872	22,546
Ojos del Salado	Argentina/Chile	6,863	22,516
Pissis	Argentina	6,779	22,241
Mercedario	Argentina/Chile	6,770	22,211
Huascaran	Peru	6,768	22,204
Llullaillaco	Argentina/Chile	6,723	22,057
Nudo de Cachi	Argentina	6,720	22,047
Yerupaja	Peru	6,632	21,758
N. de Tres Cruces	Argentina/Chile	6,620	21,719
Incahuasi	Argentina/Chile	6,601	21,654
Cerro Galan	Argentina	6,600	21,654
Tupungato	Argentina/Chile	6,570	21,555

South America (cont.)

		m	ft
Sajama	Bolivia	6,542	21,463
Illimani	Bolivia	6,485	21,276
Coropuna	Peru	6,425	21,079
Ausangate	Peru	6,384	20,945
Cerro del Toro	Argentina	6,380	20,932
Siula Grande	Peru	6,356	20,853
Chimborazo	Ecuador	6,267	20,561
Cotapaxi	Ecuador	5,896	19,344
Pico Colon	Colombia	5,800	19,029
Pico Bolivar	Venezuela	5,007	16,427

Antarctica

	m	ft
Vinson Massif	4,897	16,066
Mt Kirkpatrick	4,528	14,855
Mt Markham	4,349	14,268

Ocean Depths

Atlantic Ocean

	m	ft
Mt Kirkpatrick	4,528	14,855
Puerto Rico (Milwaukee) Deep	9,220	30,249
Cayman Trench	7,680	25,197
Gulf of Mexico	5,203	17,070
Mediterranean Sea	5,121	16,801
Black Sea	2,211	7,254
North Sea	660	2,165
Baltic Sea	463	1,519

Indian Ocean

	m	ft
Java Trench	7,450	24,442
Red Sea	2,635	8,454
Persian Gulf	73	239

Pacific Ocean

	m	ft
Mariana Trench	11,022	36,161
Tonga Trench	10,882	35,702
Japan Trench	10,554	34,626
Kuril Trench	10,542	34,587
Mindanao Trench	10,497	34,439
Kermadec Trench	10,047	32,962
Peru–Chile Trench	8,050	26,410
Aleutian Trench	7,822	25,662

Antarctica

	m	ft
Molloy Deep	5,608	18,399

Land Lows

		m	ft
Caspian Sea	Europe	−28	−92
Dead Sea	Asia	−403	−1,322
Lake Assal	Africa	−156	−512
Lake Eyre North	Oceania	−16	−52
Death Valley	North America	−86	−282
Valdés Peninsula	South America	−40	−131

Rivers

Europe

		km	miles
Volga	Caspian Sea	3,700	2,300
Danube	Black Sea	2,850	1,770
Ural	Caspian Sea	2,535	1,575
Dnepr (Dnipro)	Volga	2,285	1,420
Kama	Volga	2,030	1,260
Don	Volga	1,990	1,240
Petchora	Arctic Ocean	1,790	1,110
Oka	Volga	1,480	920

Europe (cont.)		km	miles
Belaya	Kama	1,420	880
Dnister (Dniester)	Black Sea	1,400	870
Vyatka	Kama	1,370	850
Rhine	North Sea	1,320	820
North Dvina	Arctic Ocean	1,290	800
Desna	Dnepr (Dnipro)	1,190	740
Elbe	North Sea	1,145	710
Wisla	Baltic Sea	1,090	675
Loire	Atlantic Ocean	1,020	635
West Dvina	Baltic Sea	1,019	633

Asia		km	miles
Yangtze	Pacific Ocean	6,380	3,960
Yenisey–Angara	Arctic Ocean	5,550	3,445
Huang He	Pacific Ocean	5,464	3,395
Ob–Irtysh	Arctic Ocean	5,410	3,360
Mekong	Pacific Ocean	4,500	2,795
Amur	Pacific Ocean	4,400	2,730
Lena	Arctic Ocean	4,400	2,730
Irtysh	Ob	4,250	2,640
Yenisey	Arctic Ocean	4,090	2,540
Ob	Arctic Ocean	3,680	2,285
Indus	Indian Ocean	3,100	1,925
Brahmaputra	Indian Ocean	2,900	1,800
Syrdarya	Aral Sea	2,860	1,775
Salween	Indian Ocean	2,800	1,740
Euphrates	Indian Ocean	2,700	1,675
Vilyuy	Lena	2,650	1,645
Kolyma	Arctic Ocean	2,600	1,615
Amudarya	Aral Sea	2,540	1,575
Ural	Caspian Sea	2,535	1,575
Ganges	Indian Ocean	2,510	1,560
Si Kiang	Pacific Ocean	2,100	1,305
Irrawaddy	Indian Ocean	2,010	1,250
Tarim–Yarkand	Lop Nor	2,000	1,240
Tigris	Indian Ocean	1,900	1,180
Angara	Yenisey	1,830	1,135
Godavari	Indian Ocean	1,470	915
Sutlej	Indian Ocean	1,450	900

Africa		km	miles
Nile	Mediterranean	6,670	4,140
Zaïre/Congo	Atlantic Ocean	4,670	2,900
Niger	Atlantic Ocean	4,180	2,595
Zambezi	Indian Ocean	3,540	2,200
Oubangi/Uele	Zaïre	2,250	1,400
Kasai	Zaïre	1,950	1,210
Shaballe	Indian Ocean	1,930	1,200
Orange	Atlantic Ocean	1,860	1,155
Cubango	Okavango Swamps	1,800	1,120
Limpopo	Indian Ocean	1,600	995
Senegal	Atlantic Ocean	1,600	995
Volta	Atlantic Ocean	1,500	930
Benue	Niger	1,350	840

Australia		km	miles
Murray–Darling	Indian Ocean	3,750	2,330
Darling	Murray	3,070	1,905
Murray	Indian Ocean	2,575	1,600
Murrumbidgee	Murray	1,690	1,050

North America		km	miles
Mississippi–Missouri	Gulf of Mexico	6,020	3,740
Mackenzie	Arctic Ocean	4,240	2,630
Mississippi	Gulf of Mexico	3,780	2,350
Missouri	Mississippi	3,780	2,350
Yukon	Pacific Ocean	3,185	1,980
Rio Grande	Gulf of Mexico	3,030	1,880
Arkansas	Mississippi	2,340	1,450
Colorado	Pacific Ocean	2,330	1,445
Red	Mississippi	2,040	1,270

North America (cont.)		km	miles
Columbia	Pacific Ocean	1,950	1,210
Saskatchewan	Lake Winnipeg	1,940	1,205
Snake	Columbia	1,670	1,040
Churchill	Hudson Bay	1,600	990
Ohio	Mississippi	1,580	980
Brazos	Gulf of Mexico	1,400	870
St Lawrence	Atlantic Ocean	1,170	730

South America		km	miles
Amazon	Atlantic Ocean	6,450	4,010
Paraná–Plate	Atlantic Ocean	4,500	2,800
Purus	Amazon	3,350	2,080
Madeira	Amazon	3,200	1,990
São Francisco	Atlantic Ocean	2,900	1,800
Paraná	Plate	2,800	1,740
Tocantins	Atlantic Ocean	2,750	1,710
Paraguay	Paraná	2,550	1,580
Orinoco	Atlantic Ocean	2,500	1,550
Pilcomayo	Paraná	2,500	1,550
Araguaia	Tocantins	2,250	1,400
Juruá	Amazon	2,000	1,240
Xingu	Amazon	1,980	1,230
Ucayali	Amazon	1,900	1,180
Marañón	Amazon	1,600	990
Uruguay	Plate	1,600	990
Magdalena	Caribbean Sea	1,540	960

Lakes

Europe		km²	miles²
Lake Ladoga	Russia	17,700	6,800
Lake Onega	Russia	9,700	3,700
Saimaa system	Finland	8,000	3,100
Vänern	Sweden	5,500	2,100
Rybinskoye Reservoir	Russia	4,700	1,800

Asia		km²	miles²
Caspian Sea	Asia	371,800	143,550
Aral Sea	Kazak./Uzbek.	33,640	13,000
Lake Baykal	Russia	30,500	11,780
Tonlé Sap	Cambodia	20,000	7,700
Lake Balqash	Kazakstan	18,500	7,100
Lake Dongting	China	12,000	4,600
Lake Ysyk	Kyrgyzstan	6,200	2,400
Lake Orumiyeh	Iran	5,900	2,300
Lake Koko	China	5,700	2,200
Lake Poyang	China	5,000	1,900
Lake Khanka	China/Russia	4,400	1,700
Lake Van	Turkey	3,500	1,400
Lake Ubsa	China	3,400	1,300

Africa		km²	miles²
Lake Victoria	East Africa	68,000	26,000
Lake Tanganyika	Central Africa	33,000	13,000
Lake Malawi/Nyasa	East Africa	29,600	11,430
Lake Chad	Central Africa	25,000	9,700
Lake Turkana	Ethiopia/Kenya	8,500	3,300
Lake Volta	Ghana	8,500	3,300
Lake Bangweulu	Zambia	8,000	3,100
Lake Rukwa	Tanzania	7,000	2,700
Lake Mai-Ndombe	Zaïre	6,500	2,500
Lake Kariba	Zambia/Zimbabwe	5,300	2,000
Lake Mobutu	Uganda/Zaïre	5,300	2,000
Lake Nasser	Egypt/Sudan	5,200	2,000
Lake Mweru	Zambia/Zaïre	4,900	1,900
Lake Cabora Bassa	Mozambique	4,500	1,700
Lake Kyoga	Uganda	4,400	1,700
Lake Tana	Ethiopia	3,630	1,400
Lake Kivu	Rwanda/Zaïre	2,650	1,000
Lake Edward	Uganda/Zaïre	2,200	850

Australia		km²	miles²
Lake Eyre	Australia	8,900	3,400
Lake Torrens	Australia	5,800	2,200
Lake Gairdner	Australia	4,800	1,900

North America		km²	miles²
Lake Superior	Canada/USA	82,350	31,800
Lake Huron	Canada/USA	59,600	23,010
Lake Michigan	USA	58,000	22,400
Great Bear Lake	Canada	31,800	12,280
Great Slave Lake	Canada	28,500	11,000
Lake Erie	Canada/USA	25,700	9,900
Lake Winnipeg	Canada	24,400	9,400
Lake Ontario	Canada/USA	19,500	7,500
Lake Nicaragua	Nicaragua	8,200	3,200
Lake Athabasca	Canada	8,100	3,100
Smallwood Reservoir	Canada	6,530	2,520
Reindeer Lake	Canada	6,400	2,500
Lake Winnipegosis	Canada	5,400	2,100
Nettilling Lake	Canada	5,500	2,100
Lake Nipigon	Canada	4,850	1,900
Lake Manitoba	Canada	4,700	1,800

South America		km²	miles²
Lake Titicaca	Bolivia/Peru	8,300	3,200
Lake Poopo	Peru	2,800	1,100

Islands

Europe		km²	miles²
Great Britain	UK	229,880	88,700
Iceland	Atlantic Ocean	103,000	39,800
Ireland	Ireland/UK	84,400	32,600
Novaya Zemlya (North)	Russia	48,200	18,600
West Spitzbergen	Norway	39,000	15,100
Novaya Zemlya (South)	Russia	33,200	12,800
Sicily	Italy	25,500	9,800
Sardinia	Italy	24,000	9,300
North-east Spitzbergen	Norway	15,000	5,600
Corsica	France	8,700	3,400
Crete	Greece	8,350	3,200
Zealand	Denmark	6,850	2,600

Asia		km²	miles²
Borneo	South-east Asia	744,360	287,400
Sumatra	Indonesia	473,600	182,860
Honshu	Japan	230,500	88,980
Celebes	Indonesia	189,000	73,000
Java	Indonesia	126,700	48,900
Luzon	Philippines	104,700	40,400
Mindanao	Philippines	101,500	39,200
Hokkaido	Japan	78,400	30,300
Sakhalin	Russia	74,060	28,600
Sri Lanka	Indian Ocean	65,600	25,300
Taiwan	Pacific Ocean	36,000	13,900
Kyushu	Japan	35,700	13,800
Hainan	China	34,000	13,100
Timor	Indonesia	33,600	13,000
Shikoku	Japan	18,800	7,300
Halmahera	Indonesia	18,000	6,900
Ceram	Indonesia	17,150	6,600
Sumbawa	Indonesia	15,450	6,000
Flores	Indonesia	15,200	5,900
Samar	Philippines	13,100	5,100
Negros	Philippines	12,700	4,900
Bangka	Indonesia	12,000	4,600
Palawan	Philippines	12,000	4,600
Panay	Philippines	11,500	4,400
Sumba	Indonesia	11,100	4,300
Mindoro	Philippines	9,750	3,800
Buru	Indonesia	9,500	3,700

Asia (cont.)		km²	miles²
Bali	Indonesia	5,600	2,200
Cyprus	Mediterranean	3,570	1,400

Africa		km²	miles²
Madagascar	Indian Ocean	587,040	226,660
Socotra	Indian Ocean	3,600	1,400
Réunion	Indian Ocean	2,500	965
Tenerife	Atlantic Ocean	2,350	900
Mauritius	Indian Ocean	1,865	720

Oceania		km²	miles²
New Guinea	Indon./Papua NG	821,030	317,000
New Zealand (South)	New Zealand	150,500	58,100
New Zealand (North)	New Zealand	114,700	44,300
Tasmania	Australia	67,800	26,200
New Britain	Papua NG	37,800	14,600
New Caledonia	Pacific Ocean	19,100	7,400
Viti Levu	Fiji	10,500	4,100
Hawaii	Pacific Ocean	10,450	4,000
Bougainville	Papua NG	9,600	3,700
Guadalcanal	Solomon Is.	6,500	2,500
Vanua Levu	Fiji	5,550	2,100
New Ireland	Papua NG	3,200	1,200

North America		km²	miles²
Greenland	Greenland	2,175,600	839,800
Baffin Is.	Canada	508,000	196,100
Victoria Is.	Canada	212,200	81,900
Ellesmere Is.	Canada	212,000	81,800
Cuba	Cuba	110,860	42,800
Newfoundland	Canada	110,680	42,700
Hispaniola	Atlantic Ocean	76,200	29,400
Banks Is.	Canada	67,000	25,900
Devon Is.	Canada	54,500	21,000
Melville Is.	Canada	42,400	16,400
Vancouver Is.	Canada	32,150	12,400
Somerset Is.	Canada	24,300	9,400
Jamaica	Caribbean Sea	11,400	4,400
Puerto Rico	Atlantic Ocean	8,900	3,400
Cape Breton Is.	Canada	4,000	1,500

South America		km²	miles²
Tierra del Fuego	Argentina/Chile	47,000	18,100
Falkland Is. (East)	Atlantic Ocean	6,800	2,600
South Georgia	Atlantic Ocean	4,200	1,600
Galapagos (Isabela)	Pacific Ocean	2,250	870

World Statistics – Climate

For each city, the top row of figures shows total rainfall in millimetres, the bottom row shows the average temperature in ° Celsius or centigrade. The total annual rainfall and average annual temperature are given at the end of the rows.

	Jan.	Feb.	Mar.	Apr.	May	June	July	Aug.	Sept.	Oct.	Nov.	Dec.	Total
Europe													
Berlin, Germany	46	40	33	42	49	65	73	69	68	49	46	43	603
Altitude 55 metres	1	0	4	9	14	17	19	18	15	9	5	1	9
London, UK	54	40	37	37	46	45	57	59	49	57	64	48	593
5 m	4	5	7	9	12	16	18	17	15	11	8	5	11
Málaga, Spain	61	51	62	46	26	5	1	3	29	64	64	62	474
33 m	12	13	16	17	19	29	25	26	23	20	16	13	18
Moscow, Russia	39	38	36	37	53	58	88	71	58	45	47	54	624
156 m	13	-10	-4	6	13	16	18	17	12	6	-1	-7	4
Paris, France	56	46	35	42	57	54	59	64	55	50	51	50	619
75 m	3	4	8	11	15	18	20	19	17	12	7	4	12
Rome, Italy	71	62	57	51	46	37	15	21	63	99	129	93	744
17 m	8	9	11	14	18	22	25	25	22	17	13	10	16
Asia													
Bangkok, Thailand	8	20	36	58	198	160	160	175	305	206	66	5	1,397
2 m	26	28	29	30	29	29	28	28	28	28	26	25	28
Bombay, India	3	3	3	<3	18	485	617	340	264	64	13	3	1,809
11 m	24	24	26	28	30	29	27	27	27	28	27	26	27
Ho Chi Minh, Vietnam	15	3	13	43	221	330	315	269	335	269	114	56	1,984
9 m	26	27	29	30	29	28	28	28	27	27	27	26	28
Hong Kong	33	46	74	137	292	394	381	361	257	114	43	31	2,162
33 m	16	15	18	22	26	28	28	28	27	25	21	18	23
Tokyo, Japan	48	74	107	135	147	165	142	152	234	208	97	56	1,565
6 m	3	4	7	13	17	21	25	26	23	17	11	6	14
Africa													
Cairo, Egypt	5	5	5	3	3	<3	0	0	<3	<3	3	5	28
116 m	13	15	18	21	25	28	28	28	26	24	20	15	22
Cape Town, South Africa	15	8	18	48	79	84	89	66	43	31	18	10	508
17 m	21	21	20	17	14	13	12	13	14	16	18	19	17
Lagos, Nigeria	28	46	102	150	269	460	279	64	140	206	69	25	1,836
3 m	27	28	29	28	28	26	26	25	26	26	28	28	27
Nairobi, Kenya	38	64	125	211	158	46	15	23	31	53	109	86	958
1,820 m	19	19	19	19	18	16	16	16	18	19	18	18	18
Australia, New Zealand & Antarctica													
Christchurch, New Zealand	56	43	48	48	66	66	69	48	46	43	48	56	638
10 m	16	16	14	12	9	6	6	7	9	12	14	16	11
Darwin, Australia	386	312	254	97	15	3	<3	3	13	51	119	239	1,491
30 m	29	29	29	29	28	26	25	26	28	29	30	29	28
Mawson, Antarctica	11	30	20	10	44	180	4	40	3	20	0	0	362
14 m	0	-5	-10	-14	-15	-16	-18	-18	-19	-13	-5	-1	-11
Sydney, Australia	89	102	127	135	127	117	117	76	73	71	73	73	1,181
42 m	22	22	21	18	15	13	12	13	15	18	19	21	17
North America													
Anchorage, Alaska, USA	20	18	15	10	13	18	41	66	66	56	25	23	371
40 m	-11	-8	-5	2	7	12	14	13	9	2	-5	-11	2
Kingston, Jamaica	23	15	23	31	102	89	38	91	99	180	74	36	800
34 m	25	25	25	26	26	28	28	28	27	27	26	26	26
Los Angeles, USA	79	76	71	25	10	3	<3	<3	5	15	31	66	381
95 m	13	14	14	16	17	19	21	22	21	18	16	14	17
Mexico City, Mexico	13	5	10	20	53	119	170	152	130	51	18	8	747
2,309 m	12	13	16	18	19	19	17	18	18	16	14	13	16
New York, USA	94	97	91	81	81	84	107	109	86	89	76	91	1,092
96 m	-1	-1	3	10	16	20	23	23	21	15	7	2	11
Vancouver, Canada	154	115	101	60	52	45	32	41	67	114	150	182	1,113
14 m	3	5	6	9	12	15	17	17	14	10	6	4	10
South America													
Antofagasta, Chile	0	0	0	<3	<3	3	5	3	<3	3	<3	0	13
94 m	21	21	20	18	16	15	14	14	15	16	18	19	17
Buenos Aires, Argentina	79	71	109	89	76	61	56	61	79	86	84	99	950
27 m	23	23	21	17	13	9	10	11	13	15	19	22	16
Lima, Peru	3	<3	<3	<3	5	5	8	8	8	3	3	<3	41
120 m	23	24	24	22	19	17	17	16	17	18	19	21	20
Rio de Janeiro, Brazil	125	122	130	107	79	53	41	43	66	79	104	137	1,082
61 m	26	26	25	24	22	21	21	21	21	22	23	25	23

The Earth in Focus

> Landsat image of the
San Francisco Bay area.
The narrow entrance to
the bay (crossed by the
Golden Gate Bridge)
provides an excellent
natural harbour. The
San Andreas Fault runs
parallel to the coastline.

The Universe & Solar System

BETWEEN 10 AND 20 billion (or 10,000 to 20,000 million) years ago, the Universe was created in a huge explosion known as the 'Big Bang'. In the first 10^{-24} of a second the Universe expanded rapidly and the basic forces of nature, radiation and subatomic particles, came into being. The Universe has been expanding ever since. Traces of the original 'fireball' of radiation can still be detected, and most scientists accept the Big Bang theory of the origin of the Universe.

The Nearest Stars ▾

The 20 nearest stars, excluding the Sun, with their distance from Earth in light-years.*

Proxima Centauri	4.25
Alpha Centauri A	4.3
Alpha Centauri B	4.3
Barnard's Star	6.0
Wolf 359	7.8
Lalande 21185	8.3
Sirius A	8.7
Sirius B	8.7
UV Ceti A	8.7
UV Ceti B	8.7
Ross 154	9.4
Ross 248	10.3
Epsilon Eridani	10.7
Ross 128	10.9
61 Cygni A	11.1
61 Cygni B	11.1
Epsilon Indi	11.2
Groombridge 34 A	11.2
Groombridge 34 B	11.2
L789-6	11.2

* A light-year equals approximately 9,500 billion km [5,900 billion mls].

> The Lagoon Nebula is a huge cloud of dust and gas. Hot stars inside the nebula make the gas glow red.

GALAXIES

Almost a million years passed before the Universe cooled sufficiently for atoms to form. When a billion years had passed, the atoms had begun to form proto-galaxies, which are masses of gas separated by empty space. Stars began to form within the proto-galaxies, as particles were drawn together, producing the high temperatures necessary to bring about nuclear fusion. The formation of the first stars brought about the evolution of the proto-galaxies into galaxies proper, each containing billions of stars.

Our Sun is a medium-sized star. It is

Mercury   Venus   Earth   Mars   Jupiter

PLANETARY DATA

	Mean distance from Sun (million km)	Mass (Earth = 1)	Period of orbit (Earth years)	Period of rotation (Earth days)	Equatorial diameter (km)	Escape velocity (km/sec)	Number of known satellites
Sun	–	332,946	–	25.38	1,392,000	617.5	–
Mercury	58.3	0.06	0.241	58.67	4,878	4.27	0
Venus	107.7	0.8	0.615	243.0	12,104	10.36	0
Earth	149.6	1.0	1.00	0.99	12,756	11.18	1
Mars	227.3	0.1	1.88	1.02	6,787	5.03	2
Jupiter	777.9	317.8	11.86	0.41	142,800	59.60	16
Saturn	1,427.1	95.2	29.46	0.42	120,000	35.50	20
Uranus	2,872.3	14.5	84.01	0.45	51,118	21.30	15
Neptune	4,502.7	17.2	164.79	0.67	49,528	23.3	8
Pluto	5,894.2	0.002	248.54	6.38	2,300	1.1	1

one of the billions of stars that make up the Milky Way galaxy, which is one of the millions of galaxies in the Universe.

THE SOLAR SYSTEM

The Solar System lies towards the edge of the Milky Way galaxy. It consists of the Sun and other bodies, including planets (together with their moons), asteroids, meteoroids, comets, dust and gas, which revolve around it.

The Earth moves through space in three distinct ways. First, with the rest of the Solar System, it moves around the centre of the Milky Way galaxy in an orbit that takes 200 million years.

As the Earth revolves around the Sun once every year, its axis is tilted by about 23.5 degrees. As a result, first the northern and then the southern hemisphere lean towards the Sun at different times of the year, causing the seasons experienced in the mid-latitudes.

The Earth also rotates on its axis every 24 hours, causing day and night. The movements of the Earth in the Solar System determine the calendar. The length of a year – one complete orbit of the Earth around the Sun – is 365 days, 5 hours, 48 minutes and 46 seconds. Leap years prevent the calendar from becoming out of step with the solar year.

> The diagram below shows the planets around the Sun. The sizes of the planets are relative but the distances are not to scale. Closest to the Sun are dense rocky bodies, known as the terrestrial planets. They are Mercury, Venus, Earth and Mars. Jupiter, Saturn, Uranus and Neptune are huge balls of gas. Pluto is a small, icy body.

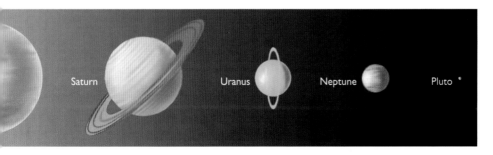

Saturn   Uranus   Neptune   Pluto

The Changing Earth

THE SOLAR SYSTEM was formed around 4.7 billion years ago, when the Sun, a glowing ball of gases, was created from a rotating disk of dust and gas. The planets were then formed from material left over after the creation of the Sun.

After the Earth formed, around 4.6 billion years ago, lighter elements rose to the hot surface, where they finally cooled to form a hard shell, or crust. Denser elements sank, forming the partly liquid mantle, the liquid outer core, and the solid inner core.

EARTH HISTORY

The oldest known rocks on Earth are around 4 billion years old. Natural processes have destroyed older rocks. Simple life forms first appeared on Earth around 3.5 billion years ago, though rocks formed in the first 4 billion years of Earth history contain little evidence of life. But

> *Fold mountains, such as the Himalayan ranges which are shown above, were formed when two plates collided and the rock layers between them were squeezed upwards into loops or folds.*

rocks formed since the start of the Cambrian period (the first period in the Paleozoic era), about 590 million years ago, are rich in fossils. The study of fossils has enabled scientists to gradually piece together the long and complex story of life on Earth.

THE PLANET EARTH

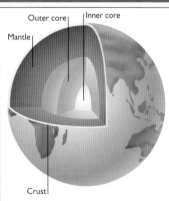

Outer core · Inner core
Mantle
Crust

CRUST The continental crust averages 35–40 km [22–25 mls]; the oceanic crust has an average thickness of 6 km [4 mls].

MANTLE 2,900 km [1,800 mls] thick. The top layer is solid, resting on a partly molten layer called the asthenosphere.

OUTER CORE 2,100 km [1,300 mls] thick. It consists mainly of molten iron and nickel.

INNER CORE (DIAMETER) 1,350 km [840 mls]. It is mainly solid iron and nickel.

ELEMENTS

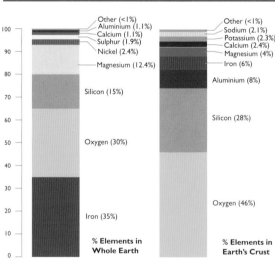

Other (<1%)
Aluminium (1.1%)
Calcium (1.1%)
Sulphur (1.9%)
Nickel (2.4%)
Magnesium (12.4%)
Silicon (15%)
Oxygen (30%)
Iron (35%)
% Elements in Whole Earth

Other (<1%)
Sodium (2.1%)
Potassium (2.3%)
Calcium (2.4%)
Magnesium (4%)
Iron (6%)
Aluminium (8%)
Silicon (28%)
Oxygen (46%)
% Elements in Earth's Crust

> *The Earth contains about 100 elements, but eight of them account for 99% of the planet's mass. Iron makes up 35% of the Earth's mass, but most of it is in the core. The most common elements in the crust, oxygen and silicon, are often combined with one or more of the other common crustal elements, to form a group of minerals called silicates. The mineral quartz, which consists only of silicon and oxygen, occurs widely in such rocks as granites and sandstones.*

PLATE BOUNDARIES

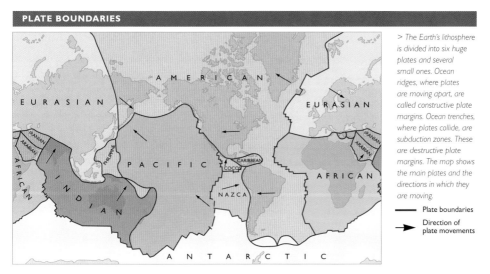

> The Earth's lithosphere is divided into six huge plates and several small ones. Ocean ridges, where plates are moving apart, are called constructive plate margins. Ocean trenches, where plates collide, are subduction zones. These are destructive plate margins. The map shows the main plates and the directions in which they are moving.

— Plate boundaries

➤ Direction of plate movements

THE DYNAMIC EARTH

The Earth's surface is always changing because of a process called plate tectonics. Plates are blocks of the solid lithosphere (the crust and outer mantle), which are moved around by currents in the partly liquid mantle. Around 250 million years ago, the Earth contained one super-continent called Pangaea. Around 180 million years ago, Pangaea split into a northern part, Laurasia, and a southern part, Gondwanaland. Later, these huge continents, in turn, also split apart and the continents drifted to their present positions. Ancient seas disappeared and mountain ranges, such as the Himalayas and Alps, were pushed upwards.

PLATE TECTONICS

In the early 1900s, two scientists suggested that the Americas were once joined to Europe and Africa. Together they proposed the theory of continental drift to explain the similarities between rock structures on both sides of the Atlantic. But no one could offer an explanation as to how the continents moved.

Evidence from the ocean floor in the 1950s and 1960s led to the theory of plate tectonics, which suggested that the lithosphere is divided into large blocks, or plates. The plates are solid, but they rest on the partly molten asthenosphere, within the mantle. Long ridges on the ocean floor were found to be the edges of plates which were moving apart, carried by currents in the asthenosphere. As the plates moved, molten material welled up from the mantle to fill the gaps. But at the ocean trenches, one plate is descending beneath another along what is called a subduction zone. The descending plate is melted and destroyed. This crustal destruction at subduction zones balances the creation of new crust along the ridges. Transform faults, where two plates are moving alongside each other, form another kind of plate edge.

GEOLOGICAL TIME SCALE

Pre-Cambrian	Lower	Paleozoic (Primary)					Upper		Mesozoic (Secondary)			Cenozoic (Tertiary, Quaternary)			Era
Pre-Cambrian	Cambrian	Ordovician	Silurian	Devonian	Carboniferous	Permian	Triassic	Jurassic	Cretaceous	Paleocene / Eocene	Oligocene / Miocene	Pliocene / Quaternary	System
			CALEDONIAN FOLDING		HERCYNIAN FOLDING					LARAMIDE FOLDING	ALPINE FOLDING		Orogeny
600	550	500	450	400	350	300	250	200	150	100	50		

Millions of years before present

5

Earthquakes & Volcanoes

PLATE TECTONICS HELPS us to understand such phenomena as earthquakes, volcanic eruptions and mountain building.

EARTHQUAKES

Earthquakes can occur anywhere, but they are most common near the edges of plates. They occur when intense pressure breaks the rocks along plate edges, making the plates lurch forward in a sudden movement.

> The earthquake that struck Kobe in January 1995 was the worst one experienced in Japan since 1923. Japan lies alongside subduction zones.

Major Earthquakes since 1900 ▾

Year	Location	Mag.	Deaths
1906	San Francisco, USA	8.3	503
1906	Valparaiso, Chile	8.6	22,000
1908	Messina, Italy	7.5	83,000
1915	Avezzano, Italy	7.5	30,000
1920	Gansu, China	8.6	180,000
1923	Yokohama, Japan	8.3	143,000
1927	Nan Shan, China	8.3	200,000
1932	Gansu, China	7.6	70,000
1934	Bihar, India/Nepal	8.4	10,700
1935	Quetta, Pakistan	7.5	60,000
1939	Chillan, Chile	8.3	28,000
1939	Erzincan, Turkey	7.9	30,000
1960	Agadir, Morocco	5.8	12,000
1964	Anchorage, Alaska	8.4	131
1968	North-east Iran	7.4	12,000
1970	North Peru	7.7	66,794
1976	Guatemala	7.5	22,778
1976	Tangshan, China	8.2	650,000
1978	Tabas, Iran	7.7	25,000
1980	El Asnam, Algeria	7.3	20,000
1980	South Italy	7.2	4,800
1985	Mexico City, Mexico	8.1	4,200
1988	North-west Armenia	6.8	55,000
1990	North Iran	7.7	36,000
1993	Maharashtra, India	6.4	30,000
1994	Los Angeles, USA	6.4	57
1995	Kobe, Japan	7.2	5,000
1996	Yunnan, China	7.0	255

> The section between the Pacific and Indian oceans shows a subduction zone under the American plate, with spreading ocean ridges in the Atlantic and Indian oceans. East Africa may one day split away from the rest of Africa as plate movements pull the Rift Valley apart.

Earthquakes are common along the mid-ocean ridges, but they are a long way from land and cause little damage. Other earthquakes occur near land in subduction zones, such as those that encircle much of the Pacific Ocean. These earthquakes often trigger off powerful sea waves, called tsunamis. Other earthquakes occur along transform faults, such as the San Andreas fault in California, a boundary between the North American and Pacific plates. Movements along this fault cause periodic disasters, such as the earthquakes in San Francisco (1906) and Los Angeles (1994).

VOLCANOES & MOUNTAINS

Volcanoes are fuelled by magma (molten rock) from the mantle. Some volcanoes, such as in Hawaii, lie above 'hot spots' (sources of heat in the mantle). But most volcanoes occur either along the ocean ridges or above subduction zones, where

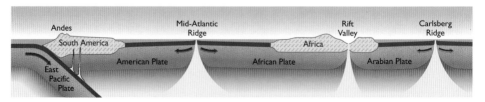

EARTHQUAKES

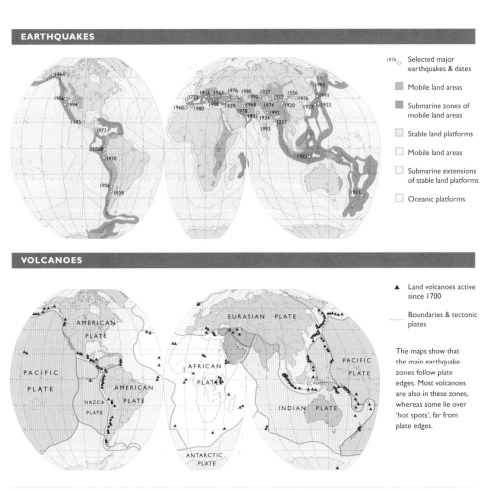

1976 ○ Selected major earthquakes & dates

▪ Mobile land areas

▪ Submarine zones of mobile land areas

▫ Stable land platforms

▫ Mobile land areas

▫ Submarine extensions of stable land platforms

▫ Oceanic platforms

VOLCANOES

▲ Land volcanoes active since 1700

— Boundaries & tectonic plates

The maps show that the main earthquake zones follow plate edges. Most volcanoes are also in these zones, whereas some lie over 'hot spots', far from plate edges.

magma is produced when the descending plate is melted.

Volcanic mountains are built up by runny lava flows or by exploded volcanic ash. Fold mountains occur when two plates bearing land areas collide and the plate edges are buckled upwards into fold mountain ranges. Plate movements also fracture rocks and block mountains are formed when areas of land are pushed upwards along faults or between parallel faults. Blocks of land sometimes sink down between faults, creating deep, steep-sided rift valleys.

> Volcanoes occur when molten magma reaches the surface under pressure through long vents. 'Quiet' volcanoes emit runny lava (called pahoehoe). Explosive eruptions occur when the magma is sticky. Explosive gases shatter the magma into ash, which is hurled upwards into the air.

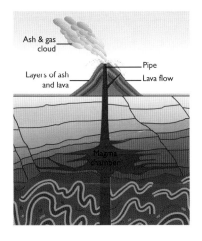

Ash & gas cloud

Layers of ash and lava

Pipe

Lava flow

Magma chamber

Water & Ice

A VISITOR FROM outer space might be forgiven for naming our planet 'Water' rather than 'Earth', because water covers more than 70% of its surface. Without water, our planet would be as lifeless as the Moon. Through the water cycle, fresh water is regularly supplied from the sea to the land. Most geographers divide the world's water into four main oceans: the Pacific, the Atlantic, the Indian and the Arctic. Together the oceans contain 97.2% of the world's water.

The water in the oceans is constantly on the move, even, albeit extremely slowly, in the deepest ocean trenches. The greatest movements of ocean water occur in the form of ocean currents. These are marked, mainly wind-blown

> Ice breaks away from the ice sheet of Antarctica, forming flat-topped icebergs. The biggest iceberg ever recorded came from Antarctica. It covered an area larger than Belgium.

EXPLANATION OF TERMS

GLACIER A body of ice that flows down valleys in mountain areas. It is usually narrow and hence smaller than ice caps or ice sheets.

ICE AGE A period of Earth history when ice sheets spread over large areas. The most recent Ice Age began about 1.8 million years ago and ended 10,000 years ago.

ICEBERG A floating body of ice in the sea. About eight-ninths of the ice is hidden beneath the surface.

ICE SHEET A large body of ice. During the last Ice Age, ice sheets covered large parts of the northern hemisphere.

OCEAN The four main oceans are the Pacific, the Atlantic, the Indian and the Arctic. Some

people classify a fifth Southern Ocean, but others regard these waters as extensions of the Pacific, Atlantic and Indian oceans.

OCEAN CURRENTS Distinct currents of water in the oceans. Winds are the main causes of surface currents.

SEA An expanse of water, but smaller than an ocean.

JANUARY TEMPERATURE AND OCEAN CURRENTS

(Northern Hemisphere – Winter)

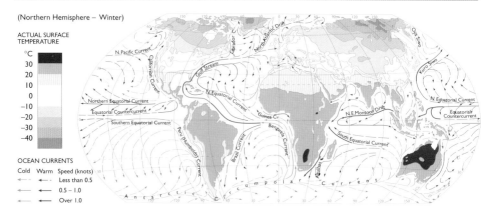

ACTUAL SURFACE TEMPERATURE

°C
30
20
10
0
−10
−20
−30
−40

OCEAN CURRENTS
Cold Warm Speed (knots)
          Less than 0.5
          0.5 – 1.0
          Over 1.0

CROSS-SECTION OF ANTARCTICA

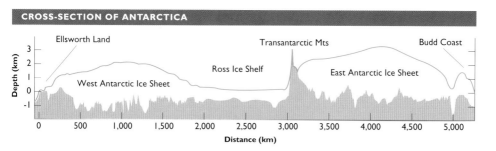

movements of water on or near the surface. Other dense, cold currents creep slowly across the ocean floor. Warm and cold ocean currents help to regulate the world's climate by transferring heat between the tropics and the poles.

ICE

About 2.15% of the world's water is locked in two large ice sheets, several smaller ice caps and glaciers. The world's largest ice sheet covers most of Antarctica. The ice is up to 4,800 m [15,750 ft] thick and it represents 70% of the world's fresh water. The volume of ice is about nine times greater than that contained in the world's other ice sheet in Greenland. Besides these two ice sheets, there are some smaller ice caps in northern Canada, Iceland, Norway and Spitzbergen, and

many valley glaciers in mountain regions throughout the world, except Australia.

If global warming was to melt the world's ice, the sea level could rise by as much as 100 m [330 ft], flooding low-lying coastal regions. Many of the world's largest cities and most fertile plains would vanish beneath the waves.

> This section across Antarctica shows the concealed land areas in brown, with the top of the ice in blue. The section is divided into the West and East Antarctic Ice Sheets. The vertical scale has been exaggerated.

Composition of Seawater ▾

The principal components of seawater, by percentage, excluding the elements of water itself:

Chloride (Cl)	55.04%	Potassium (K)	1.10%
Sodium (Na)	30.61%	Bicarbonate (HCO₃)	0.41%
Sulphate (SO₄)	7.69%	Bromide (Br)	0.19%
Magnesium (Mg)	3.69%	Strontium (Sr)	0.04%
Calcium (Ca)	1.16%	Fluorine (F)	0.003%

The oceans contain virtually every other element, the more important ones being lithium, rubidium, phosphorus, iodine and barium.

JULY TEMPERATURE AND OCEAN CURRENTS

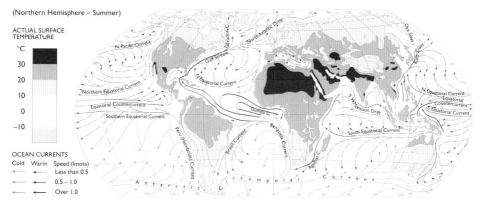

(Northern Hemisphere – Summer)

ACTUAL SURFACE TEMPERATURE

°C
30
20
10
0
−10

OCEAN CURRENTS
Cold Warm Speed (knots)
Less than 0.5
0.5 – 1.0
Over 1.0

Weather & Climate

WEATHER IS A description of the day-to-day state of the atmosphere. Climate, on the other hand, is weather in the long term: the seasonal pattern of temperature and precipitation averaged over time.

In some areas, the weather is so stable and predictable that a description of the weather is much the same as a statement of the climate. But in parts of the mid-latitudes, the weather changes from hour to hour. Changeable weather is caused mainly by low air pressure systems, called cyclones or depressions, which form along the polar front where warm subtropical air meets cold polar air.

The main elements of weather and

climate are temperature and rainfall. Temperatures vary because the Sun heats the Earth unequally, with the most intense heating around the Equator. Unequal heating is responsible for the general circulation of the atmosphere and the main wind belts.

Rainfall occurs when warm air containing invisible water vapour rises. As the rising air cools, the capacity of the air to hold water vapour decreases and so the water vapour condenses into droplets of water or ice crystals, which collect together to form raindrops or snowflakes.

> Lightning occurs in clouds and also between the base of clouds and the ground. Lightning that strikes the ground can kill people or start forest fires.

> The rainfall map shows areas affected by tropical storms, which are variously called hurricanes, tropical cyclones, willy willies and typhoons. Strong polar winds bring blizzards in winter.

LIGHTNING

Lightning is a flash of light in the sky caused by a discharge of electricity in the atmosphere. Lightning occurs within cumulonimbus clouds during thunderstorms. Positive charges build up at the top of the cloud, while negative charges build up at the base. The charges are finally discharged as an electrical spark. Sheet lightning occurs inside clouds, while cloud to ground lightning is usually forked. Thunder occurs when molecules along the lightning channel expand and collide with cool molecules.

ANNUAL RAINFALL

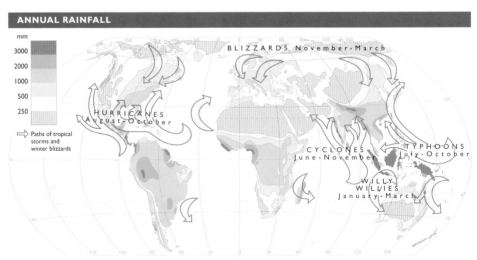

mm	
3000	
2000	
1000	
500	
250	

⇨ Paths of tropical storms and winter blizzards

BLIZZARDS November-March

HURRICANES August-October

CYCLONES June-November

TYPHOONS July-October

WILLY WILLIES January-March

GLOBAL WARMING

The Earth's climates have changed many times during its history. Around 11,000 years ago, much of the northern hemisphere was buried by ice. Some scientists believe that the last Ice Age may not be over and that ice sheets may one day return. Other scientists are concerned that air

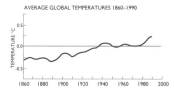

AVERAGE GLOBAL TEMPERATURES 1860–1990

pollution may be producing an opposite effect – a warming of the atmosphere. Since 1900, average world temperatures have risen by about 0.5°C [0.9°F] and increases are likely to continue. Global warming is the result of an increase in the amount of carbon dioxide in the atmosphere, caused by the burning of coal, oil and natural gas, together with deforestation. Short-wave radiation from the Sun passes easily through the atmosphere. But, as the carbon dioxide content rises, so more and more of the long-wave radiation that returns from the Earth's surface is absorbed and trapped by the carbon dioxide. This creates a 'greenhouse effect', which will change the world's climates with, perhaps, disastrous environmental consequences.

CLIMATE

The world contains six main climatic types: hot and wet tropical climates; dry climates; warm temperate climates; cold temperate climates; polar climates; and mountain climates. These regions are further divided according to the character and amount of the rainfall and special features of the temperature, notably seasonal variations. Regions with temperate climates include Mediterranean areas with hot, dry summers and mild, moist winters. The British Isles has a different type of temperate climate, with warm, rather than hot, summers and rain throughout the year.

CLIMATIC REGIONS

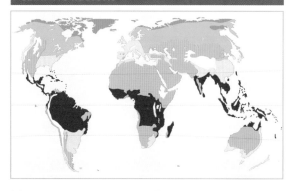

- ■ Tropical Climate (hot & wet)
- ▦ Dry Climate (desert & steppe)
- ☐ Temperate Climate (warm & wet)
- ▦ Continental Climate (cold & wet)
- ■ Polar Climate (very cold & wet)
- ☐ Mountainous Areas (where altitude affects climate types)

WORLD CLIMATIC RECORDS

Highest Recorded Temperature
Al Aziziyah, Libya: 58°C [136.4°F] on 13 September 1922

Highest Mean Annual Temperature
Dallol, Ethiopia: 34.4°C [94°F] from 1960–66

Lowest Mean Annual Temperature
Polus, Nedostupnosti, Pole of Cold, Antarctica: −57.8 °C [−72°F]

Lowest Recorded Temperature (outside poles)
Verkhoyansk, Siberia, Russia: −68°C [−90°F] on 6 February 1933

Longest Heatwave
Marble Bar, Western Australia: 162 days over 38°C [94°F], 23 October 1923 to 7 April 1924

Driest Place
Arica, northern Chile: only 0.8 mm [0.03 in] per year (60-year average)

Longest Drought
Calama, northern Chile: no recorded rainfall in 400 years to 1971

Wettest Place (average)
Tututendo, Colombia: mean annual rainfall 11,770 mm [463 in]

Wettest Place (24 hours)
Cilaos, Réunion, Indian Ocean: 1,870 mm [73.6 in] from 15–16 March 1952

Wettest Place (12 months)
Cherrapunji, Meghalaya, north-east India: 26,470 mm [1,040 in], August 1860 to 1861. Cherrapunji also holds the record for rainfall in one month: 930 mm [37 in] in July 1861

Heaviest Hailstones
Gopalganj, central Bangladesh: up to 1.02 kg [2.25 lbs] in April 1986, which killed 92 people

Heaviest Snowfall (continuous)
Bessans, Savoie, France: 1,730 mm [68 in] in 19 hours over the period 5–6 April 1969

Heaviest Snowfall (season/year)
Paradise Ranger Station, Mt Rainer, Washington, USA: 31,102 mm [1,224 in] fell from 19 February 1971 to 18 February 1972

Landscape & Vegetation

THE CLIMATE LARGELY determines the nature of soils and vegetation types throughout the world. The studies of climate and plant and animal communities are closely linked. For example, tropical climates are divided into tropical forest and tropical grassland climates. The tropical forest climate, which is hot and rainy throughout the year, is ideal for the growth of forests that contain more than half of the world's known plant and animal species. But tropical grassland, or savanna, climates have a marked dry season. As a result, the forest gives way to grassland, with scattered trees.

CLIMATE & SCENERY

The climate also helps to shape the land. Frost action in cold areas splits boulders apart, while rapid temperature changes in hot deserts make rock surfaces peel away like the layers of an onion. These are examples of mechanical weathering.

Chemical weathering usually results from the action of water on rocks. For example, rainwater containing dissolved carbon dioxide is a weak acid, which reacts with limestone. This chemical process is responsible for the erosion of the world's most spectacular caves.

Running water and glaciers play a major part in moulding scenery, while in

> The tropical broadleaf forests are rich in plant and animal species. The extinction of many species because of deforestation is one of the great natural disasters of our time.

NATURAL VEGETATION

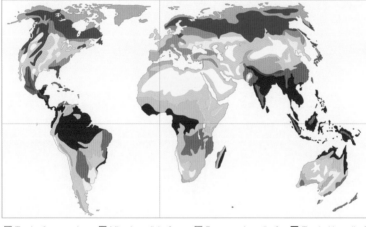

- ■ Tundra & mountain vegetation
- ■ Needleleaf evergreen forest
- ■ Broadleaf deciduous forest
- ■ Mixed needleleaf evergreen & broadleaf deciduous trees
- □ Mid-latitude grassland
- ■ Semi-desert scrub land
- ■ Evergreen broadleaf & deciduous trees & scrub
- □ Desert
- ■ Tropical grassland (savanna)
- ■ Tropical broadleaf & monsoon rainforest
- ■ Subtropical broadleaf & needleleaf forest

> Human activities, especially agriculture, have greatly modified plant and animal communities throughout the world. As a result, world vegetation maps show the natural 'climax vegetation' of regions – that is, the kind of vegetation that would grow in a particular climatic area, had that area not been affected by human activities. For example, the climax vegetation of western Europe is broadleaf, deciduous forest, but most of the original forest, together with the animals which lived in it, was destroyed long ago.

DESERTIFICATION AND DEFORESTATION

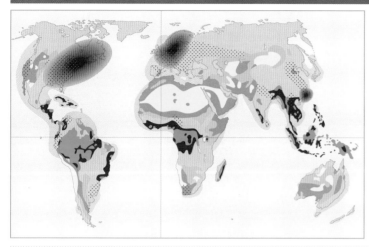

Pollution

☐ Polluted seas

▨ Main areas of sulphur & nitrogen emissions

■ Areas of acid rain

Desertification

☐ Existing deserts

■ Areas with a high risk of desertification

▨ Areas with a moderate risk of desertification

Deforestation

■ Former areas of rainforest

■ Existing rainforest

---

dry areas, wind-blown sand is a powerful agent of erosion. Most landscapes seem to alter little in one person's lifetime. But geologists estimate that natural forces remove an average of 3.5 cm [1.4 in] from land areas every 1,000 years. Over millions of years, these forces reduce mountains to flat plains.

HUMAN INTERFERENCE

Climate also affects people, though air conditioning and central heating now make it possible for us live in comfort almost anywhere in the world.

But human activities are damaging our planet. Pollution is poisoning rivers and seas, while acid rain, caused by air pollution, is killing trees and acidifying lakes. The land is also harmed by such things as nuclear accidents and the dumping of toxic wastes.

Some regions have been overgrazed or so intensively farmed that once fertile areas have been turned into barren deserts. The clearance of tropical forests means that plant and animal species are disappearing before scientists have had a chance to study them.

MOULDING THE LAND

Powerful forces inside the Earth buckle rock layers to form fold mountain ranges. But even as they rise, the forces of erosion wear them away. On mountain slopes, water freezes in cracks in rocks. Because ice occupies more space than the equivalent amount of water, this 'frost action' shatters rocks, and the fragments tumble downhill. Some end up on or inside moving glaciers. Other rocks are carried away by running water. The glaciers and streams not only transport rock fragments, but they also wear out valleys and so add to their load. The eroded material breaks down into fragments of sand, silt and mud, much of which reaches the sea, where it piles up on the sea floor in layers. These layers eventually become compacted into sedimentary rocks, such as sandstones and shales. These rocks may eventually be squeezed up again by a plate collision to form new fold mountains, so completing a natural cycle of mountain building and destruction.

MAJOR FACTORS AFFECTING WEATHERING

	WEATHERING RATE		
	SLOW ◄		► **FAST**
Mineral solubility	low *(e.g. quartz)*	moderate *(e.g. feldspar)*	high *(e.g. calcite)*
Rainfall	low	moderate	heavy
Temperature	cold	temperate	hot
Vegetation	sparse	moderate	lush
Soil cover	bare rock	thin to moderate soil	thick soil

Weathering is the breakdown and decay of rocks in situ. It may be mechanical (physical), chemical or biological.

Population

Around 10,000 years ago, the invention of agriculture had a great impact on human society. People abandoned their nomadic way of life and settled in farming villages. With plenty of food, some people were able to pursue jobs unconnected with farming. These developments eventually led to rapid social changes, including the growth of early cities and the emergence of civilization.

THE POPULATION EXPLOSION

These changes had a major effect on population. From around 8 million in 8000 BC, the world population rose to about 300 million by AD 1000. The rate of population increase then began to accelerate. The world population passed the 1 billion mark in the 19th century, the 2 billion mark in the 1920s and the 4 billion mark in the 1970s.

Today the world has a population of about 5.7 billion and experts forecast that it will reach around 11 billion by 2075. However, they then predict that it will stabilize or even decline a little towards 2100. Most of the expected increase will occur in developing countries in Africa, Asia and Latin America.

> Many cities in India, such as Bombay (also known as Mumbai), have grown so quickly that they lack sufficient jobs and homes for their populations. As a result, slums now cover large areas.

POPULATION PYRAMIDS

WORLD

	% Male									% Female					
75+
70-74
65-69
60-64
55-59
50-54
45-49
40-44
35-39
30-34
25-29
20-24
15-19
10-14
5-9
0-4

10 8 6 4 2 0 2 4 6 8 10

KENYA

75+
70-74 % Male   % Female 70-74
65-69
60-64
55-59
50-54
45-49
40-44
35-39
30-34
25-29
20-24
15-19
10-14
5-9
0-4

10 8 6 4 2 0 2 4 6 8 10

BRAZIL

75+
70-74 % Male   % Female
65-69
60-64
55-59
50-54
45-49
40-44
35-39
30-34
25-29
20-24
15-19
10-14
5-9
0-4

10 8 6 4 2 0 2 4 6 8 10

UK

75+
70-74 % Male   % Female 70-74
65-69
60-64
55-59
50-54
45-49
40-44
35-39
30-34
25-29
20-24
15-19
10-14
5-9
0-4

10 8 6 4 2 0 2 4 6 8 10

> The population pyramids compare the average age structures for the world with those of three countries at varying stages of development. Kenya, a developing country, had, until recently, one of the world's highest annual rates of population increase. As a result, a high proportion of Kenyans are aged under 15. Brazil has a much more balanced economy than Kenya's, and a lower rate of population increase. This is reflected in a higher proportion of people aged over 40. The UK is a developed country with a low rate of population growth, 0.3% per year between 1985–94, much lower than the world average of 1.6%. The UK has a far higher proportion of people over 60 years old.

The World's Largest Cities ▾

By early next century, for the first time ever, the majority of the world's population will live in cities. Below is a list of the 20 largest cities (in thousands) based on 1995 figures.

1	New York, *USA*	19,670
2	Los Angeles, *USA*	15,048
=	Mexico City, *Mexico*	15,048
4	Bombay (Mumbai), *India*	12,572
5	Tokyo, *Japan*	11,927
6	Buenos Aires, *Brazil*	11,256
7	Calcutta, *India*	10,916
8	Seoul, *South Korea*	10,628
9	São Paulo, *Brazil*	9,480
10	Paris, *France*	9,319
11	Moscow, *Russia*	8,957
12	Shanghai, *China*	8,930
13	Chicago, *USA*	8,410
14	Jakarta, *Indonesia*	8,259
15	Delhi, *India*	7,207
16	Cairo, *Egypt*	6,800
17	Manila, *Philippines*	6,720
18	Beijing, *China*	6,690
19	Istanbul, *Turkey*	6,620
20	Lima–Callao, *Peru*	6,601

This population explosion has been caused partly by better medical care, which has reduced child mortality and increased the average life expectancy at birth throughout the world. But it has also created problems. In some developing countries, nearly half of the people are children. They make no contribution to the economy, but they require costly education and health services. In richer countries, the high proportion of retired people is also a strain on the economy.

By the late 20th century, for the first time in 10,000 years, the majority of people are no longer forced to rely on farming for their livelihood. Instead, nearly half of them live in cities where many of them enjoy a high standard of living. But rapid urbanization also creates problems, especially in the developing world, with the growth of slums and an increase in crime.

POPULATION BY CONTINENT

> The cartogram shows the populations of the continents in a diagrammatic way, with each square representing 1% of the world's population. For example, North America (coloured turquoise) is represented by five squares, which means that it contains about 5% of the world's population, while Asia (in blue), the most populous continent even excluding the Asian part of the former USSR, is represented by 56 squares. By contrast, Australasia (yellow) is represented by less than half of a square because it contains only 0.45% of the world's population.

WORLD DEMOGRAPHIC EXTREMES

Fastest growing population; average annual % growth (1992–2000)		Slowest growing population; average annual % growth (1992–2000)	
1 Nigeria	5.09	1 Kuwait	-1.39
2 Afghanistan	4.21	2 Ireland	-0.24
3 Ivory Coast	3.54	3 St Kitts & Nevis	-0.22
4 Oman	3.52	4 Bulgaria	-0.13
5 Syria	3.51	5 Latvia	-0.10

Youngest populations; % aged under 15 years		Oldest populations; % aged over 65 years	
1 Kenya	49.9	1 Sweden	18.1
2 Uganda	49.6	2 Norway	16.4
= Yemen	49.6	3 Denmark	15.4
4 Botswana	49.3	= United Kingdom	15.4
5 Tanzania	49.1	5 Austria	15.0

Highest urban populations; % of population living in urban areas		Lowest urban populations; % of population living in urban areas	
1 Singapore	100.0	1 Bhutan	5.3
2 Macau	99.0	2 Burundi	5.5
3 Belgium	96.9	3 Rwanda	7.7
4 Kuwait	95.6	4 Burkina Faso	9.0
5 Hong Kong	94.1	5 Nepal	9.6

Most male populations; number of men per 100 women		Most female populations; number of men per 100 women	
1 United Arab Emirates	206.7	1 Russia	90.0
2 Qatar	167.2	2 Austria	91.2
3 Bahrain	145.3	= Somalia	91.2
4 Kuwait	128.3	4 Germany	92.0
5 Saudi Arabia	119.1	5 Barbados	92.1

Languages & Religions

All people belong to one species, *Homo sapiens*, but within that species is a great diversity of cultures. Two of the main factors that give people an identity and sense of kinship with their neighbours are language and religion.

Definitions of languages vary and as a result estimates of the total number of languages in existence range from about 3,000 to 6,000. Many languages are spoken only by a small number of people. Papua New Guinea, for example, has only 4.2 million people but 869 languages.

The world's languages are grouped into families, of which the Indo-European is the largest. Indo-European languages are spoken in a zone stretching from

> *Religion is a major force in South-east Asia. About 94% of the people in Thailand are Buddhists, and more than 40% of men over the age of 20 spend some time, if only a few weeks, serving as Buddhist monks. Confucianism, Islam, Hinduism and Christianity are also practised in Thailand.*

THE WORLD'S LANGUAGES

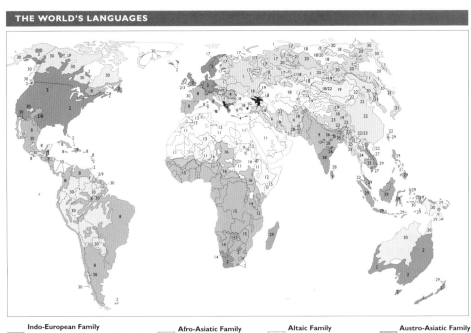

Indo-European Family

1	Balto-Slavic group (inc. Russia, Ukranian)
2	Germanic group (inc. English, German)
3	Celtic group
4	Greek
5	Albanian
6	Iranian group
7	Armenian
8	Romance group (inc. Spanish, Portuguese, French, Italian)
9	Indo-Aryan group (inc. Hindi, Bengali, Urdu, Punjabi, Marathi)
10	Caucasian Family

Afro-Asiatic Family

11	Semitic group (inc. Arabic)
12	Kushitic group
13	Berber group
14	Khoisan Family
15	Niger-Congo Family
16	Nilo-Saharan Family
17	Uralic Family

Altaic Family

18	Turkic group
19	Mongolian group
20	Tungus-Manchu group
21	Japanese & Korean

Sino-Tibetan Family

22	Sinitic (Chinese) languages
23	Tibetic-Burmic languages
24	Tai Family

Austro-Asiatic Family

25	Mon-Khmer group
26	Munda group
27	Vietnamese
28	Dravidian Family (inc. Telugu, Tamil)
29	Austronesian Family (inc. Malay-Indonesian)
30	Other Languages

NATIVE SPEAKERS

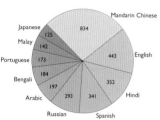

Mandarin Chinese 834
Japanese 125
Malay 142
Portuguese 173
Bengali 184
Arabic 197
Russian 293
Spanish 341
Hindi 352
English 443

> The chart shows the native speakers of major languages in millions. Mandarin Chinese is the language of 834 million, as compared with English, which has 443 million speakers. But many other people speak English as a second language because of its importance in international affairs and business.

Religious Adherents ▾	
The world's major religions, with the number of adherents in millions (latest available year)	
Christian	1,667
Roman Catholic	952
Protestant	337
Orthodox	162
Anglican	70
Other Christian	148
Muslim	881
Sunni	841
Shia	40
Hindu	663
Buddhist	312
Chinese folk	172
Ethnic/local	92
Jewish	18
Sikh	17

Europe, through south-western Asia into the Indian subcontinent. In addition, during the period of European coloniz-ation, they spread throughout North and South America and also to Australia and New Zealand. Today about two-fifths of the world's people speak an Indo-European language, as compared with one-fifth who speak a language belong-ing to the Sino-Tibetan language.

The Sino-Tibetan language family includes Chinese, which is spoken as a first language by more people than any other. English is the second most important first language, but it is more important than Chinese in international affairs and business, because so many people speak it as a second language.

RELIGIONS

Christianity is the religion of about a third of the world's population. Other major religions include Buddhism, Islam, Hinduism, Judaism, Chinese folk reli-gions and traditional tribal religions.

Religion is a powerful force in human society, establishing the ethics by which people live. It has inspired great music, painting, architecture and literature, yet at the same time religion and language have contributed to conflict between people throughout history. Even today, the cause of many of the conflicts around the world are partly the result of language and religious differences.

> Most languages have alphabetic systems of writing. The Greek alphabet uses some letters from the Roman alphabet, such as the A and B. Russians use the Cyrillic alphabet, which is based partly on Roman and partly on Greek letters. The Cyrillic alphabet is also used for Bulgarian. Serbs use either the Cyrillic or the Roman alphabet to write Serbo-Croat.

ALPHABETS

The Greek Alphabet

Α Β Γ Δ Ε Ζ Η Θ Ι Κ Λ Μ Ν Ξ Ο Π Ρ Σ Τ Υ Φ Χ Ψ Ω

A V/B G D E Z E TH I K L M N X O P R S T Y F CH PS O

The Cyrillic Alphabet

А Б В Г Д Е Ё Ж З И Й К Л М Н О П Р С Т У Ф Х Ц Ч Ш Щ Ю Я

A B V G D E YO ZH Z I Y K L M N O P R S T U F KH TS CH SH SHCH YU YA

Agriculture & Industry

BECAUSE IT SUPPLIES so many basic human needs, agriculture is the world's leading economic activity. But its relative importance varies from place to place. In most developing countries, agriculture employs more people than any other activity. For example, the diagram at the bottom of this page shows that more than 90% of the people of Nepal are employed in farming.

Many farmers in developing countries live at subsistence level, producing barely enough to supply the basic needs of their families. Alongside the subsistence sector, some developing countries produce one or two cash crops that they export. Dependence on cash crops is precarious: when world commodity prices fall, the country is plunged into financial crisis.

In developed countries, by contrast, the proportion of people engaged in agriculture has declined over the last 200

> The cultivation of rice, one of the world's most important foods, is still carried out by hand in many areas. But the introduction of new strains of rice has greatly increased yields.

years. Yet, by using farm machinery and scientific methods, notably the selective breeding of crops and animals, the production of food has soared. For example, although agriculture employs only 3% of its workers, the United States is one of the world's top food producers.

INDUSTRIALIZATION

The Industrial Revolution began in Britain in the late 18th century and soon spread to mainland Europe and other parts of the world. Industries first arose in areas with supplies of coal, iron ore and cheap water power. But later, after oil and gas came into use as industrial fuels, factories could be set up almost anywhere.

The growth of manufacturing led to an increase in the number of industrial cities. The flight from the land was accompanied by an increase in efficiency in agriculture. As a result, manufacturing replaced agriculture as the chief source of

EMPLOYMENT

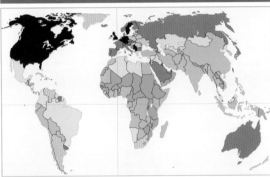

The number of workers employed in manufacturing for every 100 workers engaged in agriculture (latest available year)

■ Under 10	▢ 50 – 100	■ 200 – 500
▢ 10 – 50	▢ 100 – 200	■ Over 500

DIVISION OF EMPLOYMENT

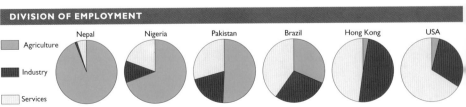

Agriculture

Industry

Services

Nepal   Nigeria   Pakistan   Brazil   Hong Kong   USA

18

PATTERNS OF PRODUCTION

> *The table shows how the economy breaks down (in terms of the Gross Domestic Product for 1993) in a selection of industrialized countries. Agriculture remains important in some countries, though its percentage share has steadily declined since the start of the Industrial Revolution. Industry, especially manufacturing, accounts for a higher proportion, but service industries account for the greatest percentage of the GDP in most developed nations. The figures for Manufacturing are shown separately from Industry because of their importance in the economy.*

Country	Agriculture	Industry (excl. manufacturing)	Manufacturing	Services
Australia	3%	14%	15%	67%
Austria	2%	9%	26%	62%
Denmark	4%	7%	20%	69%
Finland	5%	3%	28%	64%
France	3%	7%	22%	69%
Germany	1%	11%	27%	61%
Greece	18%	12%	20%	50%
Hong Kong	0%	8%	13%	79%
Hungary	6%	9%	19%	66%
Ireland	8%	7%	3%	82%
Italy	3%	7%	25%	65%
Japan	2%	17%	24%	57%
Kuwait	0%	46%	9%	45%
Mexico	8%	8%	20%	63%
Netherlands	4%	9%	19%	68%
Norway	3%	21%	14%	62%
Singapore	0%	9%	28%	63%
Sweden	2%	5%	26%	67%
UK	2%	8%	25%	65%
USA	3%	9%	25%	63%

income and employment in industrialized countries and rapidly widened the wealth gap between them and the poorer non-industrialized countries whose economies continued to rely on agriculture.

SERVICE INDUSTRIES

Eventually, the manufacturing sector became so efficient that it could supply most of the things that people wanted to buy. Trade between industrialized countries also increased, so widening the choice for consumers in the developed world. These factors led to a further change in the economies of developed countries, namely a reduction in the relative importance of manufacturing and the growth of the service sector.

Service industries include such activities as government, transport, insurance, finance, and even the writing of computer software. In the United States, service industries now account for about two-thirds of the Gross National Product (GNP), while in Japan they account for more than half. But the wealth of both countries still rests on their massive industrial production.

AGRICULTURE

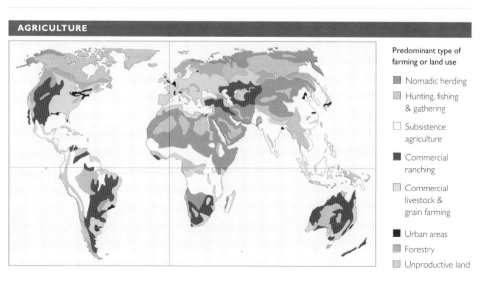

Predominant type of farming or land use

- Nomadic herding
- Hunting, fishing & gathering
- Subsistence agriculture
- Commercial ranching
- Commercial livestock & grain farming
- Urban areas
- Forestry
- Unproductive land

Trade & Commerce

Trade has always been an important human activity. It has widened the choice of goods available in any country, lowered prices and generally raised living standards. People regard any growth of world trade as a sign that the world economy is healthy, whereas a decline indicates a world recession.

Exports and imports are of two main kinds. Visible imports and exports include primary products, such as food and manufactures. Invisible imports and exports include services, such as banking, insurance, interest on loans, and money spent by tourists.

World trade, both visible and invisible, is dominated by the 25 members of the OECD (Organization for Economic Development), which includes the world's top trading nations, namely the United States, Germany, Japan, France, Italy, the United Kingdom and Canada, together with other European nations, as well as Australia, New Zealand and Mexico, which has close ties with the USA.

> The new port of the historic Italian city of Ravenna is linked to the Adriatic Sea by a canal. The port has large oil refining and petrochemical industries.

CHANGING EXPORTS

From the late 19th century to the 1950s, primary products, including farm products, minerals, natural fibres, timber and, in the latter part of this period, oil

DEBT AND AID

International debtors and the development aid they receive (1993)

The provision of aid by rich countries to developing countries is part of international politics. But the grants made to developing countries are often dwarfed by the burden of debt which the countries are expected to repay. In 1990, the debts of Mozambique, one of the world's poorest countries, were estimated to be 75 times its entire earnings from exports.

Debt, US$ per capita
Aid, US$ per capita

$4853
$279
2,750
2,500
2,250
2,000
1,750
1,500
1,250
1,000
750
500
250
0
50
100

India, Tanzania, Sierra Leone, Nigeria, Madagascar, Mozambique, Guinea Bissau, Laos, Honduras, Zambia, Egypt, Papua New Guinea, Mauritania, Ivory Coast, Jordan, Nicaragua, Jamaica, Ecuador, Congo, Panama, Israel

The World's Largest Businesses ▾

These are sales figures in billions of US$ and refer to the year ended 31 December 1992. They include sales of subsidiaries but exclude excise taxes collected by manufacturers.

1	General Motors, USA	132.8
2	Exxon, USA	103.5
3	Ford Motor, USA	100.8
4	Royal Dutch Shell, UK/Neths	98.9
5	Toyota Motor, Japan	79.1
6	IRI, Italy	67.5
7	IBM, USA	65.1
8	Daimler-Benz, Germany	63.3
9	General Electric, USA	62.2
10	Hitachi, Japan	61.5
11	British Petroleum, UK	59.2
12	Matsushita Electric Ind., Japan	57.5
13	Mobil, USA	57.4
14	Volkswagen, Germany	56.7
15	Siemens, Germany	51.4
16	Nissan Motor, Japan	50.2
=	Philip Morris, USA	50.2
18	Samsung, South Korea	49.6
19	Fiat, Italy	47.9
20	Unilever, UK/Neths	44.0

TRADED PRODUCTS

The character of world trade has greatly changed in the last 50 years. While primary products were once the leading commodities, world trade is now dominated by manufactured products. Cars are the single most valuable traded product, followed by vehicle parts and engines. The next most valuable goods are high-tech products such as data processing (computer) equipment, telecommunications equipment, and transistors. Other items include aircraft, paper and board, trucks, measuring and control instruments, and electrical machinery. Trade in most manufactured products is dominated by the OECD countries. For example, the leading car exporter is Japan, which became the world's leading car manufacturer in the 1980s. The United States, Germany, the United Kingdom, France and Japan lead in the production of data processing equipment.

and natural gas, dominated world trade.

Many developing countries still remain dependant on exporting mineral ores, fossil fuels, or farm products such as cocoa or coffee whose prices fluctuate according to demand. But today, manufactured goods are the most important commodities in world trade. The OECD nations lead the world in exporting manufactured goods, though they are being challenged by a group of nations in eastern Asia, notably Hong Kong, Singapore, South Korea and Taiwan. Other rapidly industrializing countries in Asia include Indonesia, Malaysia and Thailand. The generally cheap labour costs of these countries have enabled them to produce manufactured goods for export at prices lower than those charged for similar goods made in Western countries.

Private companies carry on most of the world's trade. The small proportion handled by governments decreased recently with the collapse of Communist regimes in eastern Europe and the former Soviet Union.

SHARE OF WORLD TRADE

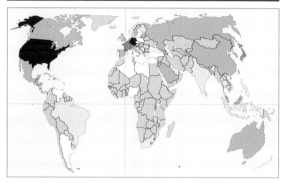

Percentage share of total world exports by value (1993)

■ Over 10%	▨ 1 – 5%	☐ 0.25 – 0.5%
■ 5 – 10%	▨ 0.5 – 1%	▨ Under 0.25%

DEPENDENCE ON TRADE

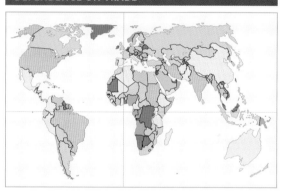

Value of exports as a percentage of Gross Domestic Product (1993)

■ Over 50% GDP	▨ 30 – 40% GDP	☐ 10 – 20% GDP
■ 40 – 50% GDP	▨ 20 – 30% GDP	▨ Under 10% GDP

Trade in Oil ▾

Major world trade in oil in millions of tonnes (1993)

Middle East (Saudi Arabia, Iran, UAE & Kuwait) to W. Europe	163	Nigeria to Western Europe	27
		Nigeria to USA	36
Middle East to Japan	126	Canada to USA	42
Middle East to Asia (not Japan)	119	Indonesia to Japan	18
Latin America (Mexico & Venezuela) to Western Europe	100	Latin America to W. Europe	24
		Western Europe (Norway & UK) to USA	21
Middle East to USA	82	
Russia to Western Europe	49	Middle East to Latin America	16
Libya to Western Europe	47	*Total world trade*	1,376

Transport & Travel

ABOUT 200 YEARS ago, most people never travelled far from their birthplace. But adventurous travellers can now reach almost any part of the world.

Transport is concerned with moving goods and people around by land, water and air. Land transport was once laborious, and was dependent on pack animals or animal-drawn vehicles. But during the Industrial Revolution, railways played a vital role in moving bulky materials and equipment required by factories. They were also important in the opening up and development of remote areas around the world in North and South America, Africa, Asia and Australia.

Today, however, motor vehicles have taken over many of the functions once served by railways. Unlike railways, motor vehicles provide a door-to-door service and, through the invention of heavy trucks, they can also carry large loads. In the mid-1990s, about 90% of inland freight in Britain was carried by road, while car and van travel accounted for 86% of passenger travel, as compared with 6% by buses and coaches, 5% by rail and less than 1% by air.

> *Traffic jams and vehicle pollution have affected cities throughout the world. Many of Bangkok's beautiful old canals have been filled in to provide extra roads to cope with the enormous volume of traffic in the city.*

TRAVEL & TOURISM

Sea transport, which now employs huge bulk grain carriers, oil tankers and container ships, still carries most of the world's trade. But since the late 1950s, fewer passengers have travelled overseas by sea, because air travel is so much faster, though many former ocean liners now operate successfully as cruise ships.

Air travel has played a major part in the rapid growth of the tourist industry,

AIR TRAVEL

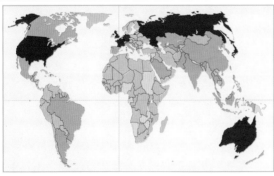

Number of passenger kilometres flown, in millions (1992). Passenger kilometres are the number of passengers (both international and domestic) multiplied by the distance flown by each passenger from airport of origin.

■ Over 100,000   ▨ 10,000 – 50,000   ▨ 500 – 1,000
■ 50,000 – 100,000   ▨ 1,000 – 10,000   ▨ Under 500

The Busiest International Airports ▾	
Number of international passengers, in thousands, (1993)	
1 Heathrow, *London*	40,848
2 Frankfurt/Main, *Frankfurt*	25,119
3 Hong Kong Intl., *Hong Kong*	24,421
4 Charles de Gaulle, *Paris*	23,336
5 Schiphol, *Amsterdam*	20,658
6 New Tokyo Intl., *Tokyo*	18,947
7 Changi, *Singapore*	18,796
8 Gatwick, *London*	16,656
9 Kennedy, *New York*	14,821
10 Bangkok Intl., *Bangkok*	12,789
11 Miami Intl., *Miami*	12,373
12 Zürich, *Zürich*	12,256
13 Los Angeles Intl., *Los Angeles*	11,945
14 Chiang Kai-Shek, *Taipei*	11,154
15 Manchester, *Manchester*	10,791

The Longest Rail Networks ▼

Extent of rail network, in thousands of kilometres, (1993)

1	USA	239.7
2	Russia	87.5
3	India	62.5
4	China	54.0
5	Germany	40.4
6	Australia	35.8
7	Argentina	34.2
8	France	32.6
9	Mexico	26.5
10	Poland	24.9

THE IMPORTANCE OF TOURISM

Nations receiving the most from tourism, millions of US$ (1993)

1	USA	53,861
2	France	25,000
3	Spain	22,181
4	Italy	21,577
5	UK	13,683
6	Austria	13,250
7	Germany	10,982
8	Switzerland	7,650
9	Hong Kong	6,037
10	Mexico	5,997

Nations spending the most on tourism, millions of US$ (1993)

1	USA	41,260
2	Germany	37,514
3	Japan	26,860
4	UK	17,244
5	Italy	13,053
6	France	12,805
7	Canada	10,629
8	Netherlands	8,974
9	Austria	8,180
10	Taiwan	7,585

Number of tourist arrivals, millions (1993)

1	France	60,100
2	USA	45,793
3	Spain	40,085
4	Italy	26,379
5	Hungary	22,804
6	UK	19,186
7	China	18,982
8	Austria	18,257
9	Poland	17,000
10	Mexico	16,534

Fastest growing tourist destinations, % change in receipts (1994–95)

1	South Korea	49%
2	Czech Republic	27%
3	India	21%
4	Russia	19%
5	Philippines	18%
6	Turkey	17%
7	Thailand	15%
8	Poland	13%
9	China	12%
10	Israel	12%

which accounted for 7.5% of world trade by the mid-1990s. Travel and tourism have greatly increased people's understanding and knowledge of the world, especially in the OECD countries, which account for about 7% of world tourism.

Some developing countries have large tourist industries which have provided employment and led to improvements in roads and other facilities. In some cases, tourism plays a vital role in the economy. For example, in Kenya, tourism provides more income than any other activity apart from the production and sale of coffee. However, too many tourists can damage fragile environments, such as the wildlife and scenery in national parks. Tourism can also harm local cultures.

THE WORLD'S VEHICLES

Proportion of the world's vehicles by region (1994)

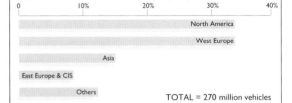

TOTAL = 270 million vehicles

CAR OWNERSHIP

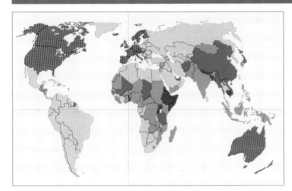

Number of people per car (latest available year)

■ Over 1,000
■ 500 – 1,000
▨ 100 – 500
☐ 25 – 100
▨ 5 – 25
▨ Under 5

Two-thirds of the world's vehicles are found in the developed countries of Europe and North America. Car ownership is also high in Australia and New Zealand, as well as in Japan, the world's leading car exporter. Car transport is the most convenient form of passenger travel, but air pollution caused by exhaust fumes is a serious problem in many large cities.

International Organizations

IN THE LATE 1980s, people rejoiced at the collapse of Communist regimes in eastern Europe and the former Soviet Union, because this brought to an end the Cold War, a long period of hostility between East and West. But hope of a new era of peace was shattered when ethnic and religious rivalries led to civil war in Yugoslavia and in parts of the former Soviet Union.

In order to help maintain peace, many governments have formed international organizations to increase co-operation. Some, such as NATO (North Atlantic

> In the early 1990s, the United Nations peacekeeping mission worked to end the civil war in Bosnia-Herzegovina and also to bring aid to civilians affected by the fighting.

UN Contributions ▾

In 1994, the top ten contributing countries to the UN budget, which was US$2,749 million, were as follows:

1	USA	25.0%
2	Japan	12.5%
3	Germany	8.9%
4	Russia	6.7%
5	France	6.0%
6	UK	5.0%
7	Italy	4.3%
8	Canada	3.1%
9	Spain	2.0%
10	Brazil	1.6%

Treaty Organization), are defence alliances, while others aim to encourage economic and social co-operation. Some organizations, such as the Red Cross, are non-governmental organizations, or NGOs.

UNITED NATIONS

The United Nations, the chief international organization, was formed in October 1945 and now has 185 member countries. The only independent nations that are not members are Kiribati, Nauru, Switzerland, Taiwan, Tonga, Tuvalu and the Vatican City.

THE UNITED NATIONS

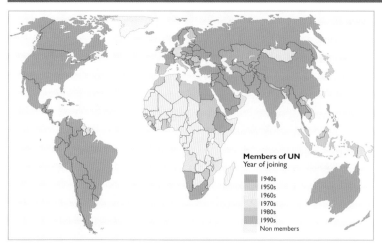

Members of UN
Year of joining

- 1940s
- 1950s
- 1960s
- 1970s
- 1980s
- 1990s
- Non members

> The membership of the UN has risen from 51 in 1945 to 185 in 1995. The first big period of expansion came in the 1960s when many former colonies achieved their independence. The membership again expanded rapidly in the 1990s when new countries were formed from the former Soviet Union and Yugoslavia. The most recent addition, Palau, is a former US trust territory in the Pacific Ocean and joined in 1994.

The United Nations was formed at the end of World War II to promote peace, international co-operation and security, and to help solve economic, social, cultural and humanitarian problems. It promotes human rights and freedom and is a forum for negotiations between nations.

The main organs of the UN are the General Assembly, the Security Council, the Economic and Social Council, the Trusteeship Council, the International Court of Justice and the the Secretariat.

The UN also operates 14 specialized agencies concerned with particular issues, such as agriculture, education, working conditions, communications and health. For example, UNICEF (the United Nations International Children's Fund), established in 1946 to deliver post-war relief to children, now aims to provide basic health care to children and mothers worldwide. The ILO (International Labour Organization) seeks to improve working conditions, while the FAO (Food and Agricultural Organization) aims at improving the production and distribution of food. The WTO (World Trade Organization) was set up as recently as January 1995 to succeed GATT (General Agreements on Tariffs and Trade).

THE UNITED NATIONS

THE GENERAL ASSEMBLY is the meeting of all member nations every September under a newly-elected president to discuss issues affecting development, peace and security.

THE SECURITY COUNCIL has 15 members, of which five are permanent. It is responsible for maintaining international peace.

THE SECRETARIAT consists of the staff and employees of the UN, including the Secretary-General (appointed for a five-year term), who is the UN's chief administrator.

THE ECONOMIC & SOCIAL COUNCIL works with the specialized agencies to implement UN policies on improving living standards, health, cultural and educational co-operation.

THE TRUSTEESHIP COUNCIL was designed to bring several dependencies to independence. This work is now complete.

THE INTERNATIONAL COURT OF JUSTICE, or World Court, deals with legal problems and helps to settle disputes. Its headquarters are at The Hague, in the Netherlands.

UN DEPARTMENTS

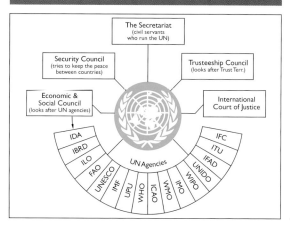

UN PEACEKEEPING MISSIONS

The United Nations tries to resolve international disputes in several ways. It sends unarmed observer missions to monitor cease-fires or supervise troop withdrawals, and the Security Council members also send peacekeeping forces.

This first of these forces was sent in 1948 to supervise the cease-fire between Arabs and Jews in disputed parts of Palestine and, since then, it has undertaken more than 30 other missions. The 'Blue Berets', as the UN troops are called, must be impartial in any dispute and they can fire only in self-defence. Hence, they can operate only with the support of both sides, which leaves them open to criticism when they are unable to prevent violence by intervening.

By the mid-1960s, the UN was involved in 15 world conflicts, was policing the boundary in partitioned Cyprus, and was seeking to enforce a peace agreement in Angola after 20 years of civil war. Other missions were in Burundi and Rwanda, El Salvador, Georgia, the Israeli-occupied Golan Heights, Haiti, Kuwait, southern Lebanon, Liberia, Mozambique, Western Sahara and the former Yugoslavia. A force known as UNPROFOR (UN Protection Force) had been operating in Bosnia-Herzegovina and, by 1995, it accounted for 60% of the total UN peacekeeping budget of US$3 billion. In 1995, under a peace agreement, the UN approved the setting up of a new force (IFOR), composed of NATO and non-NATO members, to replace UNPROFOR, though small UN forces remained in Croatia and Macedonia.

ECONOMIC ORGANIZATIONS

Over the last 40 years, many countries have joined common markets aimed at eliminating trade barriers and encouraging the free movement of workers and capital.

The best known of these is the European Union. Other organizations include ASEAN (the Association of South-east Asian Nations), which aims at reducing trade barriers between its seven members: Brunei, Indonesia, Malaysia, the Philippines, Singapore, Thailand and Vietnam.

APEC (the Asia-Pacific Co-operation Group) was founded in 1989 and in-

> The European Parliament, one of the branches of the EU, consists of 626 members. The number of members for each country is based mainly on population. It is a mainly advisory body and meets in Strasbourg, France.

cludes the countries of East and Southeast Asia, with the United States, Canada, Australia and New Zealand. APEC aims to create a free trade zone by 2020.

Together the United States, Canada and Mexico form NAFTA (North American Free Trade Agreement), which aims at eliminating trade barriers within 15 years of its foundation on 1 January 1994. Other economic groupings link the countries of Latin America.

Another economic group with more limited aims is OPEC (Organization of Petroleum Exporting Countries). It works to unify policies concerned with the sale of petroleum on world markets.

The central aim of the Colombo Plan is to provide economic development assistance for South and South-east Asia.

OTHER ORGANIZATIONS

Some organizations exist for consultation on matters of common interest. The Commonwealth of Nations grew out of the links created by the British Empire, while the OAS (Organization of American States) works to increase understanding throughout the Western Hemisphere. The OAU (Organization of

THE EUROPEAN UNION

At the end of World War II (1939–45), many Europeans wanted to end the ancient emnities that had caused such destruction and rebuild the shattered continent. It was in this mood that Belgium, France, West Germany, Italy, Luxembourg and the Netherlands signed the Treaty of Paris in 1951. This set up the European Coal and Steel Community (ECSC), the forerunner of the European Union.

In 1957, through the Treaty of Rome, the same six countries created the European Economic Community (EEC) and the European Atomic Community (EURATOM). In 1967, the ECSC, the EEC and EURATOM merged to form the

single European Community (EC).

Another economic group, the European Free Trade Association (EFTA), was set up in 1960 by seven countries: Austria, Denmark, Norway, Portugal, Sweden, Switzerland, and the United Kingdom. However, Denmark, Ireland and the UK left to become members of the EC in 1973, followed by Greece in 1981, Spain and Portugal in 1986, and Austria, Finland and Sweden in 1995. The expansion of the EC to 15 members left EFTA with just four members: Iceland, Liechtenstein, Norway and Switzerland.

In 1993, following the signing of the Maastricht Treaty, the EC was reconstituted

as the European Union (EU). The aims of the EU include economic and monetary union, a single currency for all 15 countries, and closer co-operation on foreign and security policies and also on home affairs. This step has led to a debate. Some people would like the EU to develop into a federal Europe, but others fear that this would lead to a loss of national identity. Another matter of importance is the future enlargement of the EU. By 1995, formal applications for membership had been received from Turkey, Malta, Cyprus, Poland, Hungary, Slovakia and Romania. Other possible members include the Czech Republic, Estonia, Latvia and Lithuania.

AUSTRALIA'S NEW ROLE

Most of the people who settled in Australia between 1788 and the mid-20th century came from the British Isles. However, the strong ties between Australia and Britain were weakened after Britain joined the European Community in 1973. Since 1973, many Australians have argued that their world position has changed and that they are part of a Pacific community of nations, rather than an extension of Europe. Some want closer integration with ASEAN, the increasingly powerful economic group formed by seven South-east Asian nations. But in 1995, the prime minister of Malaysia, Dr Mahathir Mohamad, argued that Australia could not be regarded as Asian until at least 70% of its people were of ethnic Asian origin.

African Unity) has a similar role in Africa, while the Arab League is made up of Arabic-speaking North African and Middle Eastern states. The recently formed CIS (Commonwealth of Independent States) aims at maintaining links between 12 of the 15 republics which made up the Soviet Union.

NORTH–SOUTH DIVIDE

The deepest division in the world today is the divide between rich and poor nations. In international terms, this is called the North–South divide, because the North contains most of the world's developed countries, while the developing countries lie mainly in the South. The European Union recognizes this division and gives special trading terms to more than 60 former European dependencies, which form the ACP (African, Caribbean and Pacific) states. One organization containing a majority of developing countries is the Non-Aligned Movement. This Movement was created in 1961 during the Cold War as a political bloc allied neither to the East nor to the West. However, the aims of the 113 members who attended the movement's 11th gathering in 1995 were concerned mainly with economic matters. The 113 countries between them produce only about 7% of the world's gross output and they can speak for the poorer South.

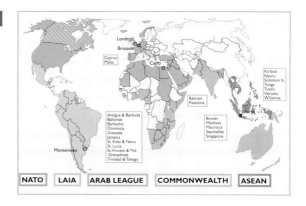

NATO   LAIA   ARAB LEAGUE   COMMONWEALTH   ASEAN

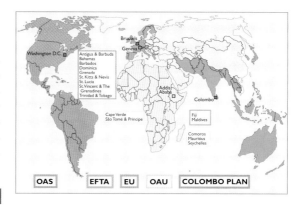

OAS   EFTA   EU   OAU   COLOMBO PLAN

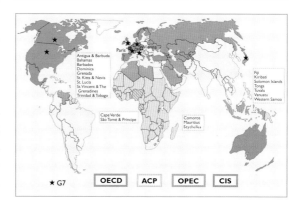

★ G7   OECD   ACP   OPEC   CIS

> The maps show the membership of major international organizations. One important grouping shown on the bottom map is the Group of Seven (often called G7), which was set up on 22 September 1985. This group of seven major industrial democracies (Canada, France, Germany, Italy, Japan, the United Kingdom and the United States) holds periodic meetings to discuss major problems, such as world recessions.

Regions in the News

> The hoped-for era of peace following the end of the Cold War in Europe in the early 1990s was not to be. Former Yugoslavia, a federation of six republics ruled by a Communist government between 1946 and 1991, became a 'region in the news' when it split apart in 1991. First, Croatia, Slovenia and Macedonia declared themselves independent nations, followed by Bosnia-Herzegovina in 1992. This left two states, Serbia and Montenegro, to continue as Yugoslavia. The presence in Croatia and Bosnia-Herzegovina of Orthodox Christian Serbs, Roman Catholic Croats, and Muslims proved an explosive mixture. Fighting broke out first in Croatia and then in Bosnia-Herzegovina. Following a bitter civil war, accompanied by 'ethnic cleansing' (the slaughter and expulsion of rival ethnic groups), the signing of a peace agreement (the Dayton Peace Accord) in 1995 ended the war and affirmed Bosnia-Herzegovina as a single state with its capital at Sarajevo. But the new country is partitioned into a Muslim–Croat Federation (51% of the country) and a Serbian Republic (49%).

Population breakdown ▾

Population totals and the proportion of ethnic groups (1995)

Yugoslavia	**10,881,000**
Serb 63%, Albanian 17%, Montenegrin 5%, Hungarian 3%, Muslim 3%	
Serbia	6,017,200
Kosovo	2,045,600
Vojvodina	2,121,800
Montenegro	696,400
Bosnia-Herzegovina	**4,400,000**
Muslim 49%, Serb 31%, Croat 17%	
Croatia	**4,900,000**
Croat 78%, Serb 12%	
Slovenia	**2,000,000**
Slovene 88%, Croat 3%, Serb 2%	
Macedonia (F.Y.R.O.M.)	**2,173,000**
Macedonian 64%, Albanian 22%, Turkish 5%, Romanian 3%, Serb 2%	

Legend:
- –·–·– International borders
- –··–··– Republic boundaries
- – – – – Province boundaries
- ——— Line of the Dayton Peace Accord
- Muslim–Croat Federation
- Serbian Republic

> Since its establishment in 1948, the State of Israel has seldom been out of the news. During wars with its Arab neighbours in 1948–49, 1956, 1967 and 1973, it occupied several areas. The largest of the occupied territories, the Sinai peninsula, was returned to Egypt in 1979 following the signing of an Egyptian–Israeli peace treaty. This left three Israeli-occupied territories: the Gaza Strip, the West Bank bordering Jordan, and the Golan Heights, a militarily strategic area overlooking south-western Syria.

Despite the peace agreement with Egypt, conflict continued in Israel with the PLO (Palestine Liberation Organization), which claimed to represent Arabs in Israel and Palestinians living in exile. Finally, on 13 September 1993 Israel officially recognized the PLO, and Yasser Arafat, leader of the PLO, renounced terrorism and recognized the State of Israel. This led to an agreement signed by both sides in Washington, DC. In May 1994, limited Palestinian self-rule was established in the Gaza Strip and in parts of the occupied West Bank. A Palestinian National Authority (PNA) was created and took over from the Israeli military administration when Israeli troops withdrew from the Gaza Strip and the city of Jericho. On 1 July 1994 the Palestinian leader, Yasser Arafat, stepped on to Palestinian land for the first time in 25 years.

Many people hoped that these developments would eventually lead to the creation of a Palestinian state, which would co-exist in peace with its neighbour Israel. But groups on both sides sought to undermine the peace process. In November 1995, a right-wing Jewish student assassinated the Israeli prime minister, Yitzhak Rabin, and then, in early 1996, Hamas, a Muslim group which aims at the overthrow of Israel and the creation of a Palestinian state in its place, launched a series of suicide bomb attacks in Israeli cities, killing many civilians and injuring hundreds more.

Meanwhile, Israel continues its long negotiations with Syria aimed at finding a resolution to the problem of the Golan Heights. This would hopefully form part of a general peace agreement between Israel and its Arab neighbours.

Population breakdown ▾

Population totals and the proportion of ethnic groups (1995)

Israel **5,696,000**
Jewish 82%, Arab Muslim 14%, Arab Christian 3%, Druse 2%

West Bank 973,500
Palestinian Arab 97% (Arab Muslim 85%, Christian 8%, Jewish 7%)

Gaza Strip 658,200
Arab Muslim 98%

Jordan **5,547,000**
Arab 99% (Palestinian Arab 50%)

Syria **14,614,000**
Arab 89%, Kurdish 6%

THE NEAR EAST

— · — · — 1949 Armistice Line

— — — — 1974 Cease-fire Lines (Golan Heights)

Efrata
● Main Jewish settlements in the West Bank and Gaza Strip

Halhul
☐ Main Palestinian Arab towns in the West Bank and Gaza Strip – under Palestinian control since May 1994 (Gaza and Jericho) and 28 September 1995 (West Bank)

World Flags

Afghanistan	Albania	Algeria	Angola	Argentina
Armenia	Australia	Austria	Azerbaijan	Bahamas
Bahrain	Bangladesh	Belarus	Belgium	Benin
Bhutan	Bolivia	Bosnia-Herzegovina	Botswana	Brazil
Bulgaria	Burkina Faso	Burma (Myanmar)	Burundi	Cambodia
Cameroon	Canada	Central African Rep.	Chad	Chile
China	Colombia	Congo	Costa Rica	Croatia
Cuba	Cyprus	Czech Republic	Denmark	Djibouti
Dominican Republic	Ecuador	Egypt	El Salvador	Equatorial Guinea
Eritrea	Estonia	Ethiopia	Finland	France
Gabon	Georgia	Germany	Ghana	Greece

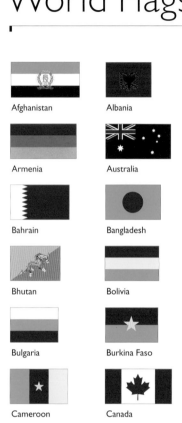

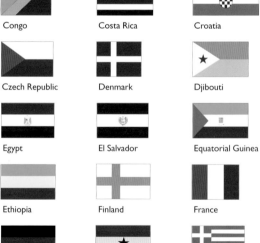

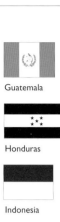 Guatemala

 Guinea

Guinea–Bissau

Guyana

Haiti

 Honduras

 Hong Kong

Hungary

Iceland

India

Indonesia

 Iran

Iraq

Ireland

Israel

Italy

Ivory Coast

Jamaica

Japan

Jordan

Kazakstan

Kenya

Korea, North

Korea, South

Kuwait

Kyrgyzstan

Laos

Latvia

Lebanon

Lesotho

Liberia

Libya

Liechtenstein

Lithuania

Luxembourg

Macedonia

Madagascar

Malawi

Malaysia

Mali

Malta

Mauritania

Mexico

Moldova

Mongolia

Morocco

Mozambique

Namibia

Nepal

Netherlands

New Zealand

Nicaragua

Niger

Nigeria

Norway

Oman
 Pakistan
 Panama
 Papua New Guinea
 Paraguay

 Peru
 Philippines
 Poland
 Portugal
 Puerto Rico

 Qatar
 Romania
 Russia
 Rwanda
 São Tomé & Príncipe

 Saudi Arabia
 Senegal
 Sierra Leone
 Singapore
 Slovak Republic

 Slovenia
 Somalia
 South Africa
 Spain
 Sri Lanka

 Sudan
 Surinam
 Swaziland
 Sweden
 Switzerland

 Syria
 Taiwan
 Tajikistan
 Tanzania
 Thailand

 Togo
 Trinidad & Tobago
 Tunisia
 Turkey
 Turkmenistan

 Uganda
 Ukraine
 UAE
 United Kingdom
 USA

 Uruguay
 Uzbekistan
 Vatican City
 Venezuela
 Vietnam

 Yemen
 Yugoslavia
 Zaïre
 Zambia
 Zimbabwe

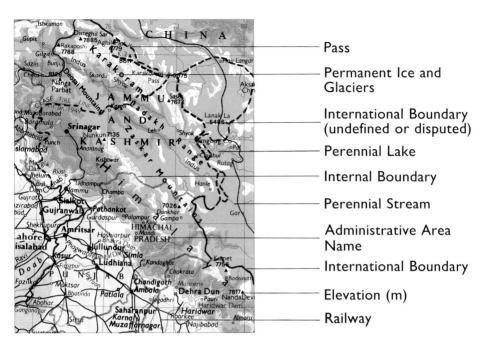

Pass

Permanent Ice and Glaciers

International Boundary (undefined or disputed)

Perennial Lake

Internal Boundary

Perennial Stream

Administrative Area Name

International Boundary

Elevation (m)

Railway

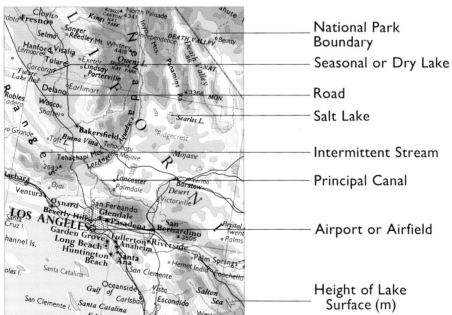

National Park Boundary

Seasonal or Dry Lake

Road

Salt Lake

Intermittent Stream

Principal Canal

Airport or Airfield

Height of Lake Surface (m)

Settlements

Settlement symbols and type styles vary according to the scale of each map and indicate the importance of towns rather than specific population figures.

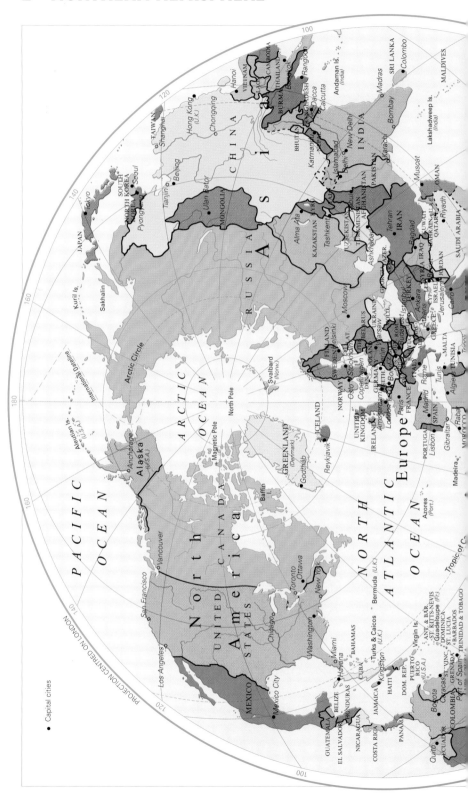

• Capital cities

PROJECTION CENTRED ON LONDON

PROJECTION CENTRED ON CAPE TOWN

+8.00
+7.00
+8.00
+5.30
+5.00
+4.30
+5.00
+6.00
+3.30
+4.00
+3.00
+2.00
+2.00
+1.00
+1.00
0.00
South Pole
-3.00
-3.00
-4.00
-3.30
-5.00

East from Greenwich

West from Greenwich

Cape Town

Greenwich

INDIAN OCEAN

Chagos Arch.
(U.K.)

SEYCHELLES

MAURITIUS

Réunion
(Fr.)
Antananarivo

MADAGASCAR

DJIBOUTI
SOMALIA
Mogadishu
ETHIOPIA
Addis Ababa
Africa
ERITREA
Asma
Khartoum
SUDAN
KENYA
Nairobi
Dar es Salaam
UGANDA
Kampala
TANZANIA
RWANDA
BURUNDI
ZAIRE
Kinshasa
Brazzaville
CAMEROON
Bangui
CENTRAL
AFRICA
Yaoundé
GABON
CONGO
Libreville
MALAWI
Lilongwe
ZAMBIA
Lusaka
Harare
ZIMBABWE
MOZAMBIQUE
Maputo
Pretoria
SWAZILAND
Johannesburg
SOUTH AFRICA
BOTSWANA
Gaborone
LESOTHO
NAMIBIA
Windhoek
ANGOLA
Luanda
Cape Town
COMOROS
Mayotte
(Fr.)

CHAD
NIGER
MALI
MAURITANIA
WESTERN
SAHARA
NIGERIA
Abuja
Lagos
BENIN
TOGO
GHANA
Accra
IVORY
COAST
Abidjan
BURKINA
FASO
Ouagadougou
LIBERIA
Monrovia
SIERRA LEONE
Freetown
GUINEA
Conakry
GUINEA BISSAU
Bissau
GAMBIA
Banjul
SENEGAL
Dakar
Nouakchott
CAPE VERDE
IS.
EQUAT.
GUINEA
SÃO TOMÉ & P.
Niamey

St. Helena
(U.K.)
Ascension
(U.K.)
Tropic of Capricorn

East from Greenwich

West from Greenwich

Equator

SOUTH ATLANTIC OCEAN

South America
GUYANA
Paramaribo
SURINAM
FRENCH GUIANA
BRAZIL
Brasília
São Paulo
Rio de Janeiro
BOLIVIA

TIME ZONES

Zones using Greenwich Mean Time

Zones fast of Greenwich Mean Time

Zones slow of Greenwich Mean Time

Standard Time not the Zone hour

No Official Time

CARTOGRAPHY BY PHILIP'S. COPYRIGHT REED INTERNATIONAL BOOKS LTD

Projection: Oblique Azimuthal Equidistant

PROJECTION CENTRED ON SAN FRANCISCO

West from Greenwich
-3.00
-3.30
-4.00
-3.00
-5.00
-4.00
-3.30
0.00
Greenwich
+1.00
+2.00
+1.00
-3.00
-4.00
-5.00
-5.00
-6.00
North Pole
-6.00
-7.00
-8.00
-9.00
+6.00
+5.00
+7.00
+8.00
+9.00
+11.00
+2.00
-10.00
-8.00
-9.00
+11.00
+10.00
+2.00
-9.00
-10.00

San Francisco

International Date Line

East from Greenwich

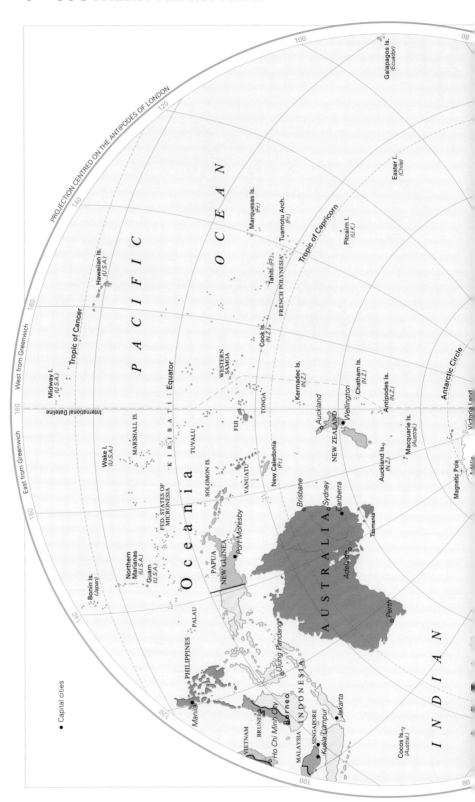

PROJECTION CENTRED ON THE ANTIPODES OF LONDON

Galapagos Is.
(Ecuador)

Easter I.
(Chile)

Tropic of Capricorn

P A C I F I C   O C E A N

Marquesas Is.
(Fr.)

Tuamotu Arch.
(Fr.)

Pitcairn I.
(U.K.)

Tahiti (Fr.)

FRENCH POLYNESIA

Cook Is.
(N.Z.)

Hawaiian Is.
(U.S.A.)

Tropic of Cancer

West from Greenwich

Antarctic Circle

Midway I.
(U.S.A.)

Equator

WESTERN
SAMOA

Kermadec Is.
(N.Z.)

Chatham Is.
(N.Z.)

International Dateline

TONGA

Auckland

Wellington

Antipodes Is.
(N.Z.)

Victoria Land

East from Greenwich

Wake I.
(U.S.A.)

MARSHALL IS

K I R I B A T I

TUVALU

FIJI

NEW ZEALAND

Macquarie Is.
(Austral.)

Magnetic Pole

O c e a n i a

SOLOMON IS.

VANUATU

New Caledonia
(Fr.)

Auckland Is.
(N.Z.)

Bonin Is.
(Japan)

Northern
Marianas
(U.S.A.)

Guam
(U.S.A.)

FED. STATES OF
MICRONESIA

PAPUA
NEW GUINEA

Port Moresby

Brisbane

Sydney

Canberra

Tasmania

PHILIPPINES

PALAU

A U S T R A L I A

Adelaide

Perth

I N D I A N

Manila

VIETNAM

BRUNEI

Borneo

Ujung Pandang

I N D O N E S I A

Jakarta

Ho Chi Minh City

MALAYSIA

SINGAPORE

Kuala Lumpur

Cocos Is.
(Austral.)

• Capital cities

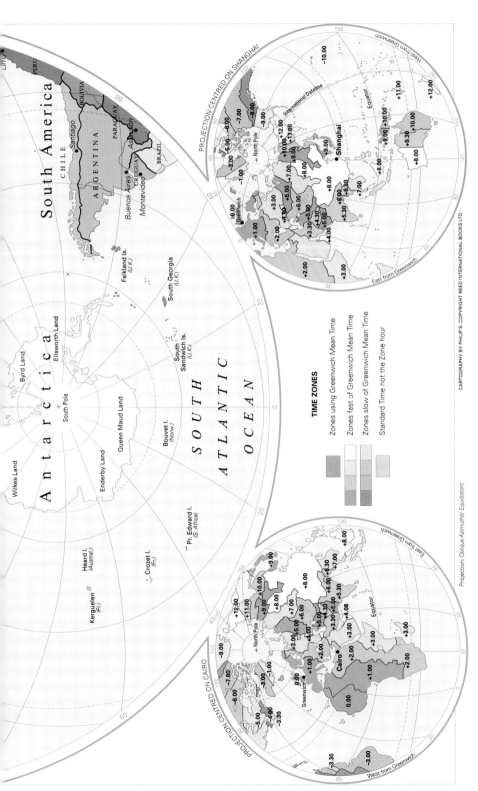

PROJECTION CENTRED ON SHANGHAI

West from Greenwich

East from Greenwich

International Dateline

North Pole

Shanghai

Greenwich

Equator

South America

CHILE

BOLIVIA

PARAGUAY

ARGENTINA

URUGUAY

Asunción

BRAZIL

Santiago

Buenos Aires

Montevideo

Falkland Is.
(U.K.)

South Georgia
(U.K.)

PERU

Antarctica

Byrd Land

Ellsworth Land

Wilkes Land

South Pole

Queen Maud Land

Enderby Land

Bouvet I.
(Norw.)

SOUTH

ATLANTIC

OCEAN

South
Sandwich Is.
(U.K.)

Pr. Edward I.
(S. Africa)

Crozet I.
(Fr.)

Heard I.
(Austral.)

Kerguelen
(Fr.)

TIME ZONES

Zones using Greenwich Mean Time

Zones fast of Greenwich Mean Time

Zones slow of Greenwich Mean Time

Standard Time not the Zone hour

PROJECTION CENTRED ON CAIRO

East from Greenwich

West from Greenwich

North Pole

Cairo

Greenwich

Equator

Projection: Oblique Azimuthal Equidistant

CARTOGRAPHY BY PHILIP'S. COPYRIGHT REED INTERNATIONAL BOOKS LTD

30 25 1 20 2 15 3 10 4 5 5 0 6 5 7 10 8 15 9

C

Arctic Circle

ICELAND

Reykjavik

Norwegian

Tromsø

60

Sea

Nor

D

Faroe Is.
(Den.)

SWEDEN

Trondheim

Shetland
Is.

NORWAY

Bergen

Oslo

Gävle

55

Orebro

Uppsala

ATLANTIC

UNITED
KINGDOM

Hebrides

Orkney
Is.

Stavanger

Vänern

Vättern
Jönköping

Baltic

E

SCOTLAND

Aberdeen

Skagerrak

Kattegat

Gothenburg

Gotlan

Glasgow

Dundee

Ålborg

DENMARK

N.
IRELAND

Edinburgh

North

Århus

Copenhagen

Malmö

Belfast

IRELAND

Newcastle-
upon-Tyne

Sea

Kiel

Gdańsk

Dublin

Manchester

Leeds

Hamburg

Szczecin

Bydgoszcz

POL

50

Cork

Liverpool

Sheffield

Bremen

Elbe

Berlin

Hannover

Magdeburg

Poznań

WALES

Birmingham

Amsterdam

NETHER-

Dortmund

GERMANY

Oder

Łódź

Cardiff

ENGLAND

The Hague

Rotterdam

LANDS

Essen

Halle

Leipzig

Dresden

Wrocław

Bristol

LONDON

Antwerp

Cologne

Bonn

Chemnitz

Katowice

F

OCEAN

Plymouth

Southampton

BELGIUM

Brussels

Wiesbaden

Frankfurt
am Main

Prague

Ostrava

English Channel

Lille

LU

CZECH REP.

Le Havre

Rouen

Maas

Luxembourg

Nuremberg

Channel Is.
(U.K.)

Seine

PARIS

Strasbourg

Rhine

Stuttgart

Munich

Vienna

SLC

Brest

Nantes

Loire

FRANCE

Dijon

Linz

Salzburg

AUSTRIA

Bratislava

Bay of
Biscay

Limoges

Lyons

St-Étienne

Zürich

Bern

LIECH.

Vaduz

Innsbruck

Graz

HUNG

La Coruña

Bordeaux

SWITZERLAND

Geneva

Milan

SLOVENIA

Ljubljana

Vigo

Garonne

Toulouse

Rhône

Grenoble

Venice

Zagreb

Trieste

CROATIA

Porto

Douro

Bilbao

Nice

Turin

Genoa

Bologna

BOSNIA-
HERZ.

40

Valladolid

Ebro

ANDORRA

Andorra-
la-Vella

Marseilles

MONACO

Florence

SAN
MARINO

Tiber

Split

Sarajevo

PORTUGAL

Zaragoza

Toulon

Corsica

Adriatic

YU

Tagus

Madrid

Barcelona

Ajaccio

ITALY

MON
NEGR

Lisbon

SPAIN

Valencia

Is.

Rome

Sea

Guadiana

Balearic

Palma

Minorca

Naples

Bari

Tirar

Seville

Córdoba

Murcia

Alicante

Ibiza

Majorca

Sardinia

Taranto

ALE

35

Cádiz

Málaga

Granada

Tyrrhenian

Co

Str. of Gibraltar

Gibraltar (U.K.)

Sea

Cagliari

Tangier

Ceuta (Sp.)

Mediterranean

Palermo

Messina

Ionian

Melilla (Sp.)

Algiers

Sicily

Catania

Sea

J

Africa

ALGERIA

Annaba

Sea

MOROCCO

Constantine

TUNISIA

Tunis

Pantelleria
(Italy)

MALTA

Valletta

1: 20 000 000

100   0   100   200   300   400   500 miles
100   0   200   400   600   800 km

10   11   12   13   14   15   16   17   18   19

Hammerfest

C

Murmansk

Ob

60

White
Sea

D

Luleå

Arkhangelsk

N. Dvina

Kotlas

Nizhniy Tagil

FINLAND

L. Onega

Perm

Yekaterinburg

55

Vaasa

Kirov

Chelyabinsk

Vyborg L.
Ladoga

Vologda

R   U   S   S   I   A

mpere

Turku

Helsinki

ST. PETERSBURG

Rybinsk
Res.

Kostroma

Kazan

Ufa

E

olm

Tallinn

ESTONIA

L.
Chudskoye

Yaroslavl

Ivanovo

Nizhniy
Novgorod

Magnitogorsk

Sea

L A T V I A

Riga

MOSCOW

Simbirsk

Samara

Orenburg

W. Dvina

Penza

E

LITHUANIA

Vitebsk

Smolensk

Tula

Volga

Uralsk

50

Kaunas

Vilnius

Mogilev

Kaliningrad

Minsk

Orel

Tambov

Saratov

Ural

K A Z A K S T A N

F

Białystok

B E L A R U S

Gomel

Kursk

Voronezh

N D

Brest

Pripet

Volgograd

Atyraū

Warsaw

Chernigov

Lubin

Zhitomir

Kiev

Dnieper

Kharkov

Don

Astrakhan

45

aków

L v o v

U   K   R   A   I   N   E

Dnepropetrovsk

Donetsk

Caspian
Sea

Dniester

Bug

Krivoy Rog

Zaporozhye

Taganrog

Rostov

REP

Miskolc

MOLDOVA

Nikolayev

Kherson

Stavropol

G

est

Debrecen

Cluj-Napoca

Kishinev

Odessa

Makhachkala

Y

Brasov

Galati

Crimea

Krasnodar

R O M A N I A

Timisoara

Ploiesti

Sevastopol

GEORGIA

Tbilisi

AZERBAIJAN

Baku

elgrade

Bucharest

Constanta

B l a c k   S e a

40

ERBIA

Danube

Varna

ARMENIA

Yerevan

Niš

SLAVIA

Sofia

B U L G A R I A

Bosporus

Samsun

Erzurum

Araks

Tabriz

H

Skopje

Plovdiv

ISTANBUL

Y

MACEDONIA

Thessaloniki

Bursa

Ankara

Kayseri

Diyarbakir

a

IRAN

AI

GREECE

Æ
g
e
a
n

T

U

R

K

E

Diyarbakir

35

Pátrai

İzmir

Konya

A

Euphrates

IRAQ

Athens

Adana

S

Aleppo

Tigris

S e a

Antalya

i

Rhodes

SYRIA

J

Crete

Nicosia

Baghdad

45

CYPRUS

10   11   12   13   14   15

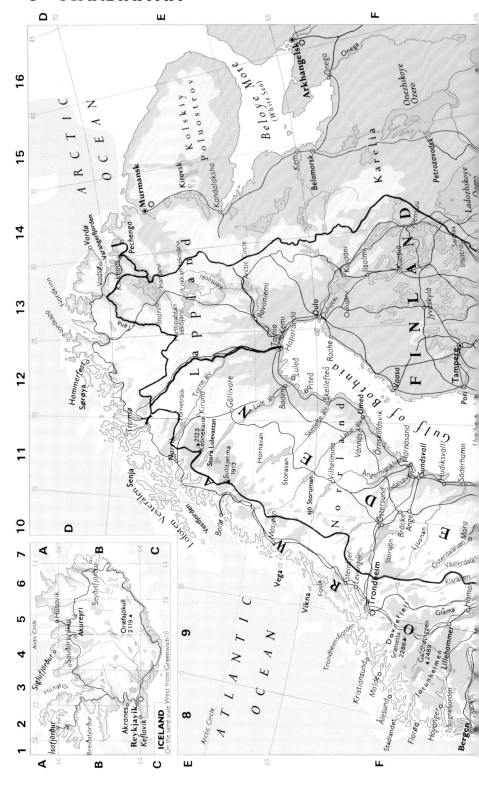

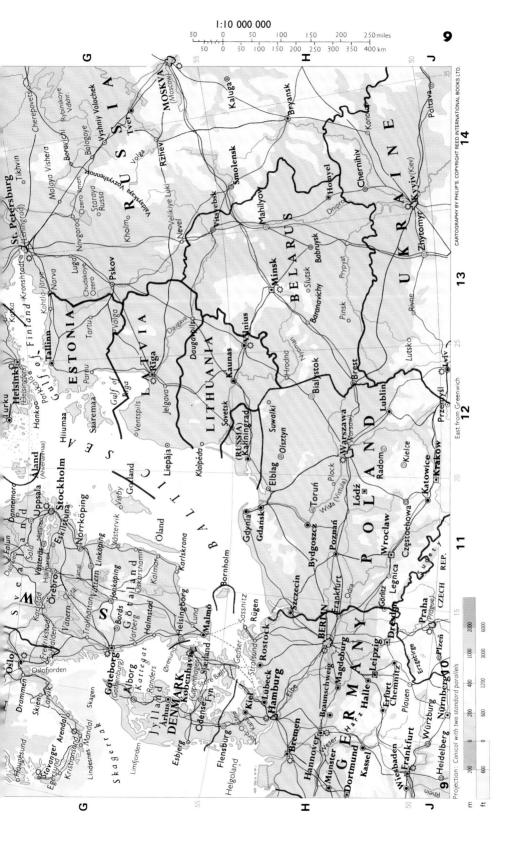

1:10 000 000

```
50      0        50       100      150      200     250 miles
50   0    50   100   150   200   250   300   350   400 km
```

Projection: Conic with two standard parallels

East from Greenwich

CARTOGRAPHY BY PHILIP'S. COPYRIGHT REED INTERNATIONAL BOOKS LTD.

RUSSIA
MOSKVA (Moscow)
Kaluga
Bryansk
Tver
Rzhev
Smolensk
Konotop
Poltava
Vyshniy Volochek
Bologoye
Rybinskoye Vdkhr.
Cherepovets
Tikhvin
Ryazhskoye Vdkhr.
Borovichi
Malaya Vishera
Staraya Russa
Ozero Ilmen
Novgorod
Velikiye Luki
Nevel
Vitebsk
Mahilyow
Homyel
Chernihiv
Kyyiv (Kiev)
Zhytomyr
Lviv
UKRAINE
Dnipro
Pryp'yat'
Rivne
Lutsko
BELARUS
Minsk
Slutsk
Baranovichy
Babruysk
Pinsk
Brest
Vilnius
Hrodna
Nyoman
Bialystok
Suwalki
Olsztyn
WARSZAWA (Warsaw)
Lublin
Przemyśl
Radom
Kielce
Kraków
Katowice
Częstochowa
POLAND
Łódź
Płock
Wisła (Vistula)
Toruń
Poznań
Bydgoszcz
Gdynia
Gdańsk
Elblag
Wrocław
Legnica
Görlitz
Sudety
CZECH REP.
Praha (Prague)
Plzeň
Dresden
Chemnitz
Erzgebirge
Leipzig
Halle
Magdeburg
Braunschweig
Hannover
BERLIN
Frankfurt
Odra
Szczecin
Rostock
Stralsund
Sassnitz
Rügen
Lübeck
Kiel
Hamburg
Bremen
Münster
Dortmund
Kassel
Wiesbaden
Frankfurt
Heidelberg
Würzburg
Nürnberg
Plauen
Harz
Erfurt
Weser
GERMANY
Rhein
Elbe
DENMARK
København (Copenhagen)
Malmö
Helsingborg
Lund
Helsingør
Sjælland
Store Bælt
Fyn
Odense
Nyborg
Nord-Ostsee Kanal
Gedser
Flensburg
Esbjerg
Århus
Jylland
Randers
Ålborg
Skagen
Kattegat
Helgoland
Limfjorden
Skagerrak
Mandal
Lindesnes
Kristiansand
Egersund
Stavanger
Haugesund
OSLO
Oslofjorden
Drammen
Skien
Larvik
Arendal
NORWAY
SWEDEN
Sverige
Göteborg (Gothenburg)
Borås
Götaland
Halmstad
Varberg
Vänern
Trollhättan
Karlstad
Vättern
Jönköping
Linköping
Norrköping
Örebro
Eskilstuna
Västerås
STOCKHOLM
Uppsala
Mälaren
Dannemora
Sala
Falun
Dalälven
Hällefors
Filipstad
Vänersborg
Frederikstad
Kalmar
Karlskrona
Oskarshamn
Västervik
Öland
Gotland
Visby
BALTIC SEA
Bornholm
Öresund
Kristianstad
ESTONIA
Tallinn
Tartu
Pärnu
Hiiumaa
Saaremaa
Gulf of Riga
LATVIA
Riga
Ventspils
Liepāja
Jelgava
Daugava
Daugavpils
LITHUANIA
Kaunas
Sovetsk
Klaipėda
(RUSSIA) Kaliningrad
St. Petersburg (Leningrad)
Kronshtadt
Narva
Pskov
Chudskoye Ozero
Kholm
Ozero
Lugo
Luga
Kohtla-Järve
Kokka
Gulf of Finland
Helsinki (Helsingfors)
Turku
Hanko
Åland (Ahvenanmaa)
FINLAND
9 10 11 12 13 14
G H J
m ft
```
```

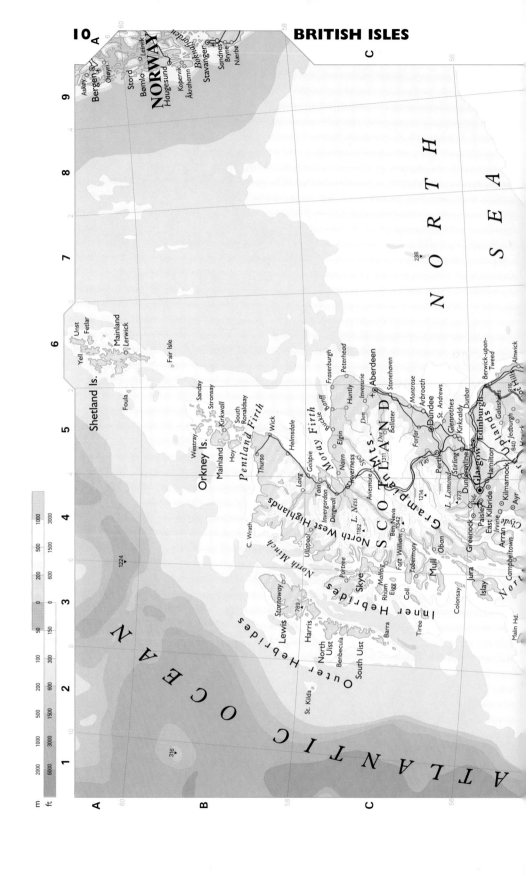

NORWAY

Askøy Bergen Osøyri Stord Bomlo Lervik Kopervik Åkrahamn Stavanger Sandnes Bryne Nærbø Haugesund Bokn Høltafjorden Sognefjorden

ATLANTIC OCEAN

NORTH SEA

SCOTLAND

Shetland Is.
Unst Fettar Yell Mainland Lerwick
Foula
Fair Isle

Orkney Is.
Westray Sanday Stronsay Sanday Kirkwall Mainland Hoy South Ronaldsay
Pentland Firth
Thurso Wick Helmsdale Golspie
Lairg
C. Wrath Ullapool Dingwall Tain Invergordon
Stornoway
Lewis
Harris
North Uist Benbecula South Uist
St. Kilda
North Minch
Outer Hebrides
Portree Skye
Rhum Eigg Coll Tiree
Barra
Inner Hebrides
Mallaig Fort William Ben Nevis 1342 Tobermory Mull Oban
Colonsay
Jura
Islay
Campbeltown
Malin Hd.
North West Highlands
Moray Firth
Inverness Nairn Elgin
Aviemore Spey
Dee 1311
Grampian Mts.
1214
L. Lomond 973 Clyde
Greenock Paisley Glasgow
East Kilbride Hamilton
Kilmarnock Irvine Arran
Ayr

Fraserburgh Peterhead
Banff Huntly Buckie
Inverurie Aberdeen
Don
Stonehaven
Ballater Montrose Arbroath
Brechin Forfar
Dundee
Perth St. Andrews
731 Stirling Glenrothes Dunfermline Kirkcaldy Dunbar
Berwick-upon-Tweed
Edinburgh 840 Galashiels 816 Jedburgh
Southern Uplands
Alnwick Cheviot Hills

1224
316
789
1182
238

m   ft
2000 6000
1000 3000
500   1500
200   600
100   300
50   150
0   0
0   0
50   150
100   300
200   600
500   1500
1000 3000

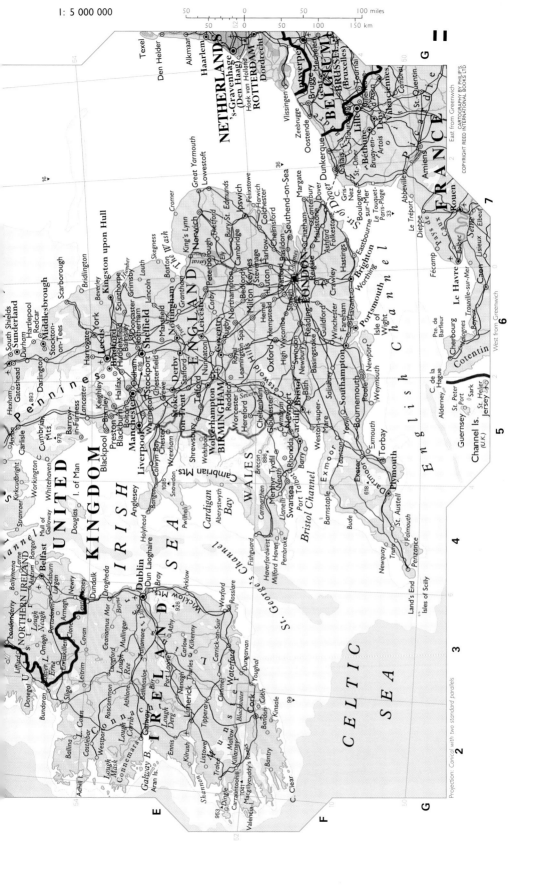

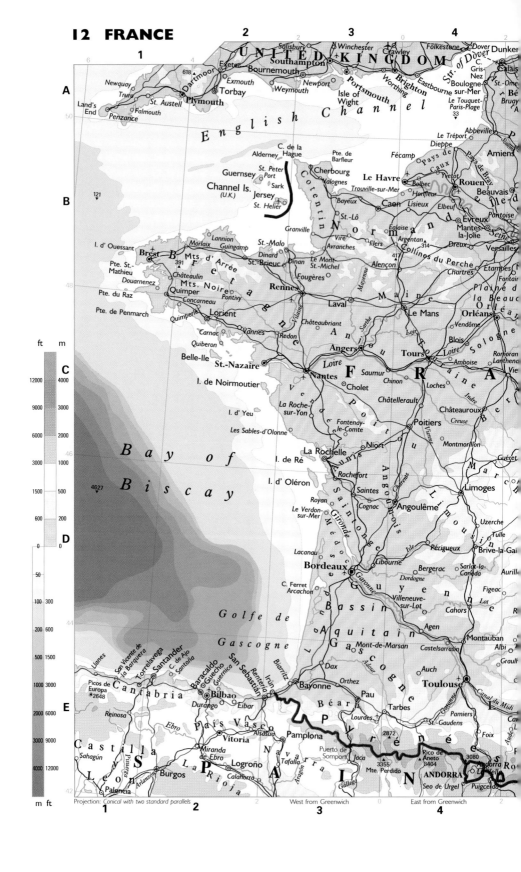

ft   m

12000 4000

9000   3000

6000   2000

3000   1000

1500   500

600   200

0   0

50

100   300

200   600

500   1500

1000   3000

2000   6000

3000   9000

4000   12000

m ft

Projection: *Conical with two standard parallels*

West from Greenwich   East from Greenwich

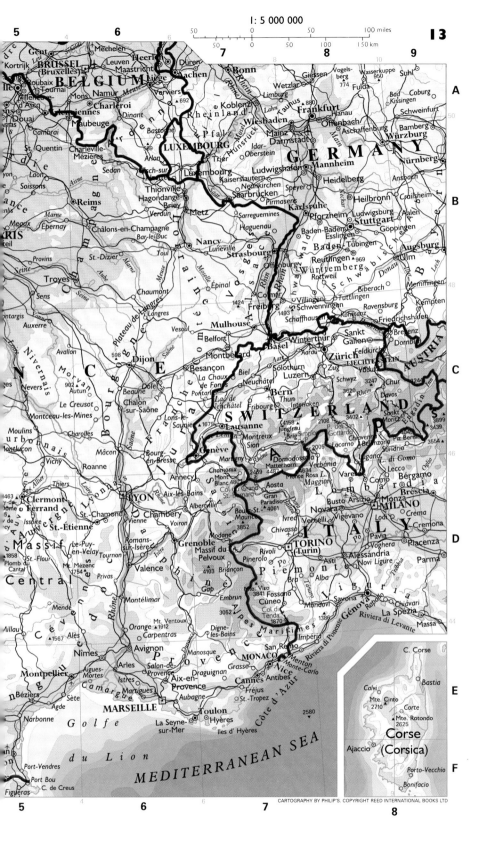

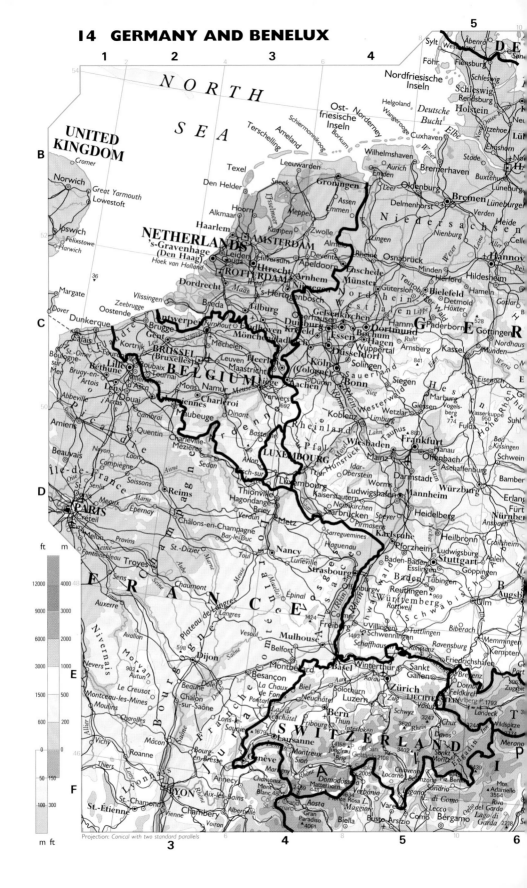

1   **2**   **3**   **4**   **5**

Kaliningrad (Russia)
Zatoka
Baltiysk
Gdańska
Gvardeysk Chernyakhovsk
Wejherowo Rumia
Bagrationovsk
Słupsk Lębork Gdynia
Darłowo Sopot Zalew
Braniewo Wiślany
Gdańsk Elbląg Kętrzyn Giżycko
Kołobrzeg 329 Tczew
Koszalin Bytów Malbork Olsztyn
Wolin Starogard Kwidzyn Pojezierze Mazurski
Świnoujście Szczecinek Gdański Iława Ostróda Szczytno
Police Chojnice Grudziądz Działdowo
Neubrandenburg Szczecin Świecie Brodnica Mława Ostrołęka
Stargard Chełmno Ciechanów
Szczeciński Piła Bydgoszcz Toruń Rypin
Schwedt Choszczno Toruń
Eberswalde Gorzów Inowrocław Włocławek Pułtusk Bug
Finow Wielkopolski Gniezno Płock Legionowo Mińsk
BERLIN Międzychód Warta Wrześnis Kutno WARSZAWA Mazowiecki
Fürstenwalde Nowy Tomyśl Poznań (Vistula) (Warsaw) Pruszków Żyrardów Otwock
Frankfurt Zielona Koło Łowicz Skierniewice Grójec
Świebodzin Góra Kościan Śrem Konin Turek Łódź
GERMANY Forst Nowa Sól Leszno Kalisz Zduńska Wola Pabianice Tomaszów Radom
Cottbus Zagań Krotoszyn Sieradz Piotrków Mazowiecki Końskie Skarżysko-
Lauchhammer Żary Głogów Ostrów Trybunalski Radomsko Kamienna
Hoyerswerda Bóbr Lubin Wielkopolski Wieluń Starachowice
Bautzen Bolesławiec Oleśnica Radomsko Kielce
Dresden Görlitz Zgorzelec Legnica Wrocław Kluczbork Częstochowa Ostrowiec-
Chemnitz Jelenia Góra Świdnica Oława Opole Świętokrzyski
Erzgebirge Děčín Liberec Wałbrzych Dzierżoniów Kłodzko Nysa Góry Myszków Jędrzejów Pińczów Tarnobrz
Teplice Jablonec Śnieżka Tarnowskie Zawiercie
Most Ústí nad 1602 Trutnov Góry Zabrze Bytom Sosnowiec
Karlovy Vary Chomutov Labem Mladá Hradec Racibórz Gliwice Chorzów Katowice Kraków Tarnów
Cheb Kladno Boleslav Králové Pardubice 1425 1492 Ostrava Karviná Oświęcim Bochnia Dębi
PRAHA Kolín Opava Havířov Bielsko-Biała
Plzeň (Prague) Sumperk Frýdek- Cieszyn Żywiec Nowy Jasło
Příbram vrchovina Olomouc Mistek 1324 Beskydy Sącz Vychodné
Beroun 836 Prostějov Přerov Považká Żilina Zakopane 2655 Poprád Prešov Hu
Klatovy Jihlava Vyškov Bystrica Ružomberok 1157
CZECH REP. Zlín Biele Karpaty Martin Nízke Tatry 2043 Košī
Písek Tábor Třebíč Brno Bielé Karpaty Trenčín Banská Bystrica 1458
České Jindřichuv Fihlava Hodonín 768 Prievidza Zvolen SLOVAK REP
Budějovice Hradec Znojmo Topoľčany Slovenské Rudohorie Sátoreljaújhely
Plöckenstein 1378 Gmünd Malé Karpaty Nitra Lučenec Ózd Miskolc
Passau Horn Stockerau Trnava Levice Salgótarján Eger
Freistadt Krems WIEN Bruck Nové Komárno Vác Gyöngyös Mezőkövesd Hajdúbö
Linz Melk Sankt (Vienna) der Leitha Zámky (Dunaj) Hatvan Debrece
Ried Wels Amstetten Pölten Baden Moson- Győr Esztergom Dunakeszi Jászberény
Steyr Wiener magyaróvár Dunaújváros BUDAPEST Karcag
Semmering P. Neustadt Sopron Tatabánya Pápa 704 Érd Székesfehérvár Kiskunfélegyháza Mezőtúr
Eisenerz Mürzzuschlag Szombathely Ajka Veszprém Nagykőrös Szolnok
Kapfenberg Bruck an der Mur Zalaegerszeg Siófok Kecskemét Csongrád Békéscsaba
Leoben Graz HUNGARY Balaton Dunaújváros Kalocsa Kiskőrös Szentes Gyula
Wolfsberg Nagykanizsa Szekszárd Kiskunhalas Orosháza
Klagenfurt Maribor Kaposvár Baja Szeged Hódmezővásárhely Makó
Triglav 2558 Celje Varaždin Koprivnica 681 Pécs Sombor Senta Sinnicolau
Kranj 2863 Ljubljana Mohács Subotica Maré Kikinda Timișo
SLOVENIA Novo Mesto Zagreb Bjelovar Drava Novi Sad
Kalce 1035 Virovitica YUG.
Trieste Postojna CROATIA Osijek Vukovar Vojvodina
1796 Sišak 989 Novi Sad Zrenjanin
Rijeka Karlovac

AUSTRIA
Böhmerwald
Kärnten
Karawanken
Steiermark

Projection: Conical with two standard parallels

ft	m
6000	2000
3000	1000
1500	500
600	200
0	0
	50
	100 300

m ft

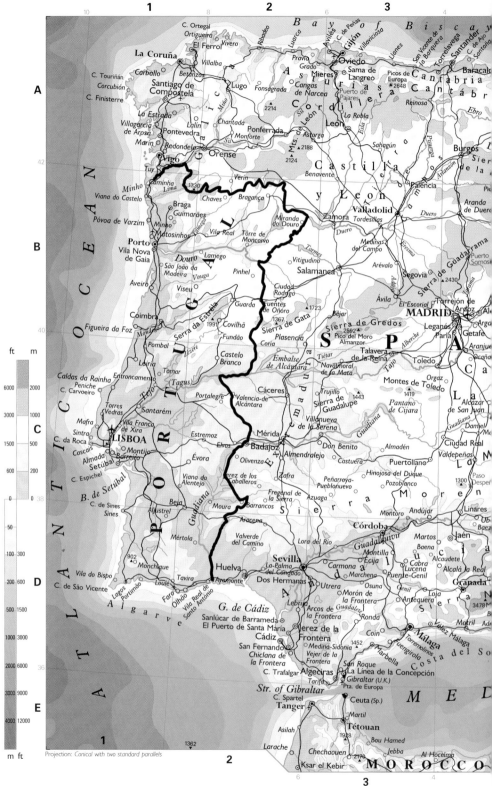

Projection: Conical with two standard parallels

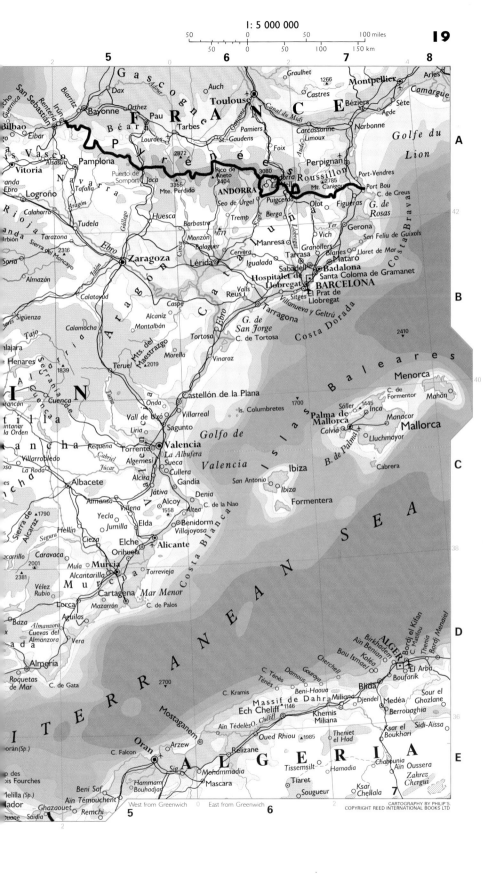

1: 5 000 000

50   0   50   100 miles
50 0   50   100 150 km

5   6   7   8

FRANCE

Graulhet   Montpellier   Arles
Auch   Castres   1266   Béziers   Sète
Dax   Orthez   Toulouse   Camargue
San Sebastián   Biarritz   Canal du Midi   Agde
Bayonne   Pau   Tarbes   Carcassonne   Narbonne   **Golfe du**
Bilbao   Eibar   Béarn   Pamiers   Limoux   **Lion**   A
   Lourdes   St-Gaudens   Foix   Aude
Vitoria   Alsasua   Pamplona   Pyrénées   Perpignan   Port-Vendres
   Navarra   Puerto de   2872   3080   2785   Roussillon
Logroño   Tafalla   Somport   Jaca   Pico de   ANDORRA   Mt Canigou   Port Bou
   Calahorra   Aragón   3355   Aneto   Seo de Urgel   La Seu   C. de Creus
Ebro   Tudela   3404   Mte. Perdido   Puigcerdá   Olot   Figueras   G. de
Rioja   Huesca   Barbastre   Tremp   Berga   Rosas
Tarazona   Monzón   1677   Vich   Gerona   San Feliu de Guixols
2316   Balaguer   Manresa   Ter   Blanes   Lloret de Mar
Sigüenza   Zaragoza   Lérida   Cervera   Tarrasa   Granollers   Mataró   42
Soria   Almazán   Calatayud   Igualada   Sabadell   Badalona   B
   Caspe   Valls   Reus   Hospitalet de   **BARCELONA**
Calamocha   Alcañiz   Llobregat   Santa Coloma de Gramanet
Montalbán   Tortosa   Sitges   El Prat de
Teruel   Morella   Vinaroz   Tarragona   Llobregat
   Mts. del   2019   G. de   Costa Dorada
   Maestrazgo   San Jorge   C. de Tortosa
Cuenca   1839   Castellón de la Plana   2410   Menorca   **Baleares**   40
   Onda   Is. Columbretes   1700   C. de   Mahón
Villarreal   Sóller   1445   Formentor
La Orden   Liria   Sagunto   **Palma de**   Inca   Manacor
   Requena   Torrente   **Valencia**   **Mallorca**   Calviá   Lluchmayor   **Mallorca**
Villarrobledo   Algemesí   La Albufera   B. de Palma   Cabrera   C
La Roda   Alcira   Sueca   **Golfo de**
Albacete   Játiva   Cullera   Gandía   San Antonio   **Ibiza**
   Almansa   Alcoy   Denia   **Valencia**   Ibiza   38
Sierra de   Yecla   1558   Altea   C. de la Nao   Formentera
Alcaraz   Villena   Benidorm
Hellín   Jumilla   Elda   Villajoyosa
Cieza   Elche   **Alicante**
Caravaca   Orihuela   **MEDITERRANEAN**
2001   Mula   **Murcia**
2381   Alcantarilla   Torrevieja   **SEA**
Vélez   Lorca   **Murcia**   D
Rubio   Mazarrón   Cartagena   Mar Menor
Baza   Aguilas   C. de Palos
Cuevas del   2700
Almanzora   Vera
Almería
Roquetas   C. de Gata
de Mar   **ALGERIA**   E

Melilla (Sp.)

West from Greenwich   0   East from Greenwich   6

5   6

CARTOGRAPHY BY PHILIP'S.
COPYRIGHT REED INTERNATIONAL BOOKS LTD

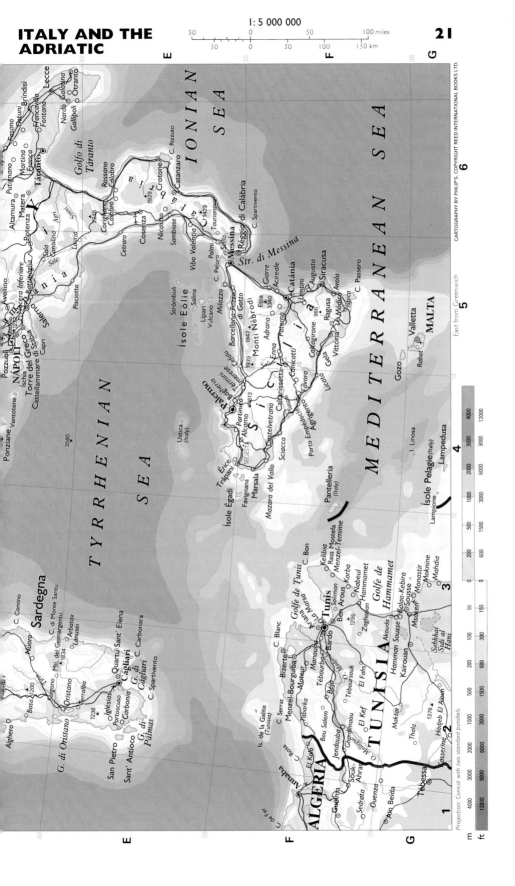

I:5 000 000

IONIAN SEA

MEDITERRANEAN SEA

TYRRHENIAN SEA

Golfo di Táranto

Sardegna

TUNISIA

ALGERIA

MALTA

East from Greenwich

Projection: Conical with two standard parallels

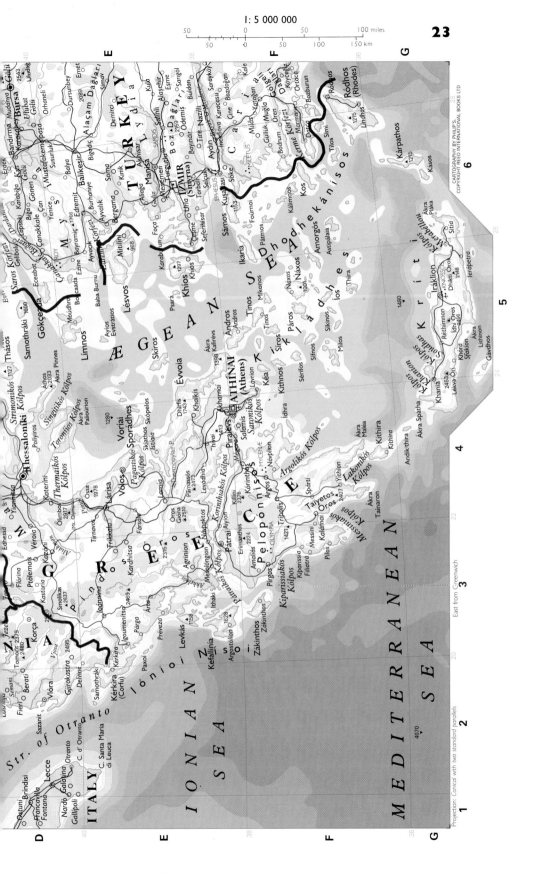

23

1: 5 000 000

CARTOGRAPHY BY PHILIP'S
COPYRIGHT REED INTERNATIONAL BOOKS LTD

Projection: Conical with two standard parallels

East from Greenwich

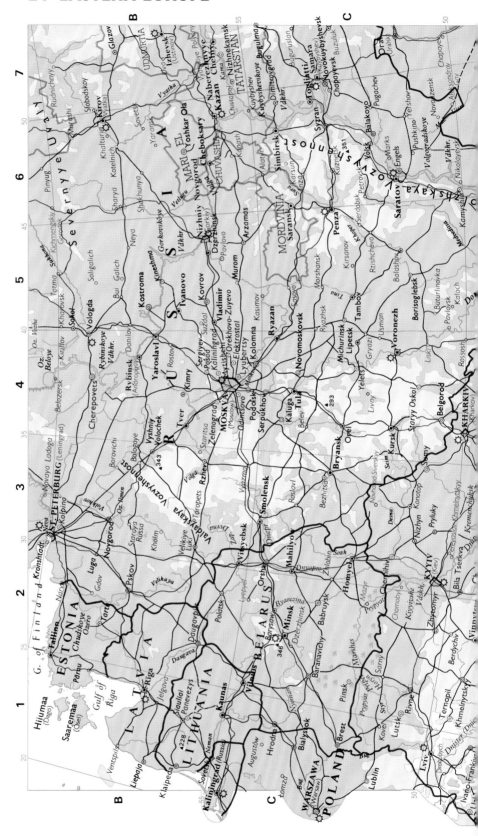

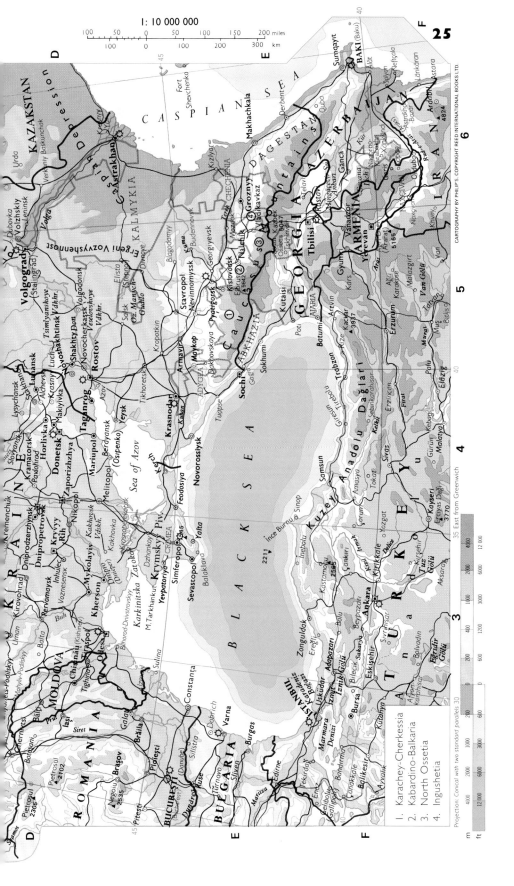

25

1:10 000 000

CARTOGRAPHY BY PHILIP'S. COPYRIGHT REED INTERNATIONAL BOOKS LTD.

1. Karachey-Cherkessia
2. Kabardino-Balkaria
3. North Ossetia
4. Ingushetia

Projection: Conical with two standard parallels 30

35 East from Greenwich

ATLANTIC OCEAN

GREENLAND

ICELAND

Svalbard

ARCTIC

Barents Sea

Novaya Zemlya

Kara Sea

Arctic Circle

Murmansk

Vorkuta

Salekhard

Ob

Yenisey

UNITED KINGDOM

NORWAY

North Sea

SWEDEN

FINLAND

White Sea

Arkhangelsk

ST. PETERSBURG

R U

LONDON

PARIS

FRANCE

GERMANY

Berlin

Europe

Warsaw

Prague

Vienna

ITALY

Rome

Belgrade

UKRAINE

Odessa

Danube

Nizhniy Novgorod

MOSCOW

Perm

Kazan

Ufa

Samara

Don

Volgograd

Astrakhan

Rostov

Yekaterinburg

Irtysh

Chelyabinsk

Omsk

Pavlodar

Sen

KAZAKSTAN

Karaganda

Mediterranean Sea

Athens

Bursa

ISTANBUL

Izmir

Ankara

TURKEY

Konya

Adana

Nicosia

CYPRUS

Beirut

LEBANON

Aleppo

SYRIA

Damascus

Black Sea

GEORGIA

Yerevan

Tbilisi

ARMENIA

AZERBAIJAN

Baku

Euphrates

Mosul

Tabriz

Caspian Sea

Aral Sea

Syrdarya

L. Balkhash

UZBEKISTAN

Tashkent

Samarkand

TURKMENISTAN

Ashkhabad

Mashhad

Bishkek

KYRGYZSTAN

Alma Ata

SI

Kashi

TAJIKISTAN

Dushanbe

U

LIBYA

Alexandria

ISRAEL

CAIRO

Jerusalem

Amman

JORDAN

Suez

EGYPT

Nile

Aswân

Red Sea

Port Sudan

SUDAN

Khartoum

ERITREA

DJIBOUTI

ETHIOPIA

Addis Ababa

UGANDA

L. Victoria

KENYA

Nairobi

ZAÏRE

TANZANIA

Mombasa

Dar es Salaam

ZAMBIA

MALAWI

Euphrates

Baghdad

IRAQ

Basra

Tigris

KUWAIT

Kuwait

SAUDI ARABIA

Medina

Jedda

Mecca

Riyadh

BAHRAIN

QATAR

Doha

Al Manāmah

UNITED ARAB EMIRATES

Abu Dhabi

G. of Oman

Muscat

OMAN

Sana

YEMEN

Aden

G. of Aden

Socotra (Yemen)

Arabian Sea

SOMALI REP.

Mogadishu

Equator

TEHRÂN

IRAN

Esfahân

Herât

Shîrâz

Zâhedân

Qandahâr

AFGHANISTAN

Kābul

Islamabad

JAMMU & KASHMIR

Hot

Faisalabad

Lahore

PAKISTAN

DELHI

New Delhi

Jaipur

Lucknow

Kanpur

Varanas

INDI

Ahmadabad

Vadodara

Indore

Bhopal

Surat

Nagp

BOMBAY

Pune

Hyderal

Bangalore

MA

Lakshadweep Is. (India)

Madurai

MALDIVES

Male

Colombo

SR

INDIAN OC

SEYCHELLES

Victoria

Aldabra Is. (Seychelles)

Amirante Is. (Seychelles)

Chagos Arch. (U.K.)

KARACHI

Indus

C   B   A
D
E
F
G
H
J
K
L

0   20   40   60   80
50
40
30
20
10
0
10

1 : 50 000 000

```
      200   0   200  400   600   800  1000  1200 miles
      200   0       400   800       1200  1600   2000 km
```

OCEAN

vernaya
Zemlya

New Siberian Is.

Laptev Sea

Wrangel I.

ALASKA (U.S.A.)

Bering Sea

Khatanga

Verkhoyansk

Gizhiga

Aleutian Is. (U.S.A.)

ilsk

S I A

Yakutsk

Okhotsk

Magadan

Petropavlovsk- Kamchatskiy

O C E A N

Lena

Angara

Sea of Okhotsk

Krasnoyarsk

Bratsk

L. Baikal

Chita

Komsomolsk

Sakhalin

Yuzhno- Sakhalinsk

sibirsk

Novokuznetsk

Irkutsk

Ulan Ude

Blagoveshchensk

Khabarovsk

Hokkaidō

Sapporo

Ürümqi

Hailar

Qiqihar

Harbin

Vladivostok

Honshū

NG

Hami

Ulan Bator

Changchun

Jilin

Sea of Japan

TŌKYŌ

Yokohama

JAPAN

Yumen

Baotou

SHENYANG Anshan

Jinzhou

Dalian

NORTH KOREA

Pyŏngyang

Nagoya

Ōsaka

Kyōto

Kitakyūshū

Lanzhou

BEIJING

TIANJIN

Taiyuan

Jinan

SEOUL

SOUTH KOREA

Pusan

Hiroshima

Bonin Is. (Japan)

M O N G O L I A

R

P A C I F I C

Hwang-ho

Xi'an

Yellow Sea

Volcano Is. (Japan)

Tropic of Cancer

C H I N A

Nanjing

SHANGHAI

East China Sea

Ryukyu Is.

Chengdu

Wuhan

HANGZHOU

Nanchang

Fuzhou

B E T

Changsha

CHONGQING

Yangtze

Taipei

TAIWAN

Lhasa

Kunming

Si Kiang

GUANGZHOU

Macau (Port.)

HONG KONG (U.K.)

GUAM (U.S.A.)

Thimphu

BHUTAN

Brahmaputra

handu

nges

BANGLADESH

BURMA

(MYANMAR)

Hanoi

Haiphong

FED. STATES OF MICRONESIA

atna

DACCA

Chittagong

LAOS

Hainan

Luzon

PHILIPPINES

PALAU

CUTTA

Irrawaddy

VIETNAM

Vientiane

MANILA

Bay of

Bengal

Salween

Rangoon

THAILAND

BANGKOK

Mekong

Cebu

Mindanao

Davao

Andaman Is. (India)

CAMBODIA

Phnom Penh

Ho Chi Minh City

South China Sea

Palawan

Sulu Sea

Zamboanga

Halmahera

Manado

IRIAN JAYA

G. of Thailand

BRUNEI

SABAH

Bandar Seri Begawan

Celebes Sea

Ceram

KA

Nicobar Is. (India)

PEN. MALAYSIA

SARAWAK

Borneo

Ambon

Celebes

Banda Sea

Str. of Malacca

Kuala Lumpur

MALAYSIA

Medan

SINGAPORE

Ujung Pandang

I N D O N E S I A

Arafura Sea

E A N

Sumatra

Banjarmasin

Java Sea

Flores

Timor

AUSTRALIA

Palembang

JAKARTA

Semarang

Surabaya

Bandung

Java

Sumba

Timor Sea

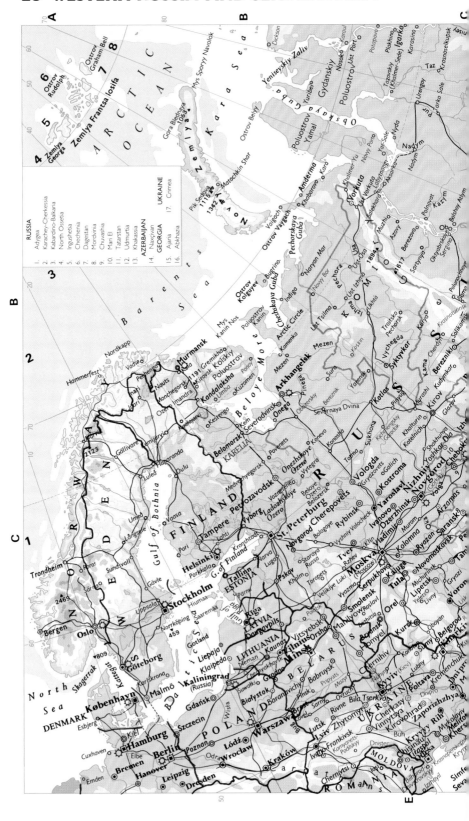

1 : 20 000 000

100   0   100   200   300   400   500 miles
100   0   200   400   600   800 km

CARTOGRAPHY BY PHILIP'S. COPYRIGHT REED INTERNATIONAL BOOKS LTD.

Projection: Conical Orthomorphic with two standard parallels

East from Greenwich

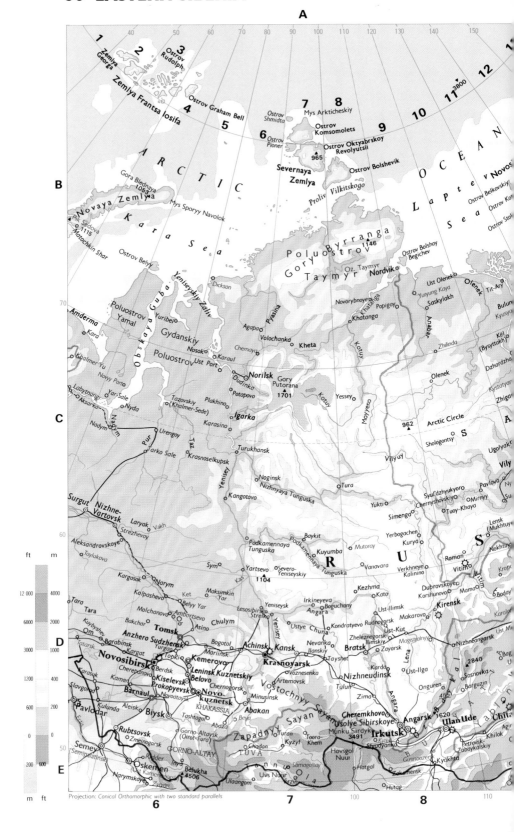

A

1 : 20 000 000

100   0   100   200   300   400   500 miles
100   0   200   400   600   800 km

14

B

15

16

C

Mys Dezhneva
(East C.)

Uelen

D

Ostrov Henrietta
Ostrov Jeanette
Ostrova Delong
Ostrov Zhokhova

Ostrova Bennett
e Ostrova
r Ostrova Faddeyevskiy

Ostrov Vrangelya

Chukchi
Sea

Amguema

Vankarem

Providenya
Lavrentiya

Anadyrskiy Zaliv

St. Lawrence I.
(U.S.A.)

60

Beringovskiy

Chukotskoye Nagorye

▲1843

Egvekinot

Iultin

Ostrov Malyy
Lyakhovskiy

Ostrov Bolshoy
Lyakhovskiy

Proliv Dmitriya Lapteva

Kiye Ostrova

Ostrova
Medezhi

Perch
Ust-Chaun

Chaunskaya

Ayon

Anadyr

Cherskiy
Ambarchik
Nizhne Kolymsk

▲1853

Bilibino

Markovo

Penzhino

Koryakskoye Nagorye

▲2562

Bering
Sea

Mys Buorkhaya

Verkhoyansk
▲2389

Chokurdakh

Indigirka

Shamanova

Druzhina

Zyryanka

Bolshoy Anyuy

▲1742

Oloy

Omolon

Evensk

Parem

Gizhiga
Penzhinskaya Guba

Slautnoye

Tilichiki

Ostrov
Karaginskiy

Komandorskiye
Ostrova

Nikolskoye Ostrov

Khonuu

Kolyma

Taskan

Omsukchan

Viliga

Gizhiginskaya
Guba

Zaliv
Shelikhova

Ust-Kamchatsk

Klyuchi
▲3621 Pushchino

Krestovskaya
Solbovoya

Oymyakon

Logashkalakh

Palatka
Debin

Magadan

Tigil

Poluostrov

Soboleva

Zhupanovo
▲3456

Petropavlovsk-
Kamchatskiy

Okhotsk

Sea
of
Okhotsk

Ust Bolsheretsk

Ozernovskiy

Ostrov
Paramushir

Olekminsk

Yakutsk

Aldan

Stanovoy Khrebet

Komsomolsk

Nikolayevsk-
na-Am.

Sakhalin

Aleksandrovsk-
Sakhalinskiy
▲1608 Gora Lopatina

Kurilskiye Ostrova

Khabarovsk

Birobidzhan

Khrebet Sikhote Alin

Sovetskaya Gavan

Yuzhno-Sakhalinsk

Korsakov

Ostrov
Kunashir

Blagoveshchensk

Qiqihar

Jiamusi

Harbin

Ussuriysk

Vladivostok
Nakhodka

Hokkaido

Sapporo

Hakodate

JAPAN

East from Greenwich   120

9

10

11

130

CARTOGRAPHY BY PHILIP'S.
COPYRIGHT REED INTERNATIONAL BOOKS LTD.

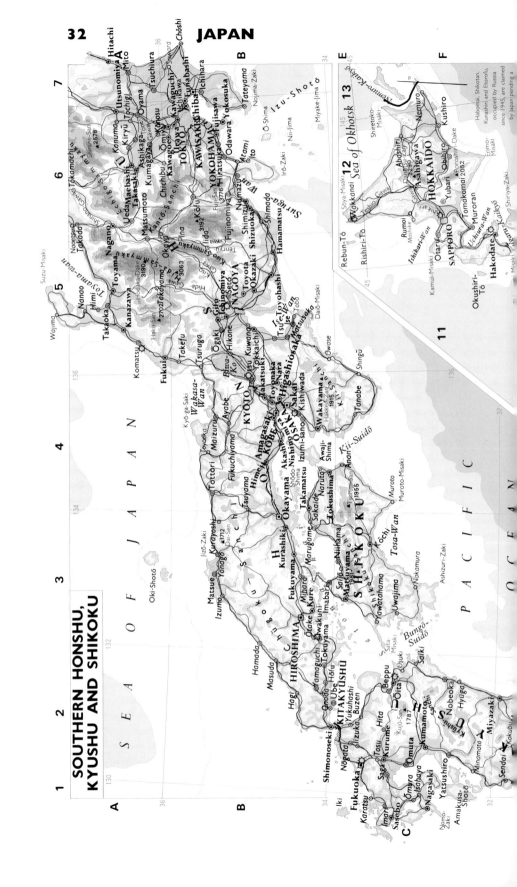

SOUTHERN HONSHU, KYUSHU AND SHIKOKU

JAPAN

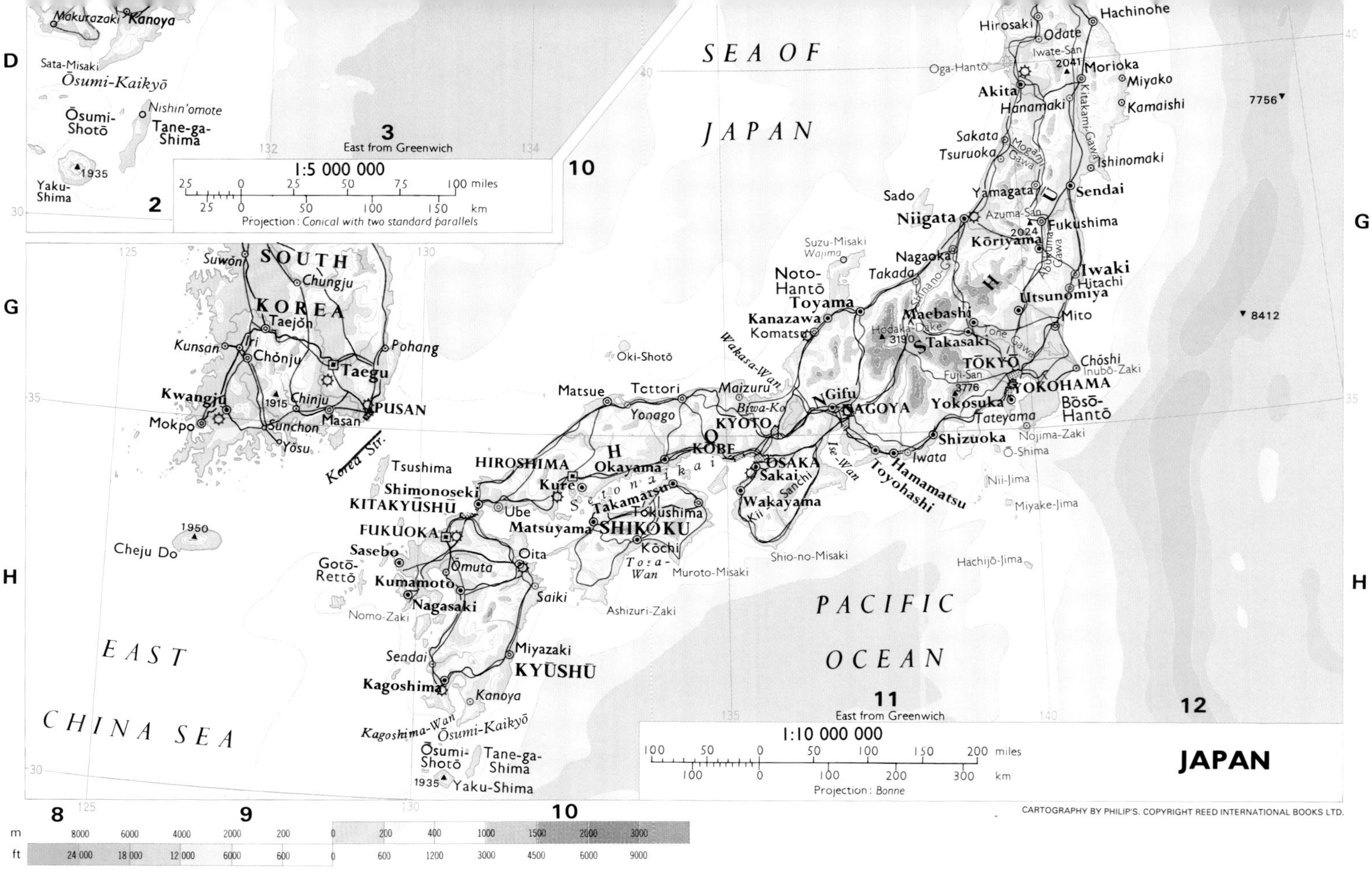

SEA OF JAPAN

PACIFIC OCEAN

EAST CHINA SEA

SOUTH KOREA

KYŪSHŪ

SHIKOKU

HONSHŪ

TŌKYŌ

YOKOHAMA

NAGOYA

KYŌTO

KŌBE

ŌSAKA

HIROSHIMA

KITAKYŪSHŪ

FUKUOKA

Sendai

Niigata

Akita

Hachinohe

Morioka

Kagoshima

Nagasaki

Kumamoto

Kōchi

Matsuyama

Okayama

Takamatsu

Tokushima

Wakayama

Shizuoka

Gifu

Kanazawa

Toyama

Nagaoka

Fukushima

Kōriyama

Utsunomiya

Mito

Yamagata

PUSAN

Taegu

Kwangju

Mokpo

Taejŏn

1:5 000 000

Projection: Conical with two standard parallels

East from Greenwich

1:10 000 000

Projection: Bonne

CARTOGRAPHY BY PHILIP'S. COPYRIGHT REED INTERNATIONAL BOOKS LTD.

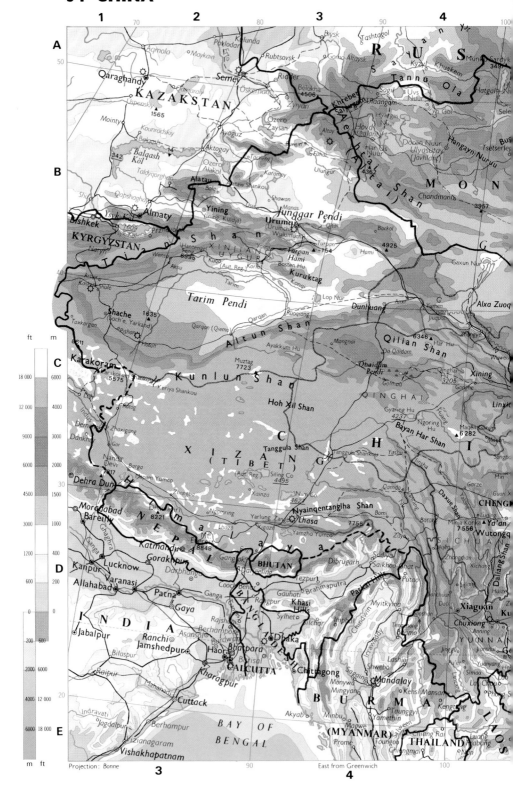

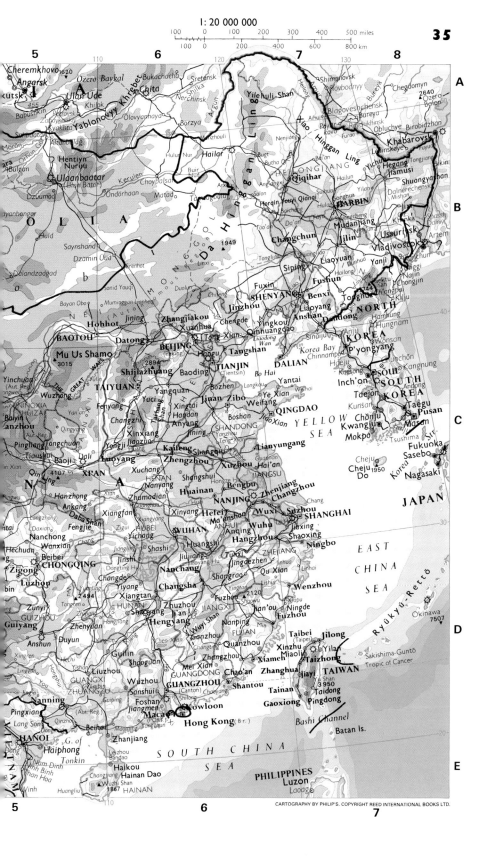

1: 20 000 000

100 0 100 200 300 400 500 miles
100 0   200   400   600   800 km

5   6   7   8

Cheremkhovo 620
Angarsk   Ozero Baykal   Bukachachi   Sretensk   Shimanovsk
kutsk   455   Khilok   Nerchinsk   Svobodnyy   Chegdomyn 2640 A
         Ulan Ude   Yilehuli Shan   Ozero Bolon
Babushkin Zabaykalsk   Olovyannaya   Xiao   Hinggan   Ling   Blagoveshchensk
Kyakhta   Altanbulag   Borzya   Ji'an   Baykit   Raikhinsk   Bureya
Suhbaatar   Manzhouli   Heihe   Furao   Obluchye   Birobidzhan
Moron Orhon   Nenjiang   Khabarovsk
Bulgan   Hentiyn   Hailar   Bugt   HEILONGJIANG   Tongliang
         Nuruu   Buir   Butha Qi   Qiqihar   Hailun   Hamusi
Ulaanbaatar   Kerulen   Choybalsan   Nur   Hulan   Da'an   Yichun   Hegang
(Ulan Bator)   Ondorhaan   Matad   Arxan   Fuyu   Suihua   Shuangyashan   B
Dzuunmod   Tamsagbulag   Salon   Harbin   Yilan   Jixi
ayanhongor   Huld   1949   Horqin Youyi Qianqi   Bei'an   Mudanjiang   Mishan
         Saynshand   Duolun   Boichang   Changchun   Jilin   Ussuriysk   Artem
Dalandzadgad   Dzamin Uud   Erenhot   Linxi   Siping   Vladivostok
                         Sonid Youqi   Liaoyuan   Yanji
       b   MONGOLIA   Duolun   Chifeng   Fuxin   Tonghua   Najin   Chongjin
   o   SHENYANG   Benxi   NORTH   Kilju   40
Yinchuan   Bayan Bobo   Mu Us Shamo   Hohhot   Jining   Zhangjiakou   Chengde   Jinzhou   Liaoyang   Anshan   Dandong   KOREA   Hamhung
(Aut Reg)   3015   BAOTOU   Datong   Xuanhua   Yingkou   Qinhuangdao   Sinuiju   Wonsan
Wuzhong   Hohhot   Tong Xian   Tangshan   Liaodong   Korea Bay   P'yongyang
NINGXIA   THE GREAT   BEIJING   Hangu   Wan   Chinnampo   Haeju
HUIZA   Baiyin   Yan'an   (Peking)   TIANJIN   Bo Hai   Kaesong   SOUL   Kangnung
anzhou   Qingyang   Shijiazhuang   Baoding   (Tientsin)   Yantai   Inch'on   SOUTH   Andong
Pinglang   Tongchuan   TAIYUAN   Yangquan   Bozhen   Longkou   Weihai   Taejon   KOREA   Taegu
Tianshui   Baoji   Fenyang   Yuci   Xingtai   Jinan   Zibo   Ye Xian   Kunsan   Pusan   C
Qing   4107   Dali   Changzhi   Handan   Boshan   Weifang   QINGDAO   Chonju   Mason
Xi'an   XIAN   Luoyang   Yongji   Liaozuo   Anyang   Jining   SHANDONG   YELLOW   Kwangju   Tsushima
         N   Xuchang   Kaifeng   Zhengzhou   Teng Xian   SEA   Mokpo   Fukuoka
                              Shangqiu   Xuzhou   Hai'an   Cheju   Sasebo
ngwu   Hanzhong   HENAN   Nanyang   Shangshui   JIANGSU   Lianyungang   Cheju   1950   Nagasaki
Langzhong   Daba Shan   Xinyang   Zhumadian   Honze Hu   Guannan   Do
Daxian   Fengjie   Xiangfan   Xiangyang   Huainan   Bengbu   Zhenjiang   Changzhou   Chang   JAPAN   30
Nanchong   Zigui   Zhumadian   Hefei   NANJING   Changzhou   Wuxi   SHANGHAI
Hechuan   Wanxian   ANHUI   Ma'anshan   Wuhu   Suzhou   Jiaxing
Beibei   Yichang   HUBEI   WUHAN   Anqing   Hangzhou   Shaoxing   Ningbo   EAST
Zigong   CHONGQING   Shashi   Huangshi   Jiujiang   ZHEJIANG   CHINA
Luzhou   Changde   Dongting Hu   NANCHANG   Tunxi   Jingdezhen   Qu Xian   Linhai   SEA
bin   Zunyi   Yiyang   Shangrao   Yingtan   Lishui   Wenzhou
GUIZHOU   Xiangtan   Changsha   HUNAN   JIANGXI   2120   Xiapu   Ryūkyū-Rettō
Guiyang   Zhenyuan   Shaoyang   Zhuzhou   Ji'an   Jian'ou   Ningde   Okinawa   7507   D
Anshun   Duyun   Hengyang   Wuyi Shan   Nanping   Fuzhou   Sakishima-Gunto
Xingren   Guilin   Chen Xian   Ganzhou   Taibei   Jilong   Tropic of Cancer
Lingyun   Hechi   Liuzhou   Shaoguan   Zhangting   Quanzhou   Xinzhu   Yilan
Nanning   Wuzhou   GUANGDONG   Chao'an   Xiamen   Miaoli   Taizhong   TAIWAN
Pingxian   GUANGXI   Sanshui   GUANGZHOU   Shantou   Zhanghua   Jiayi   3950   Taidong
Lang Son   ZHUANG   Foshan   (Canton)   Chaoyang   Tainan   Gaoxiong   Pingdong
HANOI   Qinzhou   Jiangmen   Macao   Kowloon   Bashi Channel   Batan Is.   20
Haiphong   Beihai   Maoming   Hong Kong (Br.)
VIETNAM   G. of   Zhanjiang   SOUTH   CHINA
   Tonkin   Leizhou Bandao   SEA   PHILIPPINES   E
Nam Dinh   Haikou   Hainan Dao   Luzon
han Hoa   Changhua   Wuzhi Shan   1867   HAINAN   Laoag
Vinh   Huangliu

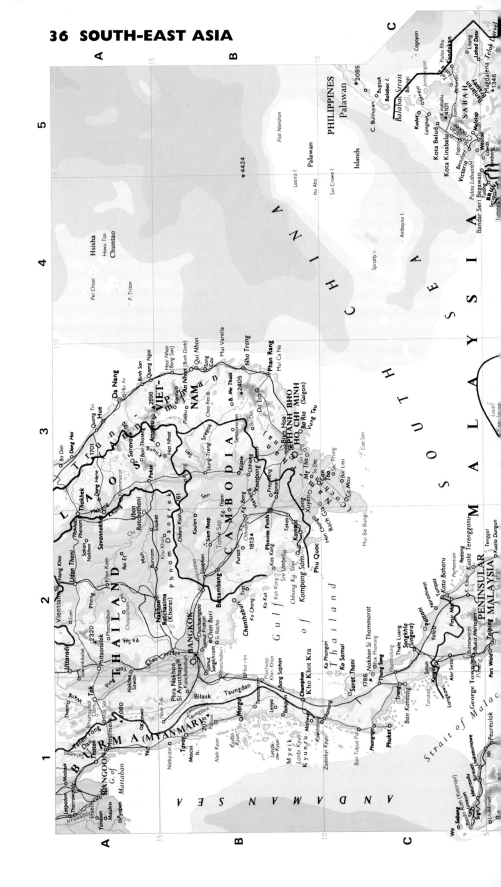

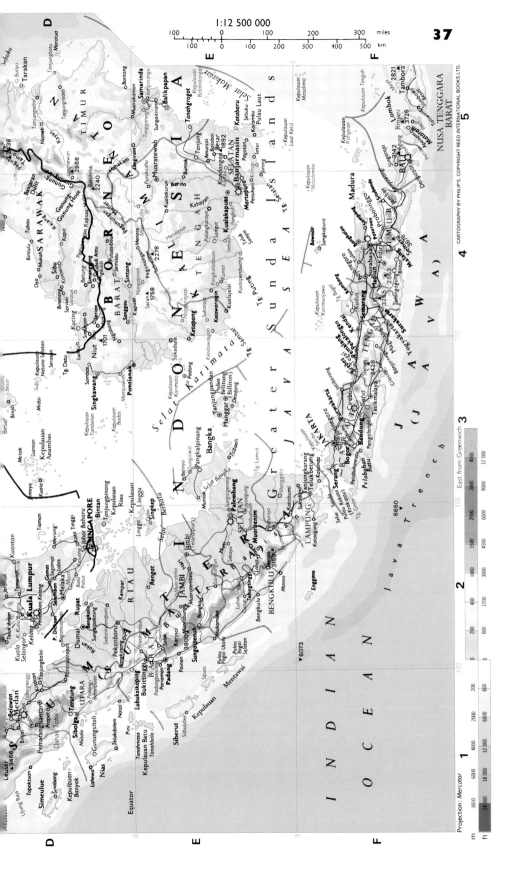

1:12 500 000

100   0   100   200   300 miles
100   0   100   200   300   400   500 km

D

Sebuku
Tarakan
Tanjungselor
Tanjungredeb
Morotua

K   A   L   I   M   A   N   T   A   N

T   I   M   U   R

Tambu 2438
Bahran Dulit
Nomeh
Keleu
Tanjungselor
Nunukan
Tubau
Bintulu
Tanjunggedeb
Bongon

Samarinda
Muarakaman
Balikpapan
Tanahgrogot

S A R A W A K
Gunung Dulit
Gunung Hose 2240
Gunung Murud 2988

B O R N E O

B A R A T

Putussibau
Semitau
Sintang
Gunung Ratu
Gunung Niut 1701
Kuching
Serian
Singkawang
Pontianak

K A L I M A N T A N

Puruk cahu
Muarateweh
Kualakurun
Barito

Kahayan
Kualakapuas
Martapura
Banjarmasin
SELAT LAUT 1892
Kotabaru
Pulau Laut

T E N G A H

Pegunungan Muller 2278
1758

Nangapinoh
Sampit
Sukadana

Kuala Kapuas
Palangkaraya
Kualapembuang
Pangkalanbuun
Buntok

S E L A T   M A K A S A R

Kepulauan Balabalangan
Kepulauan Masalima
Kepulauan Sabalana
Kepulauan Kalukaluang
Kepulauan Postilyon
Kepulauan Tengah

Kepulauan Laut Kecil

B a n d a   S e a

Kepulauan
Kangean
3142
2821
Tambora
Rinjani 3726
Lombok
Mataram
B A L I
Denpasar

NUSA TENGGARA BARAT

5

CARTOGRAPHY BY PHILIPS. COPYRIGHT REED INTERNATIONAL BOOKS LTD.

4

S u n d a   I s l a n d s

J a v a   S e a

Madura
Surabaya
Madiun
Kediri
2563
Blitar
Malang
3745
Probolinggo
Pasuruan
Jember
Banyuwangi
3428
J A W A   T I M U R

S e l a t   K a r i m a t a

P. Karimata
Kep. Karimata
Ketapang
Kotawaringin

Pulau Belitung
Pulau Bangka
Manggar
Tanjungpandan

Bangka
Pangkalpinang
Sungailiat
Mentok

Palembang
Jambi
Muarabungo
Muaraenim
Lahat

S U M A T E R A   S E L A T A N

Bengkulu
BENGKULU
Manna
Enggano

LAMPING
Tanjungkarang
Telukbetung

Teluk Semangka
Krakatau

S e l a t   S u n d a

Serang
JAKARTA
Bogor
Bandung
Cirebon
Tasikmalaya
Garut
Pekalongan
Tegal
Semarang
Kudus
3211
Yogyakarta
2911
Surakarta
J A W A   T E N G A H

J A W A   B A R A T   (J A B A R)
J A W A   (J A V A)

6650

J a v a   T r e n c h

I N D I A N   O C E A N

Equator

3800
Solok
Padang
Pariaman
Bukittinggi
Lubuksikaping
S U M A T E R A   B A R A T

Pekanbaru
R I A U
Rengat
JAMBI
Sungaipenuh
Kerinci
3805

6073

Siberut
Kepulauan Mentawai
Pulau Pagai Utara
Pulau Pagai Selatan

Medan
Tebingtinggi
Pematangsiantar
Sibolga
SUMATERA UTARA
Tarutung
Danau Toba
Gunungsitoli
Nias
Kepulauan Batu
Tanahbala

Leuser 3466
Topaktuan
Sinabang
Simeulue
Kepulauan Banyak

SINGAPORE
Johor Baharu
Gunung Tahan
Kuala Lumpur
Malaka
Kuala
R I A U
Bintan
Tanjungpinang
Kepulauan Riau
Kepulauan Lingga
Lingga
Singkep

Kepulauan Natuna Selatan
Kepulauan Anambas
Midai
Besar

Projection: Mercator

East from Greenwich

m   8000   4000   2000   1000   600   400   200   0   200   400   600
ft   24 000   12 000   6000   4500   3000   1500   600   1200   3000   6000   9000   12 000   4000   3000   2000   1500   1000

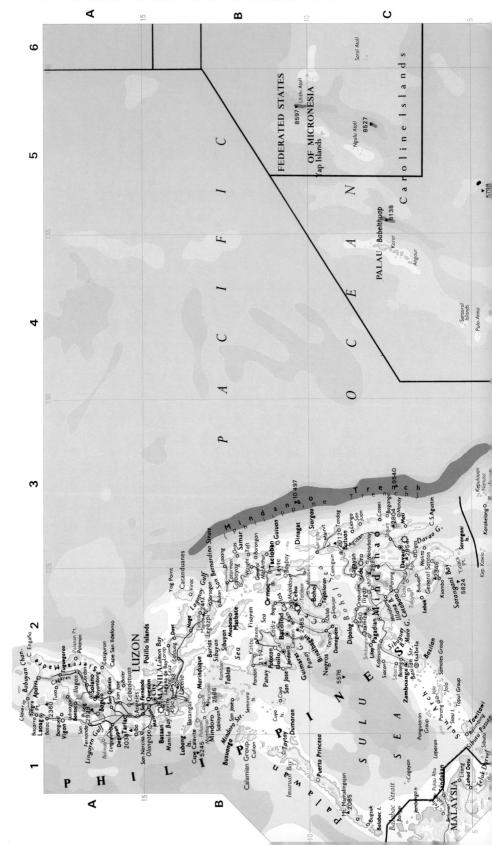

39

1:12 500 000

PAPUA NEW GUINEA

CARTOGRAPHY BY PHILIP'S. COPYRIGHT REED INTERNATIONAL BOOKS LTD.

Projection Mercator   East from Greenwich

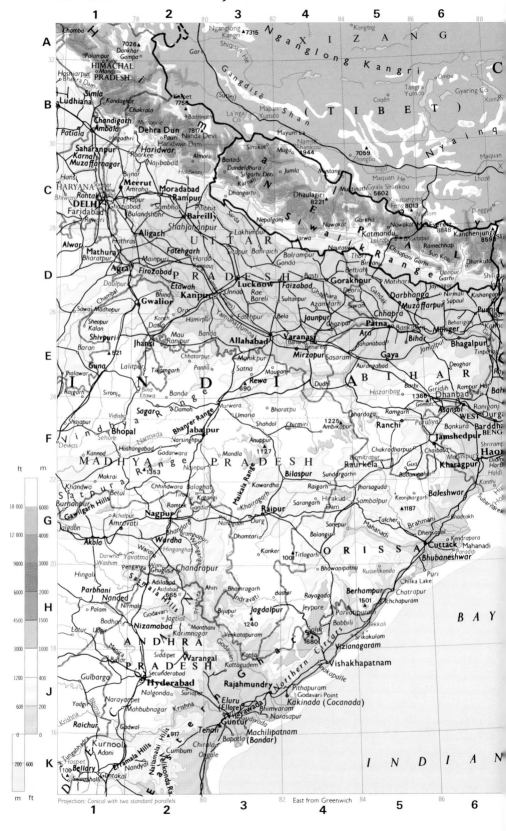

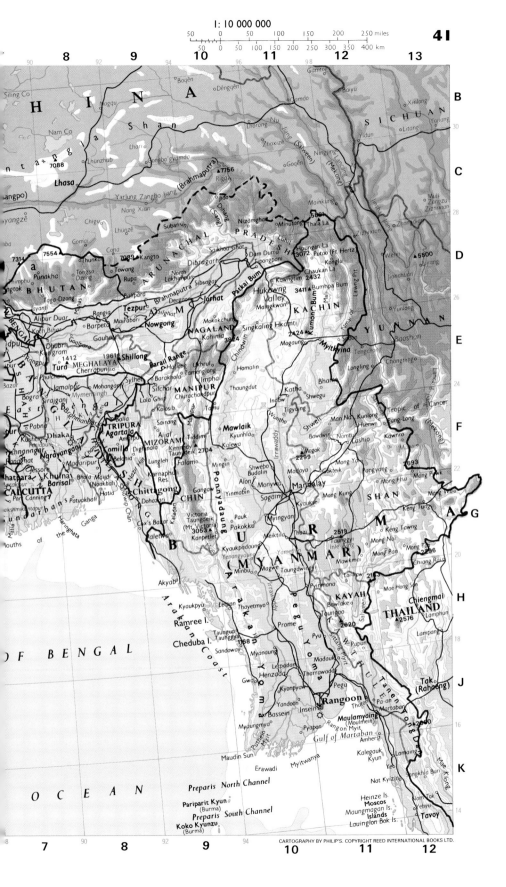

50   0   50   100   150   200   250 miles

50   0   50 100 150 200 250 300 350 400 km

B
C
D
E
F
G
H
J
K

8   **9**   **10**   **11**   **12**   **13**

90   92   94   96   98

CHINA

SICHUAN

Siling Co

H   I   N   A

Tangula Shan

Nu Jiang (Salween)

Nam Co

Lhasa

7088

Yarlung Zangbo Jiang (Brahmaputra)

Nang Xian

ARUNACHAL PRADESH

7756

KACHIN

YUNNAN

BHUTAN

7554

7314

7089

Towang

Rupa

Tezpur

Nowgong

NAGALAND

Kohima

3924

MEGHALAYA

Shillong

Barail Range

Cherrapunji

1961

1412

Tura

MANIPUR

Silchar

Imphal

Chindwin

Myitkyina

Bhamo

Katha

Shwegu

BENGAL

Dhaka

Narayanganj

Comilla

TRIPURA

Agartala

MIZORAM

Aijal

CALCUTTA

Khulna

Barisal

Chittagong

CHIN

Mawlaik

Kalewa

Shwebo

Budalin

Monywa

Mandalay

Sagaing

SHAN

Cox's Bazar

Victoria

Taungdeik

(Mt. Victoria)

3053.4

Kanpetlet

U

R

M

A

(MYANMAR)

Meiktila

Taunggyi

Akyab

Arakan Coast

Minbu

Magwe

Yamethin

KAYAH

THAILAND

Chiengmai

Ramree I.

Cheduba I.

Sandoway

Prome

Pegu Yoma

Toungoo

Pyu

Chiang Rai

Mae Hong Son

OF   BENGAL

Myanaung

Henzada

Tharrawaddy

Pegu

Rangoon

Insein

Bassein

Maulamyaing

(Moulmein)

Amherst

Gulf of Martaban

Tavoy

O   C   E   A   N

Preparis North Channel

Pariparit Kyun

(Burma)

Preparis South Channel

Koko Kyunzu

(Burma)

Heinze Is

Moscos

Maungmagan Is.

Islands

Lauinglon Bok Is.

7   **8**   **9**   **10**   **11**   **12**

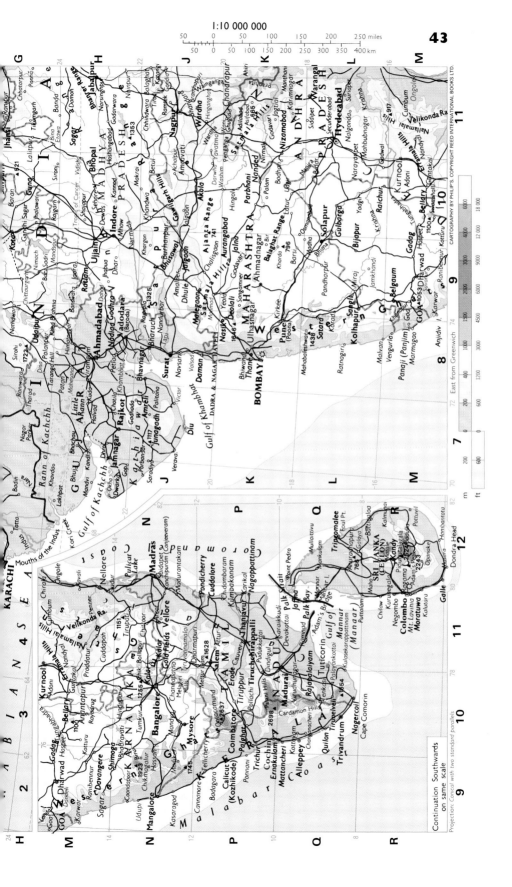

1:10 000 000

43

50 0 50 100 150 200 250 miles
50 0 50 100 150 200 250 300 350 400 km

CARTOGRAPHY BY PHILIP'S. COPYRIGHT REED INTERNATIONAL BOOKS LTD.

Continuation Southwards
on same scale

Projection: Conical with two standard parallels

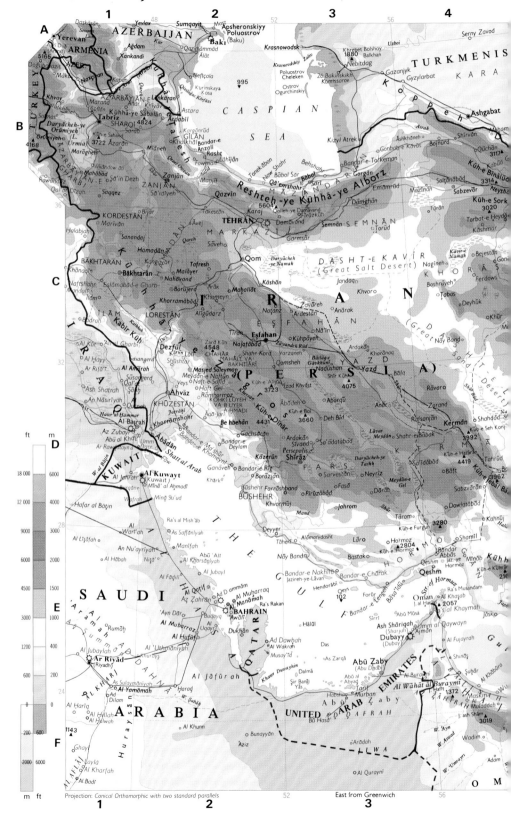

Projection: Conical Orthomorphic with two standard parallels

East from Greenwich

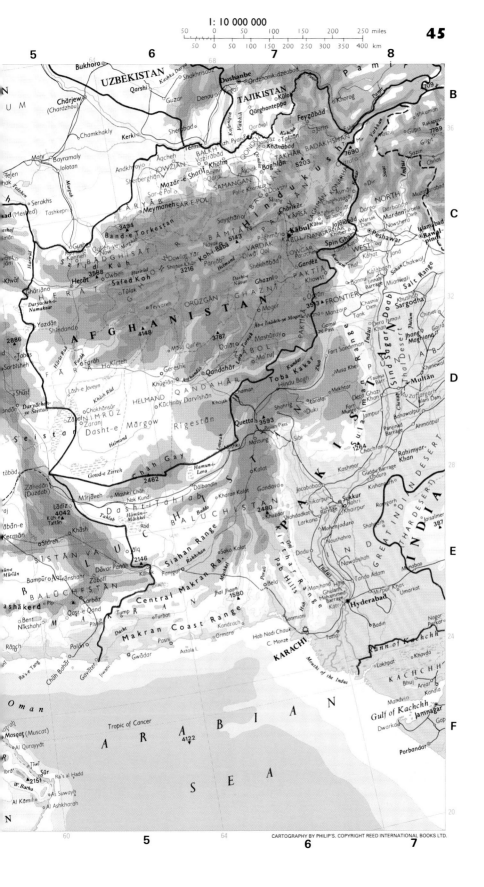

1:10 000 000

50   0   50   100   150   200   250 miles
50 0 50 100 150 200 250 300 350 400 km

UZBEKISTAN
Bukhoro
Chärjew
(Chardzhou)
Chamkhakly
Kerki
Mary
Bayramaly
Iolotan
Tejen
Serakhs
Tashkepri
Qarshi
Guzar
Denau
Shakhrisabz
Sherabad
Termez
BALKH
Andkhvoy
Aqcheh
Meymaneh
Mazar-e Sharif
Sar-e Pol
Sherberghan
DUSHANBE
TAJIKISTAN
Ordzhonikidzeabad
Külob
Qürghonteppa
Qaravpi
Feyzabad
KhorOg
Pamir
Ishkuman
7789
Gilgit
Chilas
NORTH
Peshawar
Islamabad
Rawalpindi

Bukhoro

AFGHANISTAN

Herat
Ghüriän
Daryacheh-i-Namakzar
Yazdän
Shindand
Täbas
Sarbïsheh
Shüst
Daryacheh-ye Seistan
Zäbol
Zaranj
NIMRÜZ
Dasht-e Märgow
Rigestän
HELMAND
QANDAHAR
Qandahär
Gereshk
Khügiäni
Toba Kakar
Quetta
3593
Bolan Pass

KABUL
Kabul
NANGARHAR
Jalalabad
Spin Ghar
VARDAK
Gardez
PAKTIA
Ghazni
GHAZNI
ORÜZGAN
ZABOL
Qalat
Ma'ruf

PAKISTAN

Multän
Bahawalpur
Dera Ghazi Khan
INDIA
GREAT INDIAN DESERT
THAR DESERT
Jaisalmer
387
Rahimyar Khan
Sukkur
Khairpur
Shahdadkot
Larkana
Mohenjodaro
Dadu
Nowshahro
Hyderabad
Badin
KARACHI
Mouths of the Indus
RANN OF KACHCHH
KACHCHH
Bhuj
Anjar
Kandla
Mandvi
Gulf of Kachchh
Jamnagar
Dwarka
Porbandar

Oman
Masqat (Muscat)
Al Qurayyät
Sür
Ra's al Hadd
As Suwayh
Al Ashkharah

ARABIAN
SEA

Tropic of Cancer
4122

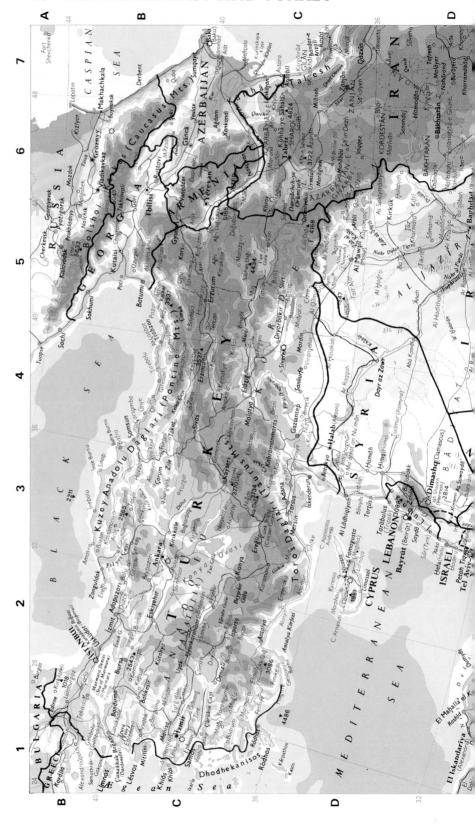

1:10 000 000

50   0   50   100   150   200   250 miles

50 0 50 100 150 200 250 300 350 400 km

47

CARTOGRAPHY BY PHILIP'S. COPYRIGHT REED INTERNATIONAL BOOKS LTD.

Division between Greeks and Turks **6**
in Cyprus; Turks to the North.

East from Greenwich

5 – – –

Projection: Conical Orthomorphic with two standard parallels

4

m 2000   0   200   400   1000   1500   2000   3000   4000
ft 6000   0   600   1200   3000   4500   6000   9000 12 000

THE GULF

THE
GULF

Zarrīn
4548
Mosjed
Soleymān

Dezfūl
Aḥvāz

KHŌZESTĀN

MESOPOTAMIA

Ābādān
Shaṭṭ al Arab

Al Baṣrah

KUWAIT
Al Kuwayt
(Kuwait)
Al Jahrah

Āz Zubayr

W. al Bāṭin

Ḥafar al Bāṭin

Raʾs al Mishʾāb
Al Qārīʿ
Az Zalfah
Al Jubayl
Al Mutkarraz
Al Hufūf
Al Khumm

DAHNĀ

Al Wafrah

An Nuʿayrīyah

Al Ḥabab

Nịṭāʿ

Ar Riyāḍ
(Riyadh)
Al Jubaylāh
Al Kharj

Al Ḥillah
Al Khariah
Al Badiʿ

JABAL ṬUWAYQ

Al ʿAramah

SUDAIR

Al Majmaʿah

Shaqrāʾ
Marāt
Al ʿUwaynid

Ad Duwādimi

1143

Al Ḥārīq
Layla
Ghoil

A R I D J I A Z

Al Hoddath
Al Homra

Buraydah
ʿUnayzah
Al Midhnab

S U D A I R

Al Qaṣīm
Ar Rass

Al Ḥinakīyah

Al Ḥadar

As Sūq
Ḥarrat Nawāṣif

Ṭābah
Fayd
Ḥāʾil

Al ʿUyūn

JABAL SHAMMAR

AN NAFŪD

Raḥḥaʾ

Ṭurabah

Al Dukhūl

Al Jalāmid

Būdanah
Sōkārah
W. Ḥawrah
Al ʿUyayzah
Buraydah

Ash Shabakah
As Salmān

S A U D I   A R A B I A

Ḥarrat Khaybar

Al Madinah

R E D   S E A

Makkah
(Mecca)
2565
al Ṭāʾif
1814

Jiddah

Al Qaḍīmah

Rabigh
Uṣfān
al Ḥamrā

Maṣtūrah

Yanbūʿ al Baḥr

Ras Bānas

AL HIJAZ

AT TIH

Gebel et Tih
2637
2884

EL QÂHIRA
(CAIRO)

EL GIZA
Pyramids

Ḥelwân

Beni Suef

Beni Mazâr
El Minyâ

Mallowi
Dairût

Asyûṭ
Sohâg
Naġ Ḥammâdi
Qenâ
Qûṣ
El Uqṣur
(Luxor)
Thebes
Isnā
Idfû
Kôm Ombo
Aswân

E G Y P T

ES SAHRÂ ESH SHARQIYA

Es Sharqiya

Nahr en Nil (Nile)

Saʿd el Aali
(Aswan High Dam)

El Khârga
Bâris
el Khârga

ES SAHRÂ NÛBIYA
(NUBIAN DESERT)

S U D A N

Buḥeiret en Naser
(Lake Nasser)

Wâdî Ḥalfa
Farîg

Dunqula

Halâib

Gebeit
Omb Mines

Muhammad Qol

2216

Mashaʾribā

Ras Abu Shagara
Ras Ḥadarba

1977
1484

Mersa Gawāsīs

G. Hamāta
2216

Bûr Ṣafâga
Ḥurghada
Abu Rudeis
2187
G. el Ṭôr

Khalîg el Suweis

El Wâḥât
el Baḥarîya

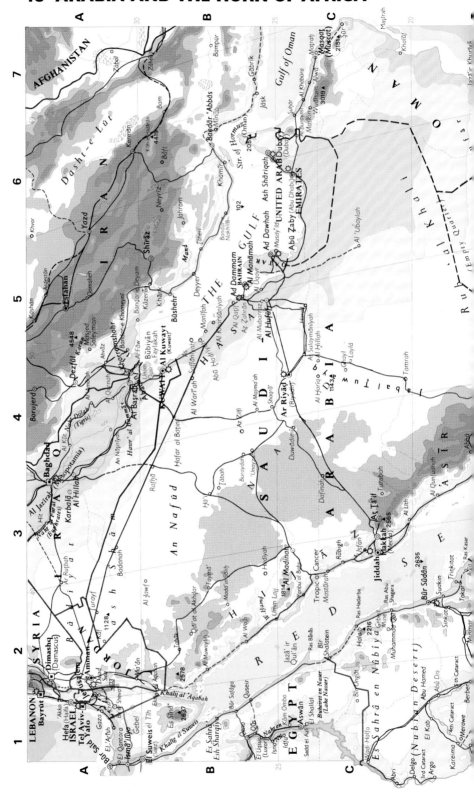

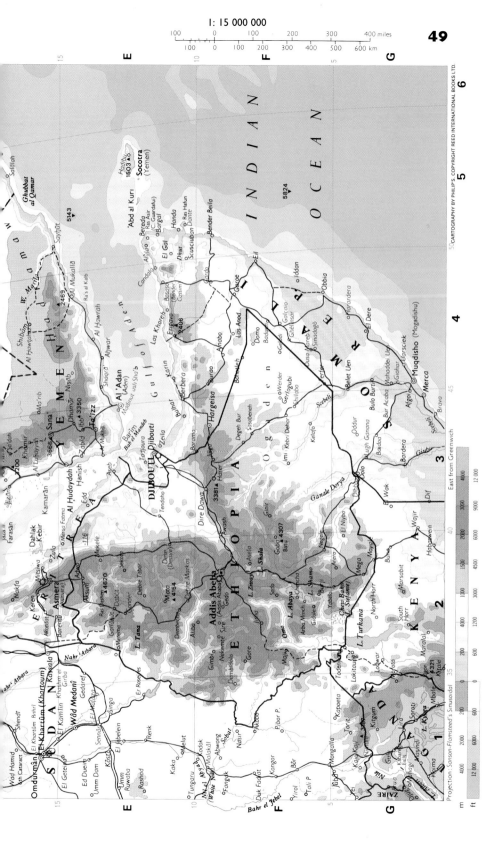

1: 15 000 000

100   0   100   200   300   400 miles

100   0   100   200   300   400   500   600 km

INDIAN   OCEAN

Socotra
(Yemen)
1503

Abd al Kuri

Ghubbat
al Qamar

YEMEN

Sana
Sa'dah
Khamir
Dhamār
Ta'izz
Al Hudaydah
Al Luḩayyah
Zabīd
Al Mukallā
Ma'rib

Dahlak
Kebir

Farasan

ERITREA
Mits̄iwa
Asmera
Keren
Makele

Al 'Adan (Aden)
Djibouti
DJIBOUTI

Gulf of Aden

Hargeisa
Berbera
Zeila

SOMALIA

Bereda
Bargal
Handa
Bender Bella

Bosaso
Gardo
El Gal
Candala

Iscia Baidoa
Galcaio
Obbia
Ferfer
Scebeli

Muqdisho (Mogadishu)
Merca
Brava

ETHIOPIA

Addis Abeba (Addis Ababa)
Dire Dawa
Harer
Nazret
Aksum
Gonder
L. Tana
Debre Markos
Dese (Dessye)

Ogaden
Werder
Shilabo
Degen Bur

KENYA

L. Turkana
Marsabit
Wajir
Moyale
Mega

SUDAN
El Khartûm (Khartoum)
Omdurmân
Wâd Medanî
Kassala
Gedaref
Kosti

White Nile
Bahr el Jebel

Ganale Dorya

ZAÏRE

East from Greenwich

Projection: Sanson-Flamsteed's Sinusoidal

CARTOGRAPHY BY PHILIP'S. COPYRIGHT REED INTERNATIONAL BOOKS LTD.

m   ft
4000 12 000
3000   9000
2000   6000
1500   4500
1000   3000
 400   1200
 200   600
   0   0
 200   600
2000   6000
4000 12 000

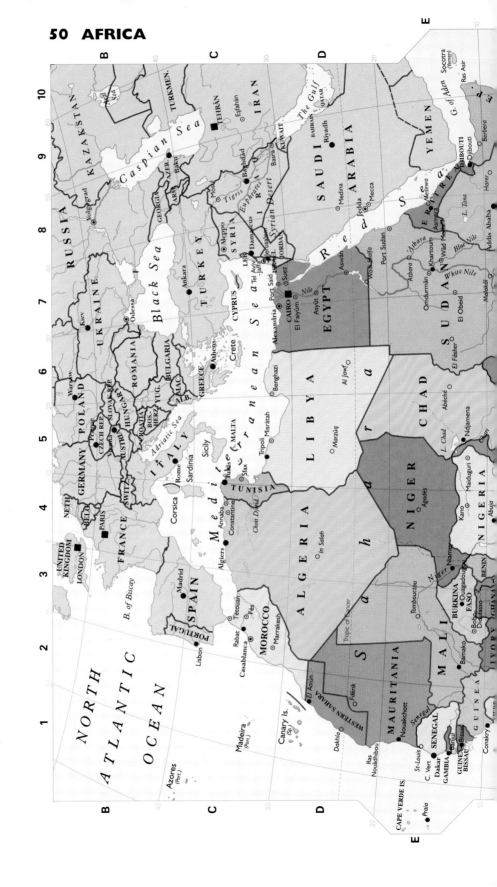

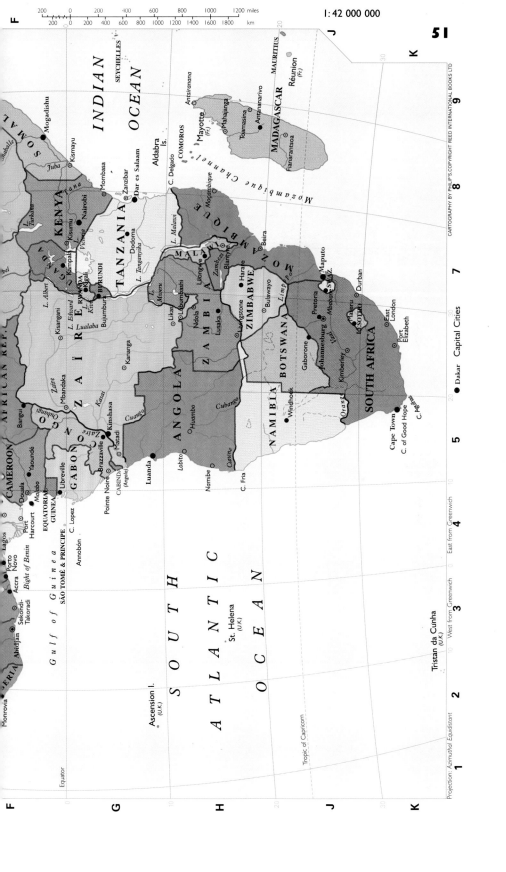

F

200   0   200   400   600   800   1000   1200 miles

200   0   200   400   600   800   1000 1200 1400 1600 1800   km

1:42 000 000

51

J   K

INDIAN

OCEAN

SEYCHELLES

MAURITIUS

Réunion
(Fr.)

MADAGASCAR

COMOROS

Antsiranana

Mayotte
(Fr.)

Mahajanga

Toamasina

Antananarivo

Fianarantsoa

SOMAL

Mogadishu

Mubello

Kismayu

Juba

Aldabra
Is.

C. Delgado

Mombasa

KENYA

Nairobi

Kisumu

Kampala

L. Victoria

Dar es Salaam

Zanzibar

TANZANIA

Dodoma

L. Tangaryika

Mozambique Channel

MOZAMBIQUE

L. Malawi

Zambezi

Beira

MALAWI

Lilongwe

Blantyre

Harare

Maputo

SWAZ.

Mbabane

Maseru

Durban

Pretoria

Johannesburg

Vaal

ZIMBABWE

Bulawayo

East
London

Port
Elizabeth

SOUTH AFRICA

BOTSWANA

Gaborone

Kimberley

Limpopo

L. Albert

Turkana

Tana

Kisangani

L. Edward

RWANDA

Kigali

BURUNDI

Bujumbura

L. Kivu

Z A Ï R E

Lualaba

Zaïre

Lubumbashi

Likasi

L. Mweru

ZAMBIA

Ndola

Lusaka

Livingstone

Ngami

Orange

C. of Good Hope

Cape Town

NAMIBIA

Windhoek

Cubango

Cubango

AFRICAN REP.

Bangui

CAMEROON

Yaoundé

Douala

Malabo

EQUATORIAL
GUINEA

Libreville

C. Lopez

GABON

Pointe Noire

Brazzaville

Kinshasa

Matadi

CABINDA
(Angola)

Luanda

Lobito

Namibe

C. Fria

Cunene

Cuanza

ANGOLA

Huambo

Kananga

Mbandaka

Kasai

Ubangi

CONGO

 Onga

Kwango

Malebo

SÃO TOMÉ & PRINCIPE

Annobón

Gulf of Guinea

Bight of Benin

Lagos

Porto
Novo

Accra

Sekondi-
Takoradi

Abidjan

Monrovia

Port
Harcourt

ERIA

Dakar Capital Cities

20

S O U T H

A T L A N T I C

O C E A N

St. Helena
(U.K.)

Ascension I.
(U.K.)

Tristan da Cunha
(U.K.)

Tropic of Capricorn

Equator

Projection: Azimuthal Equidistant

West from Greenwich 0 East from Greenwich

10   0   10   20   30

F   G   H   J   K

CARTOGRAPHY BY PHILIP'S.COPYRIGHT REED INTERNATIONAL BOOKS LTD

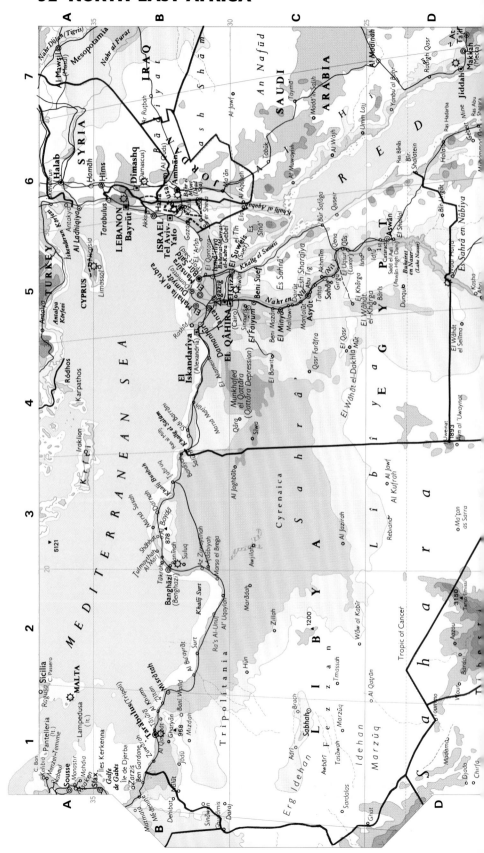

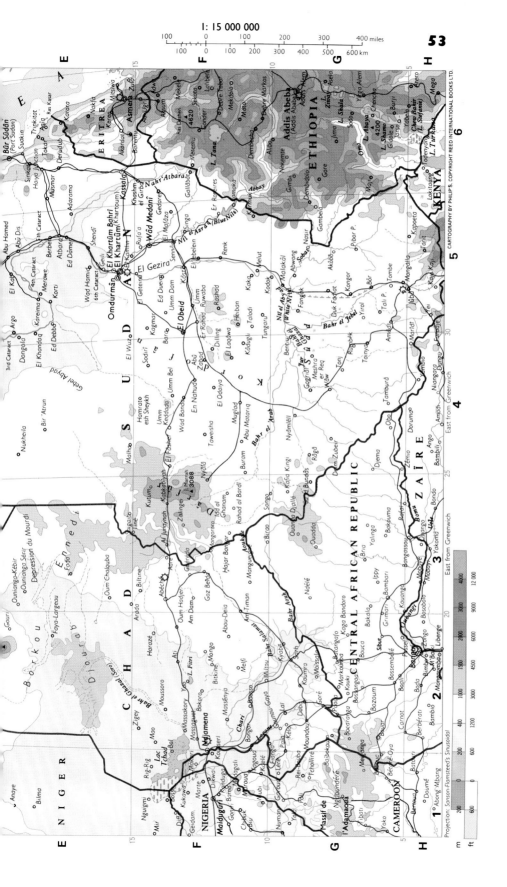

1 : 15 000 000

100   0   100   200   300   400 miles

100   0   100   200   300   400   500   600 km

53

Projection: Sanson-Flamsteed's Sinusoidal

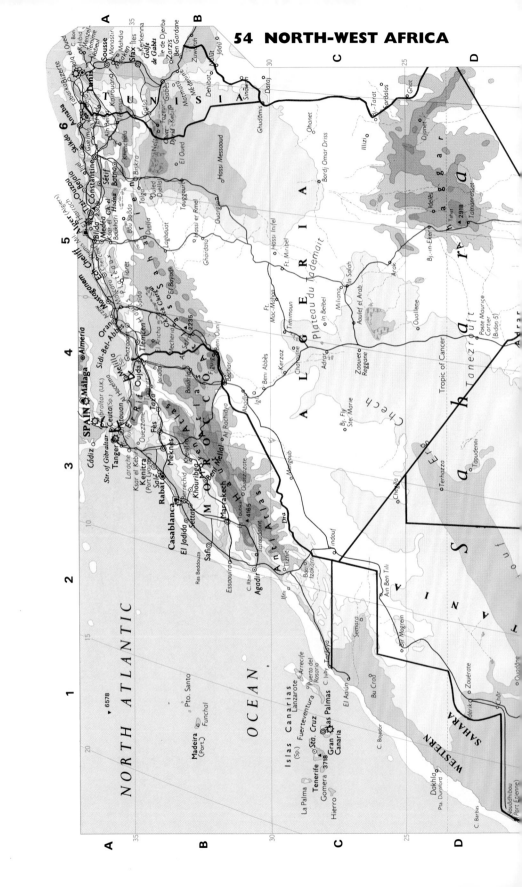

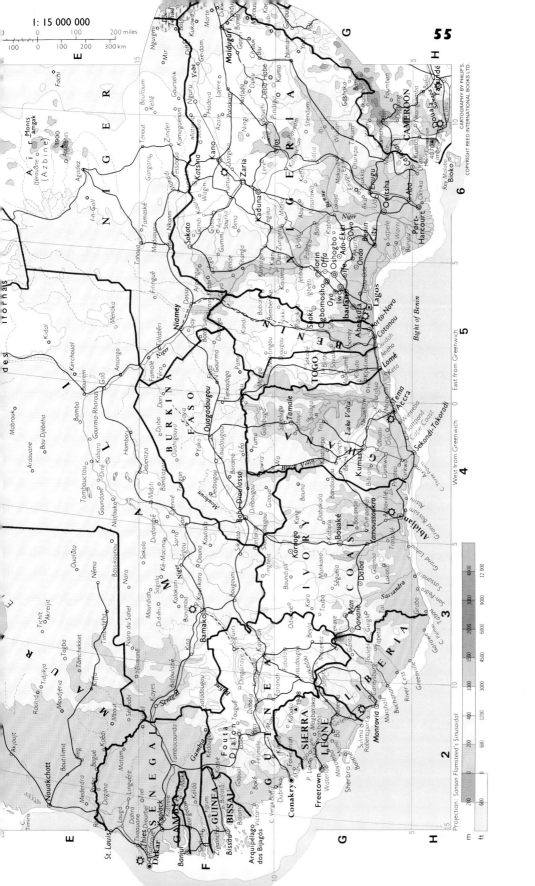

1 : 15 000 000

CARTOGRAPHY BY PHILIP'S.
COPYRIGHT REED INTERNATIONAL BOOKS LTD.

Projection: Sanson Flamsteed's Sinusoidal

1   2   3   4

A

N I G E R

Tanout
Boultoum
Kellé
Nguigmi
Rig-Rig
Zigey
Oum Chalouba
Arada
Biltine
Sigaibo
Tiné
Kutum

Gangara
Mifi
Mao
Moussoro
Harazé
Abéché
Al Junaynah
Kabi

15

Zinder
Gourselik
Lac
Tchad
Bol
C H A D
Ati
Oum Hadjer
Adre
Zalingei

Tessaoua
Kamaguenam
Diffa
Bossо
Massaguet
Massakory
Yao
L. Fitri
Am Dam
Am Guereda
Songo
30

Matsına
Nguru
Yobe
Kukawa
Marte
Massaguet
Bokoro
Mongo
Bitkine
Goz Beida
Mongororo

B

Kano
Azare
Hadejia
Geidam
Marte
Ndjamena
(Lamy)
Massenya
Abou-Deïa
Hajar Banga
Rahad el Baral

Dangora
Ningi
Potiskum
Lajere
Dikwa
Kousséri
Melfi
Am-Timan
Mangueigne

Lere
Bauchi
Nafada
Gonbi
Bama
Chari
Bongor
Bousso
Birao

Iol
Kafanchan
Biu
Deba Habe
Maroua
Magali
Gounou
Gaya
Miltou
Bahr Salamat
Ndélé
Songo

10

Jos
Pindiga
Kumo
Mubi
Kaplé
Lera
Pala
Laï
Kélo
Kyabé
Bahr Aouk
Ouanda Djallé
Soro

Bogoro
Panyam
Numan
Garoua
Rey Bouba
Moundou
Gore
Koumra
Ndélé
Ouadda

Aafia
Shendam
Yola
Poli
Tcholliré
Baïbokoum
Marcounda
Batangafo
Kaga Bandoro
Bria
Yalinga

C

Wukari
Makurdi
Gashaka
Massif de
l'Adamaoua
Bocaranga
Kouki
Bouca
Bakala
Ippy
Bambari
Bakouma

Oturkpo
Takum
Banyo
Tibati
Meiganga
Bozoum
Sibut
Grimari
Kouango
Bangassou
Raf

Mamfé
Nkambe
Bamenda
Bali
Foumban
Yoko
Bétaré-Oya
Bouar
Carnot
Boali
Zongo
Mobaye
Mobayi
Ouango
Bomu

Calabar
Kumba
Nkongsamba
Nanga Eboko
Bertoua
Berbérati
Boda
Bimbo
M'Baïki
Libenge
Busingo
Yakoma
Bonc

Oron
Bafia
Doumé
Bambio
Nola
Monduo
Gemena
Monveda
Aketi

D

4070
Douala
Yaoundé
Abong Mbang
Yokadouma
Bombma
Busu-Djanoa
Lisala
Bumba
Zaire

Limbe
Edéa
Sangmelima
Lomié
Djoum
Molounou
Dongou
Impfondo
Bomongo
Bongandango
Basankusa
Yahuma
Basok

Bioko
(Fernando
Póo)
M'Balmayo
Ambam
Bitam
Souanké
Ouesso
Bomongo
Bolomba
Djolue
Isangi

EQUATORIAL
GUINEA
Mbini

Oyem
Minvoul
Mvadh
Ousye
Belinga
Mekambo
Lulonga
Boende

Libreville
Cocobeach
Mitzic
Makokou
Kellé
Makouo
Owando
Irebu
Mbandaka
Befale
Boende
Bokungu
Opal

0

Owendo
Ndjolé
Booué
Okondja
Ewo
Lac Tumba
Z
Monkoto
A I
Ikela

Port-
Gentil
Lopez
Lambaréné
Lastoursville
Francevillle
Okoyo
Mossaka
Lukolela
Kiri
Lokolama
Loto
Lomela
Lomе

E

Omboué
Fougamou
Mouanda
L. Mai-Ndombe
Inongo
Kutu
Dekese
Kole
Lodja

Iguéla
Moabi
Mouila
Tchibanga
Zanaga
Djambala
Mushie
Tolo
Oshwe
Lusambo

Setté Cama
Nyanga
Kibangou
Mossendjo
Komono
Sibiti
Kwamouth
Bandundu
Kasai
Dibaya
Lube
Ilebo
Bena Dibele

Mayumba
Mindouli
Brazzaville
Kinshasa
Masi-Manimba
Basongo
Mweka
Kananga

F

Pointe Noire
Cacongo
Tshela
Luozi
Inkisi
Kenge
Kikwit
Idiofa
Luebo
Demba
Dimbelenge

CABINDA
Cabinda
Boma
Madimba
Mbanza Ngungu
Popokabaka
Guagu
Makumbi
Kanang
Mbuji

Matadi
Maquela
do-Zombo
Feshi
Tshikapa
Kuzumba
Dibaya
Kand

Soyo
Nqui
Mbanza
Congo
Damba
Kasongo Lunda
Luiza
Luachimo

Nzeto
Uige
Sanza Pombo
Kahemba
Camissomba
Kapanga

Ambriz
N'Gage
Camabatela
Caúngula
Lúremo
Capaia
Katai

ATLANTIC

OCEAN
Luanda
Pta. das Palmeirinhas
Caxito
Quibaxe
Ndalatando
Quela
Lubala
Chiluage
Sandoa
Kafak

Muxima
Dondo
Malanje
Saurimo

10

Gunza
Gabela
A N G O L A
Cacolo
Muconda
Luau
Dilolo
Mutshatsh

Sumbe
Andulo
Cambundi-
Catembo
Luacano

Elevation scale:
ft / m

12 000 / 4000
9000 / 3000
6000 / 2000
4500 / 1500
3000 / 1000
1200 / 400
600 / 200
0 / 0
200 / 600

m / ft

100   0   100   200   300   400 miles
100   0   100   200   300   400   500   600 km

5   30   **6**   35   **7**   40   **8**

Omdurmân   El Khartûm Bahrî   Kereng   Musiwa   Dahlak Kebir
El Khartûm   Kassala   Akordat   Asmera   Zula   **A**
(Khartoum)   Barentu   ERITREA   Mersa Fatma   15

Malhao   El Wuz   El Kamlin   Khashm   Adi Ugri
Hamrat   Sodirî   El Geteina   Rufa'a   el Girba   Edd
esh Sheykh   Gedaref   Aksum   -116
Umm   Kagmar   Ed Dueim   El Mafâza   Ras Dashen   Mekele
El Fasher   Keddada   Umm Bel   Umm Dam   Sennâr   4620   Shkota   **B**
Wad Banda   En Nahud   El Obeid   Kôstî   Singa   Metema   Gonder   Lalibela
Taweisha   Abû   Er Rahad   Umm   El Jabelein   Gallâbât   L. Tana   Debre   Tendaho
Zabad   Ruwaba   Er Roseires   Tabor   Mekdela
Buram   Abu Matariq   Dilling   Renk   Mota   Dese
shanam   Muglad   El Odaiya   Heiban   Debre Markos
Kâdugli   Talodi   Kaka   Abbay   Aliba
Kingi   SUDAN   Tungaru   Kodok   (Blue Nile)   Ankober   10
Nyâmlell   Nil el Abyad   Malakâl   Nekemte   Gedo   Addis Abeba
Gogrial   Bentiu   (White Nile)   Abwong   Gimbi   Addis Alem   Awash
Rôga   Jur   Bahr   Fangak   Gambela   Dembidolo   ETHIOPIA   **C**
lem Zuber   Meshra   Ghazal   Nasir   Gore   L. Ziway   Asela
er Req   Sûdd   Sobat   Jima   L. Shala
Wâw   Duk Fadiat   Akôbo   Sodo   Goba   Ginir
Djema   Tonj   Rumbêk   Omo   Batu   4307
Zémio   Tamburâ   Tainya   Yirol   Bôr   Môji   L. Abaya   Yirga Alem
Dorumo   Amâdi   Tali P   Tombe   L. Shamo   Chencha
Ango   Maridi   Jûbâ   Kapoeta   Gidole   Burji   Negele
Niangara   Faraje   Yei   Nimule   Jarso   Yabelo   Arero
Bambili   Amadi   Watsa   Arua   Kaja Kaji   Torit   Lokitaung   Chew Bahir   El Niybo   **5**
Poko   Dungu   Gulu   Kitgum   (L. Stefanie)   Todenyang
Banalia   Isiro   Mungbere   Kabarega   Lira   Moroto   Mega
Bomili   Wamba   Mahagi   Falls   Lodwar   L. Turkana   Moyale   El Wak   **D**
Kisangani   Bafwasende   Bunia   L. Albert   Masindi   (L. Rudolf)   Buna
Bugunu   Irumu   Soroti   L. Kyoga   Mt Elgon   South Horr   Marsabit   Wajir   Dif
Ubundu   Beni   Ruwenzori   Hoima   Mbale   4321   Kitale   KENYA   Habaswein
Butembo   5109   Fort Portal   Tororo   Eldoret   Maralal
Lubutu   Kasese   UGANDA   Mubende   Jinja   Kakamega   Meru   Isiolo   0
Kirundu   Luofu   Kampala   Kisumu   Nyahururu   Mt Kenya   5199   Garissa
Lowa   L. Edward   Entebbe   Kisii   Kericho   Nakuru   Nanyuki   Murang'a
Mbarara   George   Masaka   Naivasha   Embu   **E**
Rutshuru   Kabale   1134   Karungu   Limuru   Thika   Kitui   Tana
Kalima   Goma   Victoria   Nairobi   Garsen   Lamu
Gisenyi   RWANDA   Bukoba   Ukerewe   Musoma   Machakos
Lac Kivu   Kigali   I.   Loliondo   Konza   Magadi   Makindu   Kibwezi
Bukavu   Butare   Geita   Nyahanga   L.   Formosa
Mwenga   Mwanza   Natron   Bay   **E**
Uvira   BURUNDI   Ngudu   Kilimanjaro   5895   Malindi
Kibombo   Bujumbura   Kibondo   Kahama   Lake   Moshi   Arusha   Takaungu
Kasongo   Kasulu   Eyasi   Taveta   Voi   Mombasa
Kigoma-Ujiji   Bukene   Shinyanga   Same   Kilindini
Kabambare   Kaliua   Nzega   Lake   Tanga
Kongolo   Uvinza   Usoke   Tabora   Manyara   Mbulu   Pangani   Pemba I.
Tshofa   Kabalo   Kasanga   Singida   Kondoa   Korogwe
773   Kolwesi   Kibwesa   Manyoni   Dodoma   Handeni   Zanzibar   **F**
Kisengwa   Mpanda   Sadani   Zanzibar I.
Ankoro   Karema   Rungwa   Kibaya   Mpwapwa   Bagamoyo
Kabalo   Manono   Maba   TANZANIA   Morogoro   Dar-es-Salaam
Mwanza   Kiambi   Kipili   Iringa   Gt. Ruaha   Kisiju
Kamina   Mwaza   Molira   Sumbawanga   Mahenge   Rufiji   Utete   Mafia I.
Mitwaba   Chiengi   Pweto   L. Rukwa   Kipembawe   Mohoro
L. Upemba   Mweru   Kasanga   Njombe   Kilwa Kivinje
Bukama   Kilwa   Swamp   Chunya   Liwale
Kawambwa   Mbala   Mbeya   Tukuyu   Lindi
Kamina   Kolulwe   Rasa   Songea   Nachingwea   Mtwara   **G**
Ludub   Kasenga   ZAMBIA   Kasama   Isoka   Katenga   L. Nyasa   Manda   Masasi   Mikindani   Cabo   Delgado
wezi   Likasi   Mambilima   Luwingu   Chinsali   Livingstonia   Tunduru   Newala   Palma
Falls   Mansa   MALAWI   Mbamba Bay   Ruvuma   Moçimboa
Tenke   L. Chambeshi   Mbamba Bay   da Praia
Bangweulu   Nkata Bay

5   **6**

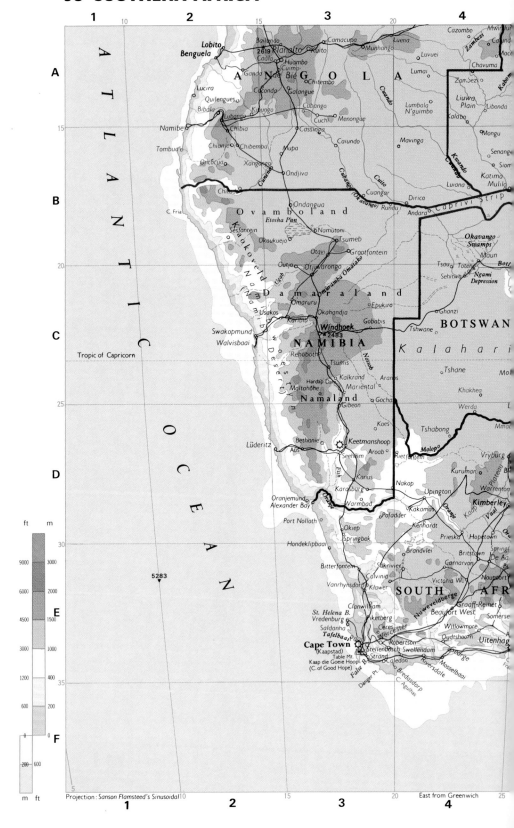

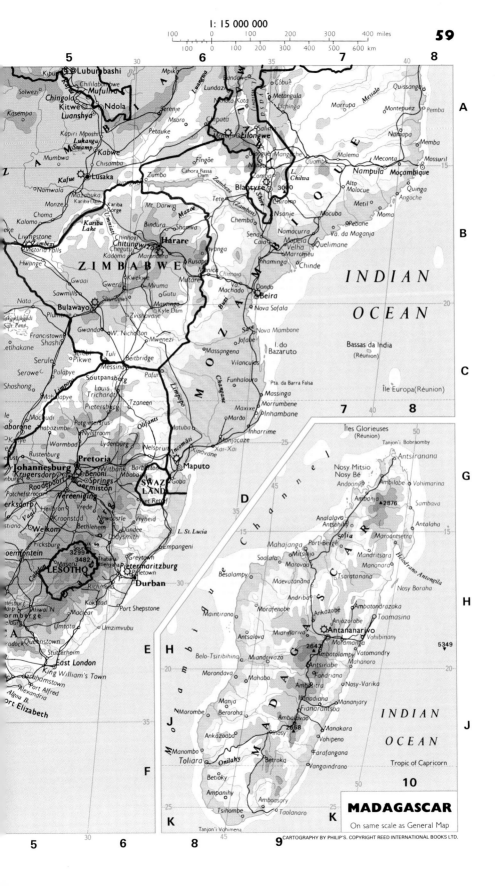

1: 15 000 000

100   0   100   200   300   400 miles

100   0   100   200   300   400   500 600 km

5   **6**   **7**   **8**

A

B

INDIAN

OCEAN

C

Kipushi ○Lubumbashi
Chililabombwe
Mufulira
Solwezi   Chingola   Kasenga
Kitwe ○Ndola
Luanshya
Kapiri Mposhi
Lukanga Kabwe
Swamp
Mumbwa   Chisamba
Kafue ○Lusaka
Namwala   Mazabuka   Kariba Dam
Monze   Kariba Lake
Choma
Kalomo
Livingstone   Zambezi
Victoria Falls
Hwange
Nata   Plumtr
okgadikgadi Salt Pans
Francistown Shashi
etlhakane
Serule
Serowe   Palapye
Shoshong
Mochudi
aborone
Kanye
Kuruman
erksdorp
stiana

Mpika
Lundazi
Nchota Kota
Chipata
Petauke
Fingoe
Cahora Bassa Dam
Zumbo
Tete (Zambezi)
Mt. Darwin
Bindura   Shamva
Chinhoyi
Kadoma Marondera
Rusape
Mutare
Chimanimani
Mvuma
Gweru Gutu
Shurugwi
Zvishavane
Gwanda
W. Nicholson
Mwenezi
Tuli
Beitbridge
Messina
Pafuri
Soutpansbe
Louis
Trichardt
Pietersburg
Olifants
Nylstroom
Warmbad
Rustenburg
Lydenburg

Cóbuè
Metangula
Lichinga
Marrupa
Montepuez Pemba
Quissanga

Nampula Moçambique
Alto Molocue Quinga
Angoche
Metil
Moma
Pebane
Va. da Maganja
Mopeia Quelimane
Velha
Marromeu
Chinde

MADAGASCAR

On same scale as General Map

CARTOGRAPHY BY PHILIP'S. COPYRIGHT REED INTERNATIONAL BOOKS LTD.

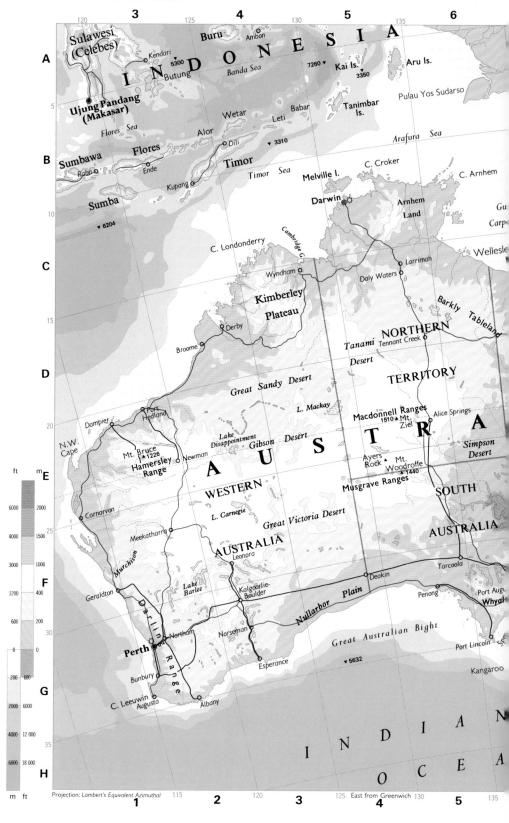

Sulawesi (Celebes)
INDONESIA
Buru
Ambon
Kendari
5300
Butung
Banda Sea
7260
Kai Is.
3350
Aru Is.
Pulau Yos Sudarso
Ujung Pandang (Makasar)
Wetar
Leti
Babar
Tanimbar Is.
Flores Sea
Alor
Arafura Sea
Sumbawa
Flores
Dili
3310
Timor
Raba
Ende
Timor Sea
Melville I.
C. Croker
C. Arnhem
Sumba
Kupang
Darwin
Arnhem Land
Gu
Carp
6204
Wellesle
C. Londonderry
Cambridge G.
Larrimah
Wyndham
Daly Waters
NORTHERN
Kimberley Plateau
Barkly Tableland
Derby
Tanami
Tennant Creek
Broome
Desert
TERRITORY
Great Sandy Desert
L. Mackay
Macdonnell Ranges
Alice Springs
Dampier
Port Hedland
1510
Mt. Ziel
N.W. Cape
Lake Disappointment
Gibson Desert
A U S T R A
Mt. Bruce
1226
Newman
Ayers Rock
Mt. Woodroffe
Simpson Desert
Hamersley Range
1440
WESTERN
AUSTRALIA
Musgrave Ranges
SOUTH
Carnarvon
L. Carnegie
Great Victoria Desert
AUSTRALIA
Meekatharra
AUSTRALIA
Tarcoola
Murchison
Leonora
Deakin
Port Augu
Geraldton
Lake Barlee
Kalgoorlie-Boulder
Penong
Whyal
Nullarbor Plain
Darling Range
Northam
Norseman
Port Lincoln
Perth
Great Australian Bight
Bunbury
Esperance
5632
Kangaroo
C. Leeuwin
Augusta
Albany
I N D I A N
O C E A

ft   m
6000 2000
4000 1500
3000 1000
1200   400
600   200
0   0
200   600
2000 6000
4000 12 000
6000 18 000
m   ft

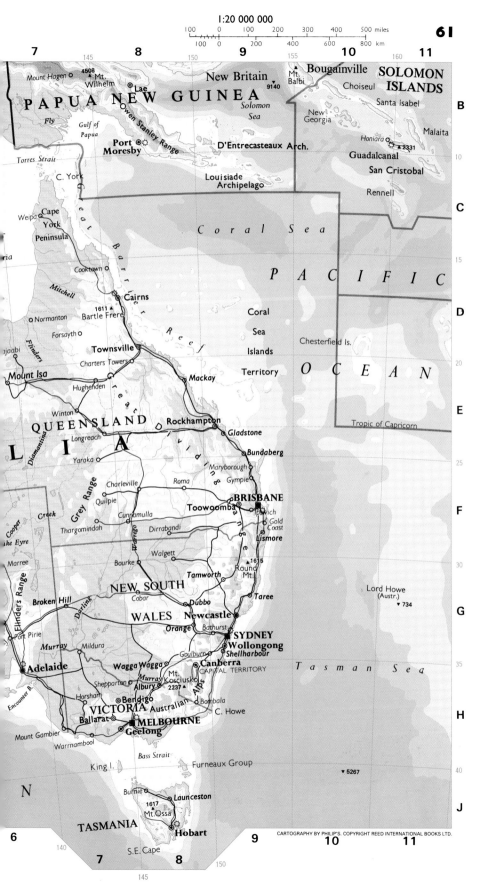

Mount Hagen 4508 Mt.
Wilhelm Lae
New Britain   Mt.   Bougainville   SOLOMON
Balbi   ISLANDS
PAPUA NEW GUINEA   9140   Choiseul   Santa Isabel   B
Owen Stanley Range   Solomon   New
Fly   Gulf of   Sea   Georgia   Malaita
Papua   Honiara ▲2331
Port ◎⊙   D'Entrecasteaux Arch.   **Guadalcanal**   10
Torres Strait   **Moresby**   **San Cristobal**
C. York   Louisiade   Rennell
Weipa   **Cape**   Archipelago   C
York
Peninsula   **C o r a l   S e a**
Cooktown   15
Mitchell   Cairns   **P A C I F I C**
1611 ▲   Coral
Normanton   Bartle Frere   Sea   D
Forsayth   Islands
Townsville   Reef   Territory   **O C E A N**
Charters Towers   20
Mount Isa   Mackay
Hughenden
Winton   E
QUEENSLAND Rockhampton   Gladstone   Tropic of Capricorn
Longreach   Bundaberg   25
Yaraka   **A**   Maryborough
Charleville   Roma   Gympie
Quilpie   Cunnamulla   **BRISBANE**   F
Toowoomba Ipswich
Thargomindah   Dirranbandi   Gold
Coast
Warrego   Lismore
Walgett   30
Bourke   Tamworth   Round   ▲1615
Mt.
NEW SOUTH   Cobar   Dubbo   Taree   Lord Howe
(Austr.)   G
WALES   **Newcastle**   ▼734
Broken Hill   Orange   Bathurst
Port Pirie   Murray   **SYDNEY**
Mildura   Goulburn   **Wollongong**   **T a s m a n   S e a**   35
Adelaide   Wagga Wagga   **Canberra**
Shepparton   Mt.   CAPITAL TERRITORY
Horsham   Albury Kosciusko
2237 ▲   H
VICTORIA Australian   Bombala
Ballarat   Alps   C. Howe
Mount Gambier **MELBOURNE**   ▼5267   40
Geelong
Warrnambool   Bass Strait
King I.   Furneaux Group
N   J
Burnie   Launceston
1617
TASMANIA Mt.Ossa   ▼5267

6   7   8   9   10   11
140   150

Hobart
S.E.Cape

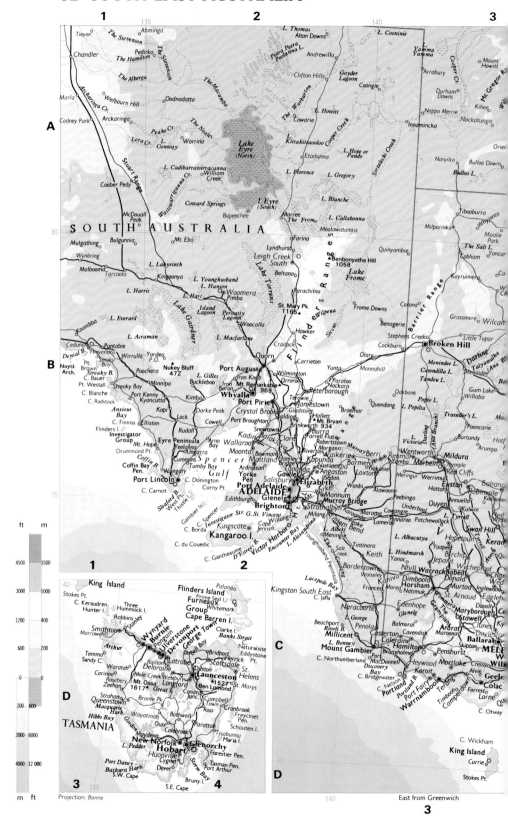

1   **2**   **3**

135   140

Tieyon
Abminga
L. Thomas
Alton Downs
L. Cooninie

The Stevenson
Pedirka
Chandler
The Hamilton
Peera Peera
Poolanna L.
Andrewilla
Yamma
Yamma

The Alberga
Welbourn Hill
Oodnadatta
Clifton Hills
Goyder
Lagoon
Arrabury
Cooper Cr.
Mount
Howitt

Marla
Arckaringa Cr.
The Neales
The Macumba
Coongie
Durham
Downs
Mc Gregor Cr.

Cadney Park
Arckaringa
Peake Cr.
The Warburton
Cowarie
Nappa Merrie
Innamincka
Kihee
Nockatunga

A
Lora Cr.
L. Conway
Warrina
Lake
Eyre
(North)
L. Kittakittaooloo Cooper Creek
L. Hope or
Pando
Etadunna
Orier
Naryilco
Bulloo Downs

Coober Pedy
Stuart Range
L. Cadibarrawirracanna Cr.
William
Creek
L. Florence
L. Gregory
Bulloo L.

McDouall
Peak
Warriguruguenna Cr.
Coward Springs
Bopeechee
L. Eyre
(South)
L. Blanche
Tibooburra
Whyjonta

30
SOUTH AUSTRALIA
Marree
The Frome
L. Callabonna
Milparinka
Moolia
Park
Yancar

Mulgathing
Bulgunnia
Mt. Eba
Farina
Moolawatana
The Salt L.
Cobham

Wynbring
L. Labyrinth
Leigh Creek
South
Benbonyathe Hill
1058
Quinyambie

Malbooma
Tarcoola
Kingoonya
Beltana
Lake
Frome
Kayrunnera
Ca

L. Harris
L. Younghusband
L. Hanson
Woomera
Pimba
Parachilna
Frome Downs
Cotona
Grassmere
Wilcan

Island
Lagoon
Perinatty
Lagoon
St Mary Pk.
1165
Flinders Range
Ilpena
Benagerie
Little Topar

L. Everard
Woocalla
Hawker
Stephens Creek
Broken Hill
Darling

Koonibba
Puntabie
L. Acraman
L. Macfarlane
Cradock
Olary
Cockburn
Menindee L.
Talyawalka
Menindee
Ana Bc

Ceduna
Yardea
P.O.
Quorn
Carrieton
Yunta
Mannahill
Cawndilla L.
Tandou L.
Ba

B
Denial B.
Thevenard
Wirrulla
Poochera
Nukey Bluff
472
Port Augusta
Wilmington
Orroroo
Parataoo
Nackara
Oakbank
Popio L.
Gum Lake
Willaba

Nuyts
Arch.
Pt.
Brown
Smoky
Bay
L. Gilles
Iron Knob
Iron
Baron
Mt. Remarkable
969
Peterborough
Terowie
Quondong
L. Popilta

Streaky B.
C. Bauer
Minnipa
Buckleboo
Whyalla
Jamestown
Braemar
Traveller's L.

Pt. Westall
C. Blanche
Port Kenny
Kyancutta
Kimba
Port Pirie
Gladstone
Hallett
Mt. Bryan 934
Riverton
L.
Victoria
Darling
Ana Branch
Pooncarie

Anxious
Bay
C. Finniss
Elliston
Kopi
Darke Peak
Crystal Brook
Spalding
Burra
Wentworth
Burtundy
Hatf

Flinders I.
Rudall
Cowell
Port Broughton
Snowtown
Farrell Flat
Morgan
Berri
Merbein
Mildura
Arumpo

**Investigator
Group**
Mt. Hope
Eyre Peninsula
Arno
Bay
Kadina
Balaklava
Clare
Robertstown
Riverton
Murray Berri
Renmark
Yamba
Red Cliffs
Belrang

Drummond Pt.
Cummins
Ungarra
Wallaroo
Moonta
Bowmans
Kapunda
Barmera
Nangiloc
Meringur
Werrimul
Iraymple
Euston

Coffin B.
Coffin Bay
Pen.
Wangary
Tumby Bay
Ardrossan
Spencer
Maitland
Hamley
Bridge
Angaston
Nuriootpa
Loxton
Karoonda
Robinvale

Port Lincoln
C. Donington
Gulf
Yorke
Pen.
Salisbury
Sedan
Wanbi
Alawoona
Hattah

C. Carnot
Corny Pt.
Port Adelaide
Elizabeth
Mannum
Peebinga
Annuello
Ouyen

35
Sleaford B.
West Pt.
Thistle
ADELAIDE
Glenelg
Murray Bridge
Marama
Cowangie
Underbool
Kulwin

C. Spencer
Edithburgh
Brighton
Strathalbyn
Karoonda
Lameroo
Pinnaroo
Patchewollock
L. Tyrrell

C. Borda
Investigator Str.
G. St. Vincent
Willunga
Milang
Albert
L. Alexandrina
Meningie
Keith
Tintinara
Younghusband Peninsula
Salt
Creek
L. Albacutya
Hopetoun
Swan Hill
Keran

Gambier Is.
Kingscote
Cape
Jervis
Victor Harbor
Encounter Bay
Birchip
Jeparit
Witchcliffe
V

Kangaroo I.
C. Gantheaume
D'Estrees B.
C. du Couedic
Lacepede Bay
Kingston South East
C. Jaffa
Bordertown
Wolseley
Kaniva
Nhill
Warracknabeal
Donald
Char

ft   m
Naracoorte
Edenhope
Dimboola
Horsham
Murtoa
St. Arnaud
Eaglehawk

4500 1500
Beachport
L. George
Penola
Balmoral
Natimuk
Maryborough
Stawell
Dunolly
Ky

3000 1000
Riswli B.
Millicent
Casterton
Cavendish
Coleraine
Hamilton
Clunes
Daylesford

1200 400
Mount Gambier
C. Northumberland
Port
MacDonnell
Discovery
Bay
C. Bridgewater
Penshurst
Skipton
Ararat
Maroona
Ballarat
MELB
Wil

600   200
C. Nelson
Heywood
Mortlake
Cressy
Geele

0   0
Portland
Port Fairy
Warrnambool
Keroit
Alvie
Colac

200   600
Lorne

2000 6000
C. Wickham
King Island
Currie

4000 12 000
C. Otway
Stokes Pt.

m   ft

1
King Island
Palana
Flinders Island
Prime Seal I.
Whitemark

40
Stokes Pt.
Furneaux
Group
Cape Barren I.
Clarke I.
Banks Strait

C. Keraudren
Hunter I.
Three
Hummock I.
Robbins I.
Naturaliste
Pt.
Eddystone
Pt.

Smithton
Marrawah
Stanley
Wynyard
Burnie
Penguin
Ulverstone
Devonport
George Town
Bell Bay
Bridport
Herrick
St.
Helens

Temma
Sandy C.
Arthur
Waratah
Railton
Latrobe
Deloraine
Dilston
Longford
Conara
Junc.
St. Marys

Corinna
Zeehan
Rosebery
Mt. Ossa
1617
Mole Creek
Westbury
Launceston
Ben Lomond
1527
Campbell
Town
Ross

Strahan
Queenstown
Macquarie
Harb.
Bronte Pk.
Cranbrook
Freycinet
Pen.
Schouten I.

Hibbs Bay
TASMANIA
Wayatinah
Bothwell
Ouse
Triabunna
Maria I.

Port Davey
Maydena
Colebrook
Parattah
Forestier Pen.

Bathurst Harb.
L. Pedder
Huopville
Cygnet
New Norfolk
Glenorchy
Hobart
Tasman Pen.
Port Arthur

S.W. Cape
Dover
Bruny I.
Storm Bay

3
S.E. Cape
4
Stokes Pt.

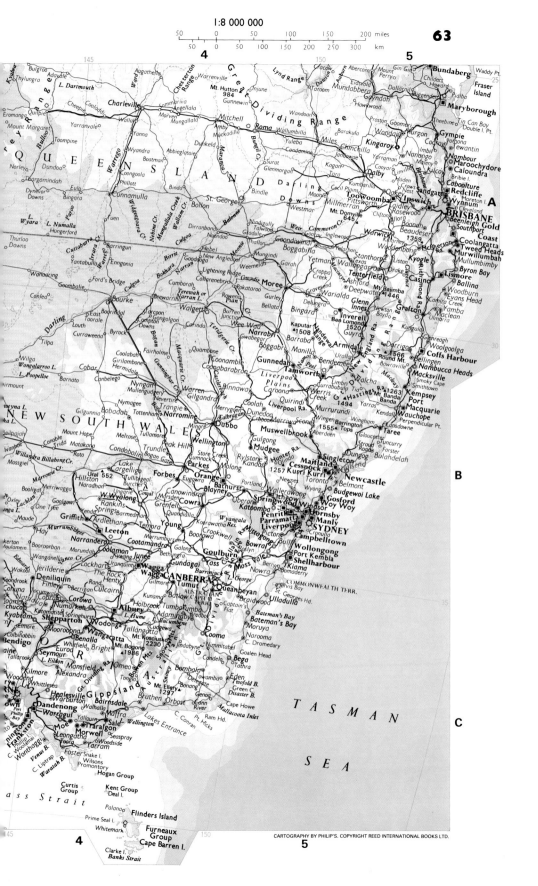

1:8 000 000

50   0   50   100   150   200 miles

50   0   50   100   150   200   250   300 km

CARTOGRAPHY BY PHILIP'S. COPYRIGHT REED INTERNATIONAL BOOKS LTD.

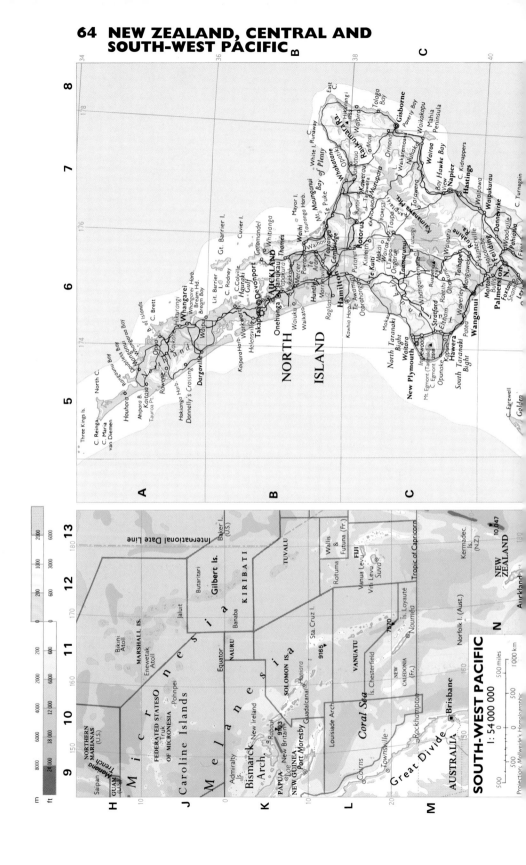

NORTH ISLAND

Three Kings Is.
C. Reinga
C. Maria van Diemen
North C.
Parengarenga Harb.
Houhora
Ahipara B.
Kaitaia
Tauroa Pt.
Hokianga Harb.
Donnelly's Crossing
Kaikohe
Rawene
Dargaville
Opua
Waipu
Bay of Islands
C. Brett
Kaipara Harb.
Workworth
Helensville
Takapuna
Devonport
AUCKLAND
Whangarei
Whangarei Harb.
Bream Hd.
Bream Bay
C. Rodney
Lt. Barrier I.
Gt. Barrier I.
Cuvier I.
C. Colville
Hauraki Gulf
Coromandel
Whitianga
Onehunga
Manukau
Waiuku
Waikato
Mercer
Huntly
Hamilton
Thames
Waihi
Te Aroha
Te Awamutu
Cambridge
Putaruru
Tauranga
Mt. Maunganui
Mayor I.
Bay of Plenty
Whakatane
Opotiki
East C.
Hicks Bay
Te Araroa
Tologa Bay
Gisborne
Poverty Bay
Waipa
Tokomaru
Whakatane
Rotorua
Te Kuiti
Kinleith
Roglan
Kawhia Harb.
North Taranaki Bight
New Plymouth
Mt. Egmont (Taranaki)
C. Egmont
Opunake
South Taranaki Bight
Hawera
Patea
Wanganui
Taihape
Taumarunui
Ongarue
Ruapehu
Ohakune
Raetihi
Waiouru
Stratford
Eltham
Inglewood
Waverley
Marton
Bulls
Feilding
Palmerston N.
Foxton
Levin
Waikanae
Martinborough
Masterton
Woodville
Dannevirke
Waipukurau
Waipawa
Hastings
Napier
Hawke Bay
C. Kidnappers
Wairoa
Mahia Peninsula
Nuhaka
Wairakei
Taupo
L. Taupo
Turangi
Kaimanawa Mts.
Kaweka Ra.
Ruahine Ra.

NORTH
ISLAND

C. Farewell
Golden

SOUTH-WEST PACIFIC
1:54 000 000

Projection: Mollweide's Homolographic

m ft
8000 24 000
6000 18 000
4000 12 000
2000 6000
200 600
0 0

Mariana Trench
Saipan
GUAM (U.S.)
NORTHERN MARIANAS (U.S.)

M i c r o n e s i a

Caroline Islands
FEDERATED STATES OF MICRONESIA
Truk
Pohnpei

Bikini Atoll
Enewetak Atoll
MARSHALL IS.

Jaluit
Banaba
NAURU
Equator
Butaritari
Gilbert Is.
KIRIBATI
Baker I. (U.S.)
International Date Line

M e l a n e s i a

Admiralty Is.
Bismarck Arch.
New Ireland
Rabaul
New Britain
PAPUA NEW GUINEA
Lae
Port Moresby
Louisiade Arch.
SOLOMON IS.
Honiara
Guadalcanal
9165
Sta Cruz I.
TUVALU
Rotuma
Wallis & Futuna (Fr.)
FIJI
Vanua Levu
Viti Levu
Suva

Coral Sea
Is. Chesterfield
VANUATU
7520
Is. Loyauté
NEW CALEDONIA (Fr.)
Nouméa

AUSTRALIA
Great Divide
Cairns
Townsville
Rockhampton
Brisbane
Norfolk I. (Aust.)
Tropic of Capricorn
Kermadec Is. (N.Z.)
10 047
NEW ZEALAND
Auckland

0 500 1000 km
0 500 miles

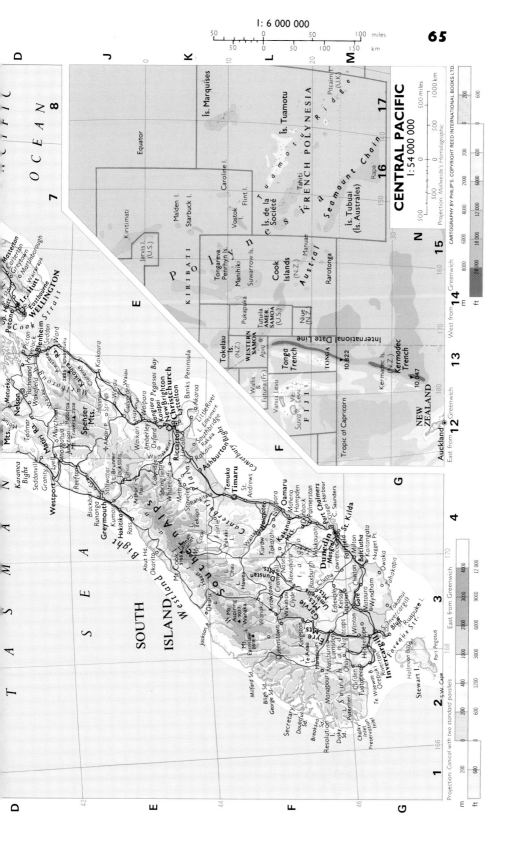

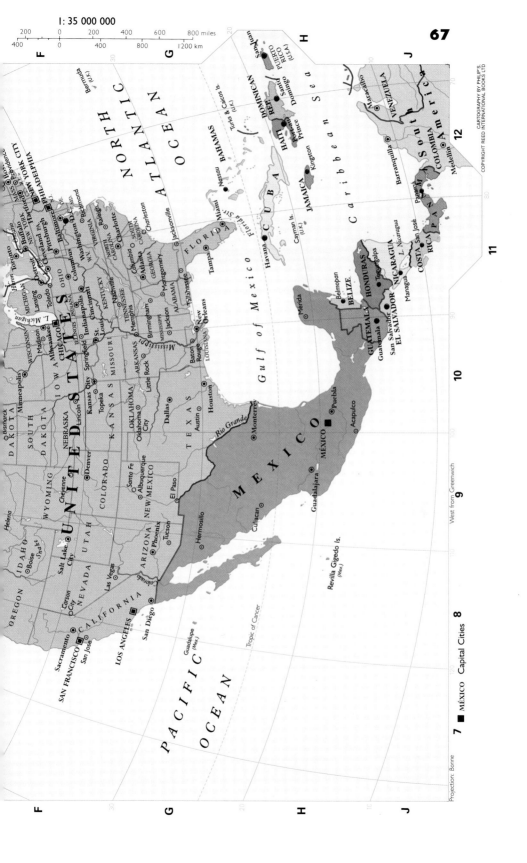

1 : 35 000 000

200   0   200   400   600   800 miles
400   0   400   800   1200 km

Projection: Bonne

7 ■ MÉXICO Capital Cities 8

West from Greenwich

NORTH ATLANTIC OCEAN

PACIFIC OCEAN

UNITED STATES

MEXICO

Gulf of Mexico

Caribbean Sea

CUBA

BAHAMAS

JAMAICA

HAITI
DOMINICAN REP.
PUERTO RICO (U.S.A)

GUATEMALA
BELIZE
HONDURAS
EL SALVADOR
NICARAGUA
COSTA RICA
PANAMA

COLOMBIA
VENEZUELA

SOUTH America

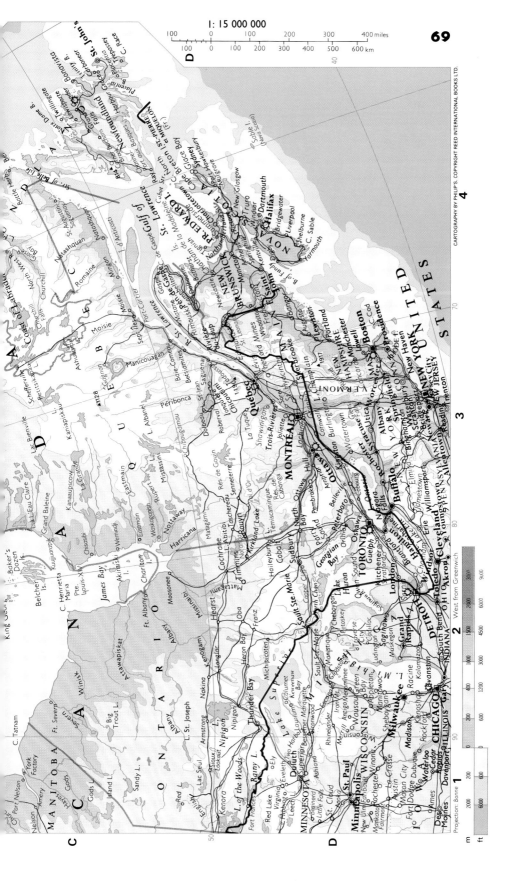

1 : 15 000 000

| 100 | 0 | 100 | 200 | 300 | 400 miles |

| 100 | 0 | 100 | 200 | 300 | 400 | 500 | 600 km |

D

40

4

CARTOGRAPHY BY PHILIP'S. COPYRIGHT REED INTERNATIONAL BOOKS LTD.

Projection: Bonne

West from Greenwich

C A N A D A

MANITOBA

ONTARIO

QUÉBEC

NEWFOUNDLAND

NOVA SCOTIA

NEW BRUNSWICK

PRINCE EDWARD I.

Gulf of St. Lawrence

Coast of Labrador

James Bay

MAINE

NEW HAMPSHIRE

VERMONT

NEW YORK

MASS.

CONN.

NEW JERSEY

PENNSYLVANIA

OHIO

MICHIGAN

INDIANA

ILLINOIS

WISCONSIN

MINNESOTA

IOWA

U N I T E D S T A T E S

Lake Superior

L. Huron

L. Michigan

Lake Erie

L. Ontario

MONTRÉAL

Québec

TORONTO

DETROIT

CHICAGO

Milwaukee

Minneapolis

St. Paul

Boston

NEW YORK

Buffalo

Cleveland

Halifax

Dartmouth

St. John's

Ottawa

Hull

Georgian Bay

C. Cod

L. Champlain

Thunder Bay

Duluth

Green Bay

Madison

Grand Rapids

Toledo

Akron

Youngstown

Allentown

Reading

Trenton

Newark

New Haven

Providence

Worcester

Springfield

Hartford

Albany

Syracuse

Rochester

Utica

Schenectady

Elmira

Scranton

Williamsport

Binghamton

Kingston

Watertown

Portland

Manchester

Lowell

Lawrence

Bangor

Fredericton

Saint John

Moncton

Charlottetown

Sydney

Truro

New Glasgow

Yarmouth

Shelburne

Liverpool

Bridgewater

Windsor

Kentville

Amherst

Sherbrooke

Trois-Rivières

Shawinigan

Joliette

Drummondville

Chicoutimi

Rimouski

Gaspé

Bay of Fundy

C. Sable

Sable I. (Nova Scotia)

ST. PIERRE & MIQUELON (Fr.)

Cape Breton I.

Notre Dame B.

Placentia B.

Trinity B.

C. Race

Bonavista B.

Str. of Belle Isle

Anticosti I.

Îs. de la Madeleine

Manicouagan

Baie-Comeau

Sept-Îles

Port-Cartier

Schefferville

Churchill Falls

Moisie

Romaine

Natashquan

Goose B.

Battle Harbour B.

Hamilton Inlet

Churchill

L. Bienville

Kanaaupscow

La Grande

Eastmain

Rupert

Nottaway

Harricana

Moosonee

Ft. Albany

Attawapiskat

Winisk

Fort Severn

C. Tatnam

C. Henrietta Maria

Akimiski I.

Belcher Is.

King George Is.

Charlton I.

Ft. George

Kuujjuarapik

Great Whale

Grand Baleine

Ft. l'Eau Claire

Baker's Dozen

Pte. Louis-XIV

York Factory

Port Nelson

Nelson

Hayes

Gods

Island L.

Sandy L.

Big Trout L.

L. St. Joseph

Winisk L.

Kenogami

Albany

Ogoki

Nipigon

L. Nipigon

Sioux Lookout

Kenora

Red Lake

English R.

Rainy L.

L. of the Woods

Fort Frances

International Falls

Virginia

Hibbing

Ely

Mesabi Range

Ashland

Eau Claire

Wausau

Oshkosh

Appleton

Sheboygan

Racine

Kenosha

Evanston

Gary

South Bend

Kalamazoo

Lansing

Flint

Saginaw

Muskegon

Traverse City

Cheboygan

Sault Ste. Marie

Sudbury

North Bay

Sturgeon Falls

Timmins

Cochrane

Kapuskasing

Hearst

Geraldton

Kirkland Lake

Noranda

Val d'Or

Senneterre

Matagami

Moosonee

Rés. de Gouin

Rés. Cabonga

La Tuque

Roberval

L. St-Jean

Dolbeau

Péribonca

Mistassini

L. Mistassini

Chibougamau

Pembroke

Peterborough

Belleville

Cobourg

Owen Sound

Kitchener

Guelph

Brantford

London

Windsor

Hamilton

Niagara Falls

St. Catharines

Oshawa

Cornwall

Burlington

Montpelier

Concord

St-Hyacinthe

St-Jean

St. Laurent

Rivière-du-Loup

Edmundston

Mingan

Betsiamites

Tadoussac

Saguenay

St-Siméon

C N S

C A N A D A

D

C

1

2

3

4

50

90

80

70

60

m   ft

2000 6000

600   1200

200   600

0   0

3000	9000
2000	6000
1500	4500
1000	3000
600	1200
200	600

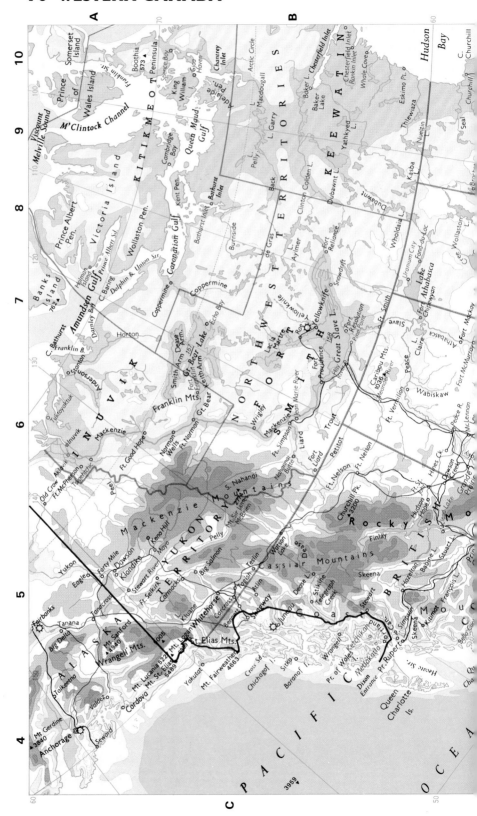

1: 15 000 000

100   0   100   200   300   400 miles
100   0   100 200 300 400 500 600 km

CARTOGRAPHY BY PHILLIPS. COPYRIGHT REED INTERNATIONAL BOOKS LTD.

ALASKA
1:30 000 000
100   200   300 miles
100 0 100 200 300 400 km

Projection: Bonne

West from Greenwich

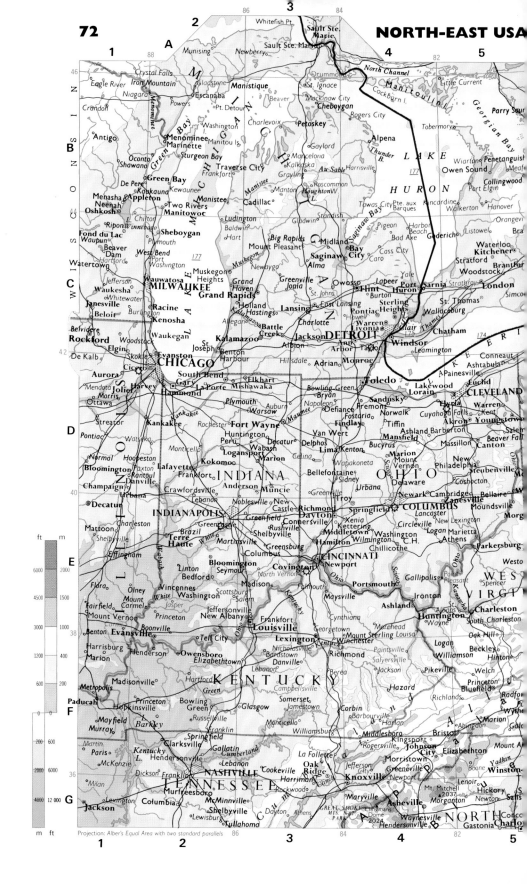

Projection: Alber's Equal Area with two standard parallels

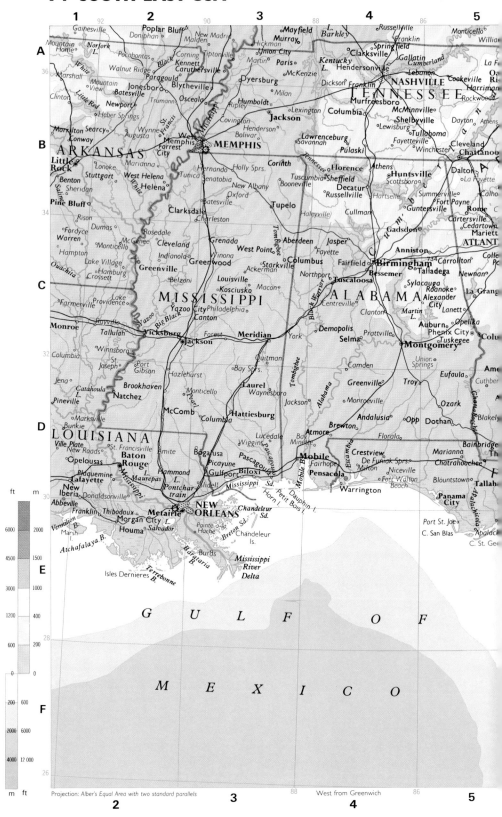

Projection: Alber's Equal Area with two standard parallels

1: 6 000 000

50   0   50   100 miles
50   0   50   100   150 km

A

Currituck Sd.

Harlan Abingdon S. Galax Martinsville Eden Danville Emporia Winton Elizabeth
Middlesboro Bristol City
Kingsport Roxboro Oxford Roanoke Rapids Roanoke City
Rogersville Johnson Mount Airy Reidsville Henderson Edenton Manteo
Morristown City Elizabethton Yadkin Greensboro Burlington Durham Roanoke I.
Jefferson Greeneville Boone Winston-Salem Thomasville High Graham Chapel Hill Rocky Mount Albemarle Sd. 36°
City Newport Lenoir Hickory Point Lexington Haw Raleigh Wilson Williamston
Knoxville Mt. Mitchell Statesville Salisbury Asheboro Smithfield Washington Greenville Pamlico
Maryville Asheville 2037 Morganton Newton Kannapolis Sanford Dunn Goldsboro Kinston New Bern
EAT. SMOKY Clingmans Concord CAROLINA Neuse Pamlico Sound
MTS. NAT. Dome NORTH Shelby Gastonia Southern Clinton Jacksonville Hatteras
PARK 2024 Waynesville Hendersonville Charlotte Pines Fayetteville Beaufort Raleigh
Murphy Brevard Monroe Clinton Jacksonville B.
Brasstown Bald Spartanburg Gaffney C. Lookout
1458 Greenville Rock Hill Laurinburg Cape Onslow
Seneca Easley Union Chester Lancaster Bennettsville Lumberton Fear B.
Toccoa Anderson Laurens Hartsville Dillon Whiteville Wilmington
Hartwell Bolton Greenwood Newberry Camden Darlington Mullins
Gainesville Abbeville Saluda Columbia Florence Marion
Elberton L. Sumter Lake City Conway Southport 34°
Athens Clark Murray Monning Kingstree Myrtle Beach C. Fear
Lawrenceville Hill L. SOUTH CAROLINA Manning
Decatur East Point Orangeburg L. Georgetown
Covington Augusta Aiken Bamberg Marion L.
Griffin Sparta Waynesboro Summerville Moultrie Cooper
Thomaston Milledgeville Walterboro North Charleston 32°
t Valley Macon Millen Ridgeland Charleston Mt. Pleasant
Warner Swainsboro Ogeechee Hampton Combahee
Perry Robins Statesboro Beaufort
Cochran Dublin Parris I.
Eastman Vidalia Savannah
Cardele Hazlehurst Altamaha Hinesville Ossabaw I.
Fitzgerald Baxley St. Catherines I.
Sylvester Douglas Jesup Sapelo I.
Tifton Satilla Brunswick 30°
Moultrie Adel Waycross
Cairo Okefenokee Folkston Cumberland I.
sville Valdosta Swamp Fernandina Beach
Quitman Jasper St. Johns
Monticello Madison Jacksonville
Apalachee Live Oak JACKSONVILLE Beach
abelle B. Perry Lake Green Cove Springs
City Starke St. Augustine
High Springs Palatka
Cross City Gainesville Bunnell
L. Ormond
Ocala George Beach 28°
Crystal River Daytona Beach New
Inverness De Land Smyrna
Eustis Sanford Beach
Leesburg Titusville
Brooksville Winter Park
Dade City Orlando C. Canaveral
Tarpon Springs Kissimmee Cocoa Merritt Island
Lakeland Haines City Melbourne
Clearwater Winter Haven
Largo TAMPA Bartow Vero Beach
St. Petersburg Indian 26°
Bradenton Sebring Fort Pierce River
Sarasota Istokpoga Stuart Grand Cays
Arcadia Kissimmee Okeechobee
L. Little Abaco I. Gt. Guana Cay
Punta Gorda Okeechobee Pahokee Settlement Hope
La Belle Belle Freeport Pt. Grand Town
Charlotte Harb. Fort Glade West Palm Bahama I. Great
Cape Myers Beach Abaco I.
Coral Immokalee Boca Raton Delray Beach BAHAMAS
Naples Big Cypress Swamp Pompano Beach
Everglades Carol City Fort Lauderdale Hollywood
Hialeah Miami Beach
EVERGLADES MIAMI Biscayne
NAT. PARK B.
Homestead

B **C** **D** **E** **F** **G**

GEORGIA FLORIDA ATLANTIC OCEAN

6   7   8   9   10

84° 82° 80° 78° 76°

1 2 3 4 5 6

104 102 100 98

A

Scobey Plentywood Crosby Bowbells Mohall Bottineau Rolla Cavalier Emers
Kenmare Souris Langdon
Williston Stanley Minot Towner Rugby Cando Grafton
48 Wolf Point Missouri Park River Warren Ec
Fairview New Town Velva Lakota Larimore Grand Gra
Fort Peck L. Sidney Watford City Garrison Fessenden New Rockford Sheyenne Northwood Cooperstown Moyville
Circle L. Sakakawea McClusky Carrington Cooperstown Hillsbor

B

N O R T H   D A K O T A

Manning Stanton Washburn Harvey Jamestown Valley City Fargo
Glendive Center
Terry Wibaux Beach Dickinson Hebron Mandan Steele Lisbon
Miles City Little Missouri Bismarck La Moure Wahpet
Baker Heart Carson Napoleon Ellendale Forman
46 Mott Cannonball Linton Ashley
Ekalaka Bowman Fort Yates Lake Ellendale Britton Sisseton
Broadus Hettinger Lemmon McIntosh Eureka Leola Webster
Buffalo Grand Mound City Selby Ipswich Aberdeen
Bison Timber Lake Mobridge Oahe Webster Milb

C

Moreau Dupree Eagle Butte Gettysburg Faulkton Clark Wa
Little Belle Fourche **S O U T H   D A K O T A** Redfield La
Sundance Spearfish Cheyenne Onida Highmore Miller De Smet Coteau
44 Gillette Sturgis Deadwood Oahe Dam Pierre Huron Big
Belle Lead Black Rapid City Fort Pierre Wessington Sprs. Howard
Newcastle Hills 2207 Philip Bad Woonsocket Madison De
Custer Harney Pk. Kadoka Murdo Kennebec Chamberlain Mitchell Salem Rapic
Hot Springs White White River Alexandria Siou
Edgemont B a d l a n d s L. Francis Case Parker Falls
Martin Winner Armour Lake Andes

D

Douglas Chadron Niobrara Butte Yankton
Lusk Harrison Valentine Bassett Por
Crawford Rushville Ainsworth Atkinson O'Neill South Siou
Hemingford North Loup 1036 Plainview Dako
42 Torrington Alliance Sand Hills Mullen Neligh Wayne
Wheatland Hyannis Thedford Madison Norfolk
Scottsbluff Middle Taylor Burwell West Po
Gering Bridgeport **N E B R A S K A** Albion
Harrisburg Oshkosh Stapleton Broken Bow Greeley Columb
Laramie Lodgepole Cr. Kimball L. McConaughy Loup City Fullerton Schuyl
Cheyenne Sidney Ogallala North Platte St. Paul Central David City

E

6000 2000 Julesburg Gothenburg Cozad Grand Island York Seward
Fort Collins Sterling South Platte Grant Lexington Kearney Aurora
Loveland Greeley Holyoke Imperial Curtis Elwood Hastings Geneva Wilt
Boulder Longmont Fort Morgan Akron Frenchman Cr. Trenton McCook Holdrege Red Cloud Hebron Fairbury
Lafayette Wray Benkelman Beaver City Alma Franklin Belleville
Golden Brighton Byers Atwood Oberlin Norton Smith Center Mankato Concordia Republi
DENVER Aurora Englewood St. Francis Phillipsburg Solomon Beloit Cla
Lakewood Castle Rock Colby Stockton Solomon Osborne Minneapolis Manha
Limon Burlington Oakley Hill City S. Fork Lincoln Junction

F

C O L O R A D O Hugo Goodland S. Fork Smoky Hills Saline Abilene City
Pikes Pk. Colorado Springs Big Sandy Cr. Cheyenne Wells Smoky Hill Hays Russell Salina
Canon City Fountain Sharon Springs Ellsworth
Pueblo Eads Leoti Scott City La Crosse Lyons McPherson
Ordway **K A N S A S**
Las Animas Lamar Tribune Dighton Great Bend Larned

38

ft m
12 000 4000
9000 3000
6000 2000
4500 1500
3000 1000
1200 400
600 200
0 0
200 600
m ft

Projection: Alber's Equal Area with two standard parallels

West from Greenwich

2 3 4 5 6
102 100 98

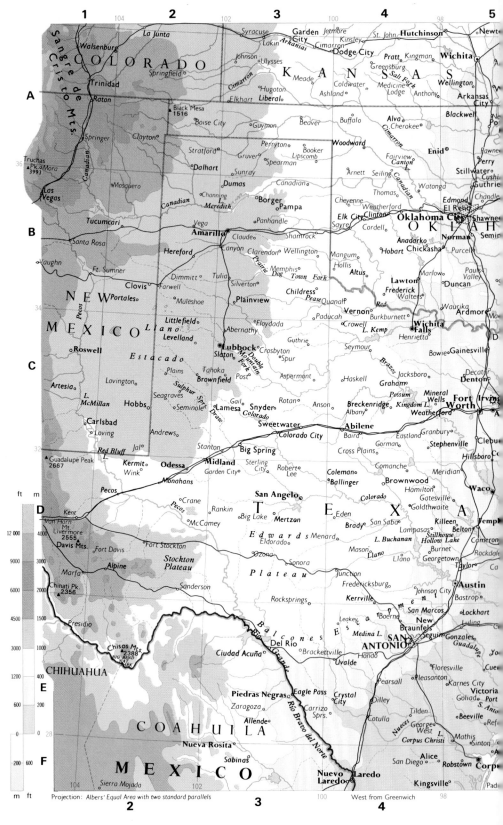

Projection: Albers' Equal Area with two standard parallels

West from Greenwich

1: 6 000 000

50   0   50   100 miles
50   0   50   100   150 km

6   7   8   9

A

MISSOURI

Steelville Murphysboro Marion
Camdenton Rolla Perryville Carbondale
Nevada Salem Ironton Fredericktown Anna
Yates Center Bolivar Jackson Metropolis
Iola Lebanon Houston Cape Girardeau Paducah
Fort Scott Buffalo Cairo
Chanute Stockton Marshfield Cabool Charleston Sikeston
Fredonia Lamar Greenfield Van Buren Poplar Bluff Dexter New Madrid Mayfield
Girard Pittsburg Springfield Ozark Hickman Union City
Parsons Carthage Aurora West Plains Doniphan Malden Tiptonville
Independence Joplin Monett Gainesville Corning Kennett Caruthersville McKenzie
Coffeyville Sedan Miami Neosho Cassville Norfork Pocahontas Black Blytheville Dyersburg
Bartlesville Vinita Bull Shoals Mountain Home Walnut Ridge Paragould Osceola Humboldt
Lake O' The Cherokees Rogers White Berryville Jonesboro Ripley Jackson
Claremore Springdale Siloam Springs Harrison Marshall Trumann Covington Henderson
Tulsa Fayetteville Mountain View Batesville Forrest City Bolivar
Wagoner Stilwell Clinton Heber Springs Wynne St. Francis Memphis MEMPHIS
Haskell Muskogee Ozark Clarksville Searcy Augusta Hernando Holly Sprs.
Okmulgee Sallisaw Van Buren Russellville Morrilton Marianna Tunica Senatobia
Henryetta Eufaula Stigler Ft. Smith Arkansas Conway Lonoke Helena New Albany
Wewaka Holdenville Poteau Booneville Little Rock Stuttgart West Helena Oxford
McAlester Eufaula L. Ouachita Mts. Benton Batesville Tupelo
Ada Wilburton Waldron Ouachita L. Sheridan White Clarksdale
Heavener Mena Hot Springs Pine Bluff Rison Charleston
Coalgate Broken Bow Lake Malvern Saline Dumas Rosedale Grenada Aberdeen
Atoka Antlers Arkadelphia De Queen Nashville Fordyce McGehee Cleveland Winona West Point
Tishomingo Millwood L. Prescott Warren Monticello Indianola Greenwood Columbus Starkville
Durant Idabel Hope Camden Lake Village Greenville Louisville Ackerman
Lake Texoma Red Hampton Ouachita Hamburg Belzoni Kosciusko Macon
Sherman Paris Clarksville Texarkana Magnolia Crossett MISSISSIPPI Philadelphia
Bonham Commerce El Dorado Providence Yazoo City Canton Meridian
McKinney Sulphur Springs Mount Pleasant Atlanta Linden Haynesville Homer Farmerville Lake Big Black Forest Quitman
Greenville Pittsburg Gilmer Jefferson Rayville Vicksburg Jackson
DALLAS Garland Quitman Minden Tallulah Port Gibson Bay Sprs.
Terrell Marshall Shreveport Ruston Monroe Winnsboro St. Joseph Hazlehurst Laurel
Waxahachie Cedar Creek Res. Longview Bossier City Jonesboro Columbia Waynesboro
Ennis Tyler Kilgore Coushatta Brookhaven Monticello Hattiesburg
Athens Henderson Carthage Mansfield Winnfield Jena Colfax Natchez McComb Columbia Lucedale
Jacksonville Tenaha Natchitoches Catahoula L. Pineville Wiggins
Fairfield Palestine Center Toledo Bend Reservoir Many Alexandria Marksville Bogalusa
Groesbeck Nacogdoches San Augustine Leesville Bunkie Picayune
Crockett Lufkin Sam Rayburn Res. LOUISIANA St. Francisville Amite Gulfport Biloxi
Centerville Groveton Jasper Oakdale Ville Platte New Roads Baton Rouge Slidell Mississippi Sd.
Hearne Madisonville Woodville Newton De Ridder Opelousas Hammond
Bryan Livingston Huntsville Livingston Eunice Plaquemine Lafayette Maurepas Pontchartrain
Navasota Conroe L. Kountze Silsbee Sulphur Lake Charles Crowley New Iberia Donaldsonville Metairie NEW ORLEANS Chandeleur Sd.
Brenham Hempstead Cleveland Orange Calcasieu L. Abbeville Franklin Thibodaux Morgan City L. Salvador Pointe-a-la-Hache Chandeleur Is.
HOUSTON Pasadena Liberty Port Arthur Sabine Grand L. Cameron Houma Breton Sd.
Rosenberg Richmond Baytown White L. Vermilion B.
Wharton Texas City Galveston B. Marsh Atchafalaya B. Burds Mississippi River Delta
Bellville Angleton Galveston Isles Dernieres B. Terrebonne
Bay City Freeport
Matagorda B.

GULF OF

MEXICO

Matagorda I.
Pass
Christi

B

ARKANSAS

TENNESSEE

C

D

E

Kingsville
Hebbronville Falfurrias Sarita
Salado Zapata Padre I.
Falcon L. Laguna Madre
Rio Grande City Raymondville
MEXICO McAllen Edinburg Harlingen San Benito
Brownsville

F

Continuation
Southwards
on same scale

6   7   4   5

VANCOUVER • Port Coquitlam
Duncan New • Chilliwack
C. Flattery *Juan de Fuca Strait* Westminster
C. Alava Victoria Anacortes • Lynden Oliver B R I T I S H C O
Port Townsend Bellingham NORTH Grand Forks Trail Creston
Port Angeles Sedro Woolley CASCADES Oroville Kootenay L.
Forks Sequim Mt. Vernon Mt. Baker NAT. PARK Okanogan Settle Pend Bonners
Olympic Oak 3285 Okanogan Republic Franklin D Falls Ferry
Mt. Mts. Harbor Darrington Omak Roosevelt L Colville Priest
Olympus OLYMPIC Arlington L Chelan Brewster Chewelah Newport Sandpoint
2428 PARK Edmonds Glacier Pk Priest
Everett 3213 Grand L.
Bremerton Leavenworth Waterville Coulee Davenport Deer Park
A SEATTLE Chelan Grand
Port Renton Cashmere Coulee Spokane
Orchard Tacoma Wenatchee Dam Post Falls
B Hoquiam Shelton WASHINGTON Odessa Cœur
Aberdeen Olympia Puyallup Ephrata d'Alene L.
Grays Harb Montesano Tumwater Cle Elum Quincy Moses Lake Ritzville St. Maries
Westport Columbia Othello Wallace
Willapa B. Raymond Centralia Ellensburg Basin Colfax Palouse
Long Beach Winlock Chehalis Mt. Rainier Yakima Connell Pullman Moscow
C. Disappointment 4392 Selah Pomeroy
Warrenton Castle Rock MT. RAINIER Union Gap Snake Clarkston Lewiston
Astoria Kelso 2550 NAT. PARK Mt. Toppenish Dayton Waitsburg Nezperce Kami
Seaside St. Longview Adams Sunnyside Richland Walla Walla Salmon
Helens Kalama 3751 Grandview Pasco Milton-Freewater Clangley
Tillamook PORTLAND Vancouver Prosser Kennewick
Hillsboro Goldendale Hermiston Enterprise
C Newberg Gresham The Dalles Columbia La Grande Wallowa
McMinnville Milwaukie Mt. Hood Pendleton 3011 Mts.
Lincoln Oregon City 3427 Maupin Pilot Rock
City Dallas Mount Angel Heppner Blue
Newport Salem Condon
Waldport Independence Mill City Fossil Elgin
Corvallis Albany Mt. John Day Bakers
Albany Lebanon Jefferson Madras Mountains Council
Florence Junction City 3200 Mitchell New
Sweet Home Prineville Meadows
Eugene Springfield Three Sisters Crooked Redmond John Brogan Weiser
Cottage Grove 3156 Bend Day 2755 Seneca Vale Payette
North Bend Oakridge OREGON Silvies Ontario New Plymouth
Coos Bay Drain Malheur Nyssa Emmett
Coquille Sutherlin Caldwell Idaho
Myrtle Point Roseburg Great Harney Basin Burns Nampa Boise Home
C.Blanco Myrtle Creek Sandy Harney L. Juntura Boise Arrowrock
D Port Orford Canyonville Desert Malheur Murphy Owyhee Res.
Grants Pass Crater L. Summer L. Mountain
Gold Beach CRATER LAKE Home
Brookings Medford NAT. PARK Owyhee
Jacksonville Upper Lake Abert 2962
Crescent Ashland Klamath Klamath Falls Alvord
City L. Desert
Yreka Montague Clear Lake Goose L. Lakeview
Klamath Res. McDermit
E Arcata Mt. Shasta Alturas Alkali
Eureka Mts. Dunsmuir 4317 Lake
Fortuna Mount Shasta Black Rock Ra. Santa Rosa Ra. Independence Mts.
Ferndale Thompson McCloud Burney Winnemucca
Cape Pk 2724 CALIFORNIA Eagle L. Wells
Mendocino Redding Shasta Lake Lassen Peak Elko 3437
Anderson 3187 Susanville Carlin Humboldt
Red Bluff Chester Westwood Rye Patch Battle Ruby Mts.
Corning Almanor Res. Mountain 3235 Franklin
Fort Bragg Orland Honey L. Winnemucca L. 2997 Ruby
Willits Chico Lovelock L.
Willows Quincy Portola Pyramid Trinity Range Eureka
Ukiah Clear Oroville L. Carson McGill
Lakeport Colusa Downieville Reno Sink Fallon Toiyabe Ra. Ely
Cloverdale L. Marysville Nevada City Sparks Shoshone Mountains Austin Schell Creek
Healdsburg Yuba City Grass Valley Truckee Virginia City NEVADA
Calistoga Arbuckle Auburn Tahoe Carson Yerington Mt. Jefferson McGill
Santa Rosa Woodland Citrus Heights City Walker L. 3599 Monitor Ra.
Sebastopol Davis Placerville Gardnerville Mt. Grant
Petaluma Napa Sacramento Jackson 3426
San Rafael Fairfield Arden S. Andreas Hawthorne
G Berkeley Concord Antioch Lodi
Golden Gate Vallejo Richmond

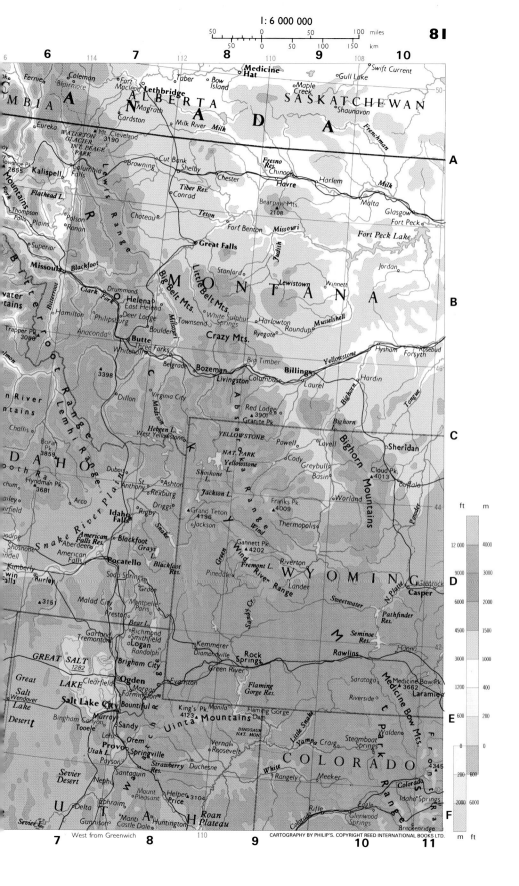

50   0   50   100 miles

50   0   50   100   150   km

Fernie○   Coleman○   ○Taber   Bow   Medicine   ○Gull Lake   ○Swift Current
       Blairmore   ○Fort   Island   Hat   Maple
       Lethbridge   Macleod   Creek
                                            SASKATCHEWAN
A L B E R T A   ○Magrath   Shaunavon○
       ○Cardston   Milk River Milk   Frenchman
Eureka○ Mt. Cleveland   50
       WATERTON 3190   A
       GLACIER INT. PEACE
       PARK ○Browning Cut Bank ○Shelby Chinook   Harlem   Milk
Kalispell○ Columbia   Chester   Havre   Malta   Glasgow○
       Falls   ○Conrad Tiber Res.   Bearpaw Mts.   Fort Peck
Flathead L. Choteau   Teton   Fort Benton 2108   Fort Peck Lake
       ○Polson   ○   Fort Benton Missouri Judith   B
Superior○ ○Ronan   Great Falls   Lewistown○   Jordan○
Missoula○ Blackfoot   Stanford○   ○Winnett
       Clark Fork Drummond Helena White Sulphur Harlowton○ Musselshell
       Hamilton○ ○East Helena Townsend Springs ○Roundup
       Philipsburg Deer Lodge Boulder○   Ryegate○
Trapper Pk Anaconda○ Butte○ Crazy Mts.   B
3098   Whitehall○ Three Forks Big Timber Yellowstone Hysham○ Rosebud○
       ○3398   Belgrade Bozeman○ Columbus○ Billings Forsyth
       ○Dillon Livingston○   Laurel   Hardin○
                Virginia City○ Red Lodge○ Bighorn   Tongue
I D A H O Borah Pk Madison 3901 Granite Pk   Bighorn
       3859 Hebgen L. YELLOWSTONE Powell○ Lovell○ Mountains Sheridan○   C
       ○Hyndman Pk West Yellowstone NAT. PARK Cody○   Greybull○ Cloud Pk
       3681   St. Ashton Yellowstone Shoshone Basin   4013   Buffalo○
       Arco○ ○Anthony L. L. Franks Pk ○Worland
Snake River Plain Rexburg Jackson L. 4009 Thermopolis○
       Idaho ○Driggs   Grand Teton Wind Riverton○
Shoshone○ Falls ○Rigby   4196 ○Jackson Gannett Pk Fremont L. Lander○   D
       ○Blackfoot Grays 4202   Green Pinedale○ WYOMING Casper○
American L.   Wind River Range   Sweetwater N. Platte Glenrock○
Falls Res. Pocatello○ Blackfoot   Seminoe   Pathfinder
Kimberly○ Soda Springs○ Res.   Res.   Res.
Burley○ ○3151   Grace○   Rawlins○ Hanna
Twin Malad City○ Montpelier Kemmerer○   Saratoga○ Medicine Bow Pk
Falls   Preston○ Paris   Diamondville○ Rock Green River Riverside○ 3662 Laramie○
       Garland○ Bear L. Richmond Springs   Laramie
GREAT SALT Tremonton○ Smithfield Green River Evanston○ Flaming
       Logan○ Randolph○   Gorge Res.   Medicine Bow Mts
Great 1282 Brigham City○   King's Pk Manila○ Flaming Gorge   Walden○
Salt LAKE Clearfield○ Morgan Uinta Mountains Dam DINOSAUR Steamboat Park   E
Lake Ogden○ Farmington○ 4123 NAT. MON. Springs○ 4345
Salt Lake City○ Murray○ Vernal○ Yampa Craig○
Desert Bingham Canyon Sandy○ Roosevelt○ Little Snake   COLORADO
Tooele○ Lehi○ Provo○ Strawberry Duchesne○ White ○Meeker Colorado○
Sevier Orem○ Springville Res. ○Rangely   Idaho○ Springs
Desert Payson○ Santaquin○ 3104   Rifle○ Eagle○ Glenwood
U T A H Nephi○ Mount Helper○ Roan Colorado Springs○ Breckenridge○   F
Delta○ Pleasant○ Price○ Plateau
Sevier L. Ephraim○ Manti Castle Dale○ Huntington○
       Gunnison○

ft   m

12 000   4000
9000   3000
6000   2000
4500   1500
3000   1000
1200   400
600   200
0   0
200   600
2000   6000

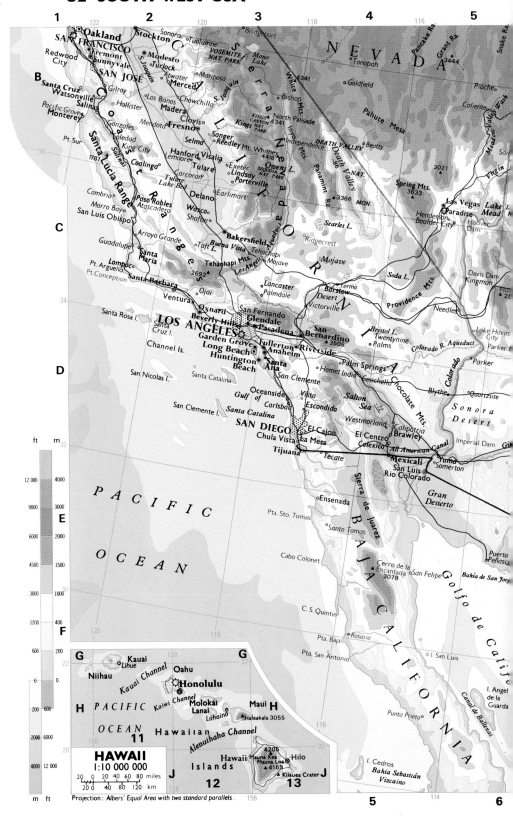

1 122 **2** 120 **3** 118 **4** 116 **5**

N E V A D A

Oakland
SAN FRANCISCO
Stockton
Sonora
Tuolumne
Bridgeport
Tonopah
3444
Grant Ra.
Snake R.
Pancake Ra.

A

Redwood City
Fremont
Sunnyvale
Modesto
Turlock
YOSEMITE NAT. PARK
Mono Lake
Mono L.

Goldfield
Pioche

B

Santa Cruz
Watsonville
SAN JOSE
Gilroy
Atwater
Merced
Mariposa
White Mts.
4341
Calciente
Valley Wash

Pacific Grove
Salinas
Hollister
Los Banos
Chowchilla
Bishop
North Palisade
4341
Virgin

Monterey
Gonzales
Madera
Clovis
Mendota
KINGS CANYON NAT. PARK
Kings Nat. Park
DEATH VALLEY
NAT. MON.
3021

Pt. Sur
Soledad
King City
Fresno
Selma
Sanger
Reedley
Mt. Whitney 4418
SEQUOIA NAT. MON.
Death Valley 86
Pahute Mesa
Beatty

C

Santa Lucia Range
1787
Coalinga
Hanford
Visalia
Lemoore
Tulare
Exeter
Lindsay
Porterville
Inyo Mts.
Spring Mts. 3633
Las Vegas
Paradise
Lake Mead

Cambria
Paso Robles
Atascadero
Corcoran
Earlimart
Tulare Lake Bed
Delano
3366 MON.
Henderson
Boulder City
Hoover Dam

Morro Bay
San Luis Obispo
Arroyo Grande
Wasco
Shafter
Taft L.
Searles L.
Mojave
Soda L.
Davis Dam
Kingman

C

Guadalupe
Santa Maria
Bakersfield
Buena Vista
Tehachapi
Ridgecrest
Providence Mts.
Needles
25

Lompoc
Pt. Arguello
Pt. Conception
Santa Barbara
Tehachapi Mts. 2692
Taft L.
Mojave
Barstow
Termo
Desert
Victorville
Lake Havasu City
Parker

Ventura
Ojai
Lancaster
Palmdale
Soda L.
Colorado R. Aqueduct
Parker

Santa Rosa I.
Oxnard
Beverly Hills
San Fernando
Glendale
Pasadena
San Bernardino 3505
Twentynine Palms
Bristol L.
Colorado

D

LOS ANGELES
Garden Grove
Long Beach
Huntington Beach
Fullerton
Anaheim
Santa Ana
Riverside
Hemet
Indio
Palm Springs
Coachella
Blythe
Quartzsite
Parker

Santa Cruz I.
Channel Is.
San Clemente
Palm Springs
Chocolate Mts.
Sonora Desert

San Nicolas I.
Santa Catalina
San Clemente I.
Oceanside
Carlsbad
Vista
Escondido
Salton Sea
Westmorland
Calipatria
Brawley
Blythe

SAN DIEGO
Chula Vista
La Mesa
El Cajon
El Centro
Calexico
All American Canal
Imperial Dam
Gi

Tijuana
Tecate
Mexicali
San Luis
Rio Colorado
Somerton
Yuma

E

P A C I F I C
Ensenada
Pta. Sto. Tomas
Santa Tomas
Sierra de Juarez
Gran Desierto

O C E A N
Cabo Colonet
Cerro de la Encantada 3078
San Felipe
Bahía de San Jor
Puerto Peñasco

F

C. S. Quintin
I. San Luis

120 118 Pta. Baja
Rosario
Pta. San Antonio
116
I. Angel de la Guarda

G 158 **G**
Kauai
Lihue
Oahu
Honolulu
Punta Prieta
Canal de Ballenas

Niihau
Kauai Channel

H PACIFIC
Kaiwi Channel
Molokai
Lanai
Lahaina
Maui **H**
Haleakala 3055
I. Cedros

O C E A N
11 H a w a i i a n
Alenuihaha Channel
116

160
20
Hawaii
Mauna Kea 4205
Mauna Loa
Hilo
5 114 **6**

HAWAII
1:10 000 000
20 0 20 40 60 80 miles
20 0 40 80 120 km
Islands
12 158 **J** Kilauea Crater 4169 **13**
Bahía Sebastián Vizcaino

Projection: Albers' Equal Area with two standard parallels. 156

ft m
12 000 — 4000
9000 — 3000
6000 — 2000
E
4500 — 1500
3000 — 1000
1200 — 400
F
600 — 200
200 — 0
0 —
200 — 600
2000 — 6000
4000 — 12 000
m ft

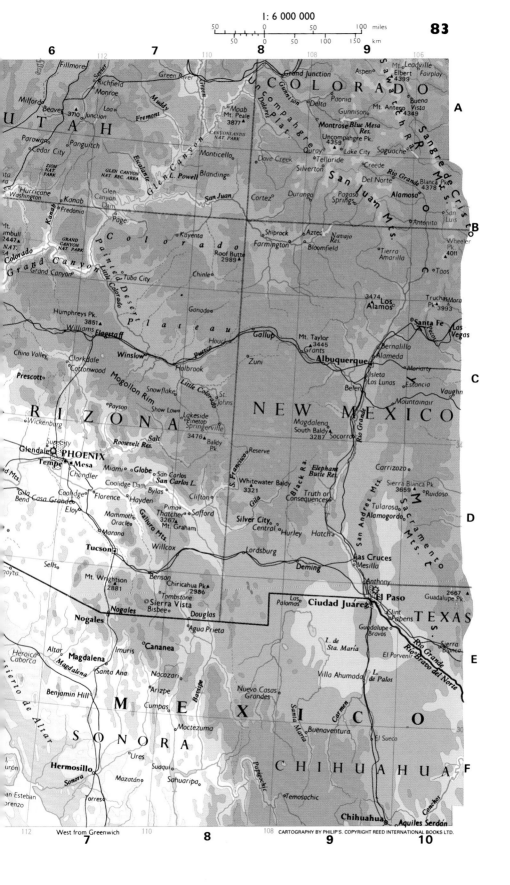

1: 6 000 000

50   0   100 miles
50   0   50   100   150 km

6   112   **7**   110   **8**   108   **9**   106

UTAH

Fillmore
Richfield
Monroe
Milford
Beaver   Sevier   Green River
3710   Junction   Loa
Parowan   Panguitch   Fremont   Muddy
Cedar City
ZION NAT. PARK   ESCALANTE   GLEN CANYON NAT. REC. AREA
Hurricane   Washington   Kanab   Fredonia
Mt.   Page
umbull   Glen   Canyon Dam
2447   Kayenta
NAT.   GRAND CANYON NAT. PARK
Colorado   Grand Canyon   Little Colorado
Grand Canyon   Tuba City
Chinle
Humphreys Pk.   Ganado
3851   Plateau
Williams   Flagstaff   Houck
Chino Valley   Clarkdale   Winslow   Holbrook   Little Colorado   Puerco
Prescott   Cottonwood   Snowflake   St. Johns
ARIZONA   Mogollon Rim   Show Low   Lakeside   Pinetop   Springerville
Payson   3476   Baldy Pk.
Wickenburg
Sun City   Salt   Roosevelt Res.
Glendale   PHOENIX   Miami   Globe
Tempe   Mesa   San Carlos
Chandler   San Carlos L.
Mts.   Coolidge   Bylas
Gila   Coolidge Dam   Hayden   Clifton
Bend   Casa Grande   Florence   Pima   Safford
Eloy   Thatcher   3267   Mt. Graham
Mammoth   Oracle   Galiuro Mts.
Marana   Willcox
Tucson   Benson
Sells   Mt. Wrightson   Chiricahua Pk.
2881   Tombstone   2986
Sierra Vista   Bisbee   Douglas
Nogales   Nogales   Agua Prieta
oyta

COLORADO

Grand Junction   Aspen   Mt. Leadville   Fairplay
Elbert   4399
Gunnison   Paonia   Mt. Antero   Buena
Delta   4349   Vista
Montrose   Blue Mesa   Res.
Uncompahgre Pk.   Saguache
Ouray   Lake City
Telluride   Creede   Rio Grande   Blanca
Silverton   Del Norte   4378   Pk.
San Juan Mts.   Alamosa
Durango   Pagosa Springs   San Luis
Dove Creek   Antonito
Shiprock   Aztec   Navajo   Wheeler Pk.
Farmington   Bloomfield   Res.   4011
Roof Butte   Tierra   Taos
2989   Amarilla
3474   Los   Truchas Mora
Alamos   Pk. 3993
Santa Fe   Las Vegas
Mt. Taylor   Bernalillo
Gallup   3445   Grants   Alameda
Zuni   Albuquerque   Moriarty
Isleta   Estancia
Belen   Los Lunas   Vaughn
Mountainair
NEW MEXICO
Magdalena   Socorro
South Baldy   Rio Grande
3287
Reserve   Carrizozo
Whitewater Baldy   Elephant   Sierra Blanca Pk.
3321   Butte Res.   3659   Ruidoso
Gila   Truth or   Tularosa
Silver City   Consequences   Alamogordo
Central   Hurley   Hatch   San Andres Mts.   Sacramento Mts.
Lordsburg
Deming   Las Cruces
Mesilla
Anthony   2667
El Paso   Guadalupe Pk.
Las   Ciudad Juárez   Clint   **TEXAS**
Palomas   Fabens
Guadalupe   Sierra
Bravos   Blanca
I. de   El Porvenir   Rio Grande   Rio Bravo del Norte
Cananea   Sta. María
Magdalena   Villa Ahumada   L. de Palos
Altar   Santa Ana   Nacozari
Heroica   Magdalena   Arizpe
Caborca   Benjamin Hill   Cumpas   Nuevo Casas   Carmen   Buenaventura   El Sueco
MEXICO   Grandes
sierto de Altar   **SONORA**   Moctezuma   **CHIHUAHUA**
Ures   Baviope   **CHIHUAHUA**
Hermosillo   Suaqui   Papigochic
Sonora   Mazatán   Sahuaripa   Temosachic
Torres   Chihuahua   Aquiles Serdán
an Esteban
orenzo

112   West from Greenwich   110   **8**   108   CARTOGRAPHY BY PHILIP'S. COPYRIGHT REED INTERNATIONAL BOOKS LTD.
7   **8**   **9**   **10**

A
B
C
D
E
F

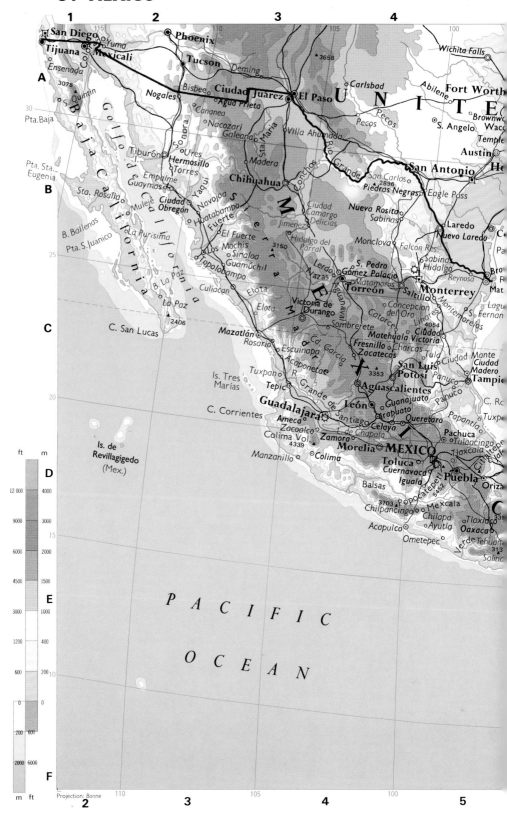

San Diego
Tijuana
Ensenada
Pta.Baja
A
3078
Quintin
Pta. Sta.
Eugenia
B. Ballenas
B
Sta. Rosalia
Pta.S.Juanico
C
C. San Lucas
2406
La Paz
B.La Paz

Yuma
Mexicali
Nogales
Bisbee
Cananea
Nacozari
Galeana
Ures
Hermosillo
Torres
Empalme
Guaymas
Muleje
Ciudad
Obregon
La Purisima
Navojoa
Huatabampo
Fuerte
El Fuerte
Los
Mochis
Sinaloa
Topolobampo
Guamuchil
Culiacan
Elota
Elota
Rosario

Phoenix
Tucson
Deming
Ciudad
Juarez
El Paso
Agua Prieta
Sta. Maria
Villa Ahumada
Madera
Conchos
Rio
Chihuahua
Ciudad
Camargo
Delicias
Jimenez
Hidalgo del
Parral
3150
Lerda
Nazas
Gomez Palacio
Matamoros
Torreon
Victoria de
Durango
Sombrerete
Cd. Garcia
Fresnillo
Charcas
Zacatecas
3353
Escuinapa
Acaponeta
R.Grande de
Santiago
Tuxpan
Tepic
Ameca
Guadalajara
Zacoalco
Colima Vol.
4339
Zamora
Manzanillo
Colima

Wichita Falls
Carlsbad
Abilene
Fort Worth
S. Angelo
Brownw
Waco
Pecos
Temple
Austin
San Antonio
He
Piedras Negras
Eagle Pass
Nueva Rosita
Sabinas
Monclova
Falcon Res
Laredo
Nuevo Laredo
Sabinas
Hidalgo
Reynosa
S. Pedro
Monterrey
Saltillo
Concepcion
del Oro
Catorce
Matehuala
Linares
4054
Ciudad
Victoria
Tula
San Luis
Potosi
Panuco
Ciudad
Mante
Ciudad
Madero
Tampic
Aguascalientes
Leon
Guanajuato
Panuco
C. Rc
Irapuato
Celaya
Queretaro
Papantla
Tuxp
Chapala
Pachuca
Morelia
MEXICO
Tlaxcala
Toluca
Cuernavaca
Puebla
Oriza
Iguala
5452
Balsas
Popocatepetl
C
3703
Mexcala
Chilpancingo
Chilapa
Tlaxiac
Acapulco
Ayutla
Oaxaca
Ometepec
Verde
313
Saline

U N I T E
D E

S I E R R A
M
A
D
R
E

O C C I D E N T A L

Golfo de California
Baja California
Sonora
Sierra Madre

Is. Tres
Marias
C. Corrientes

Is. de
Revillagigedo
(Mex.)
D

P A C I F I C
E

O C E A N

ft m
12 000 4000
9000 3000
6000 2000
4500 1500
3000 1000
1200 400
600 200
0 0
200 600
2000 6000
m ft

F
Projection: Bonne

110
2   3   105   4   100   5

1: 15 000 000

100   0   100   200   300   400 miles

100   0   100   200   300   400   500   600 km

6   7   8   9

UNITED STATES

Gainesville

Dallas   Marshall   Birmingham   Columbia   C. Royal   A
Shreveport   Jackson   Atlanta   Augusta   Charleston
Tyler   Monroe   Vicksburg   Meridian   Montgomery   Macon
Natchez   Hattiesburg   Montgomery   Columbus   Savannah
Alexandria   Baton Rouge   Dothan   Albany   Altamaha
Beaumont   Lake Charles   Mobile   Pensacola
Port Arthur   Lafayette   New Orleans   Tallahassee   Jacksonville   30
Galveston   Mississippi   C. San   Apalachee B.   Daytona Beach
Matagorda I.   Delta   Blas
Christi   Orlando   C. Canaveral
Tampa   Lakeland   B
Grande del Norte   St. Petersburg   W. Palm Beach
Sarasota   L. Okeechobee   Grand
adre   Bahama
Fort   I.
Lauderdale
Miami

GULF OF MEXICO   C. Sable   25

Key West   Andros I.

Tropic of Cancer

Canal de   Matanzas   Sagua la Grande
La Habana   Cárdenas   Sta. Clara   C
(Havana)   Colón
C. Catoche   Marianao   Batabanó   Caibarién
El Cuyo   Pinar del Rio   CUBA
Progreso   C. San   G. de   Cienfuegos   Trinidad   Sancti Spiritus
Temax   Antonio   Guane   Batabanó   I. de Juventud   Jucaro   Ciego de Avila
El Diaz   Puerto   Ciego de Avila
Mérida   Valladolid   Morelos   Yucatan
Peto   I. de   I. de Juventud
Golfo de   Campeche   Cozumel
acruz   Vigía Chico   Grand Cayman   20
Campeche   Felipe   (U.K.)
arado   Ciudad del Carmen   Carillo Puerto   Yucatan
acotalpan   Laguna   Ciudad Chetumal
Coatzacoalcos   de Terminos   Corozal
me de   Villahermosa   Ambergris Cay
uantepec   Tuxtla   Belize   Turneffe Is.
Judith   Gutierrez   Belmopan   BELIZE   Golfo de Hondu   D
Chiapa   San Cristobal   Middlesex   Pto. Barrios   Gulfo de Hondu
Tonala   Chiapa   GUATEMALA   Pto. Cortés
uantepec   4217   Tela   Trujillo   Iriona
G. de   Huixtla   Guatemala   La Ceiba   L. Caratasca
Acapa   San Pedro Sula   15
Sta. Ana   Sta. Rosa   HONDURAS   C. Gracias á Dios
San José   Comayagua   Wanks or Coco
Sonsonate   San   Tegucigalpa   Puerto Cabezas
San Salvador   Vincente   Jinotega   Providencia
EL SALVADOR   S. Miguel   Nacaome   (Col.)
G. de Fonseca   Choluteca   Matagalpa   San Andrés
Chinandega   El Gallo   (Col.)
León   NICARAGUA
Managua   Granada   Bluefields   E
Masaya   L. Nicaragua
S. Juan
Pen. de Nicoya   Vol. Irazú   Limón
COSTA RICA   Colón   10
Puntarenas   Aldijue   San José   PANAMA   La
Coiba   San José   Cartago   3374   Panama   Palma
3391   Chitré   El
Pen. de   Arch. de   Real
Azuero   las Perlas   F
G. de   Coiba   Panama

CARTOGRAPHY BY PHILIP'S.
COPYRIGHT REED INTERNATIONAL BOOKS LTD.

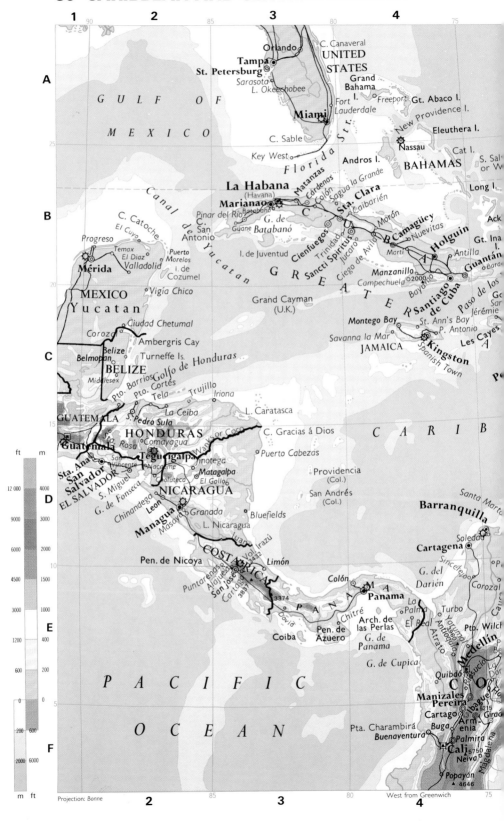

1   2   3   4

90   85   80   75

A

GULF   OF

MEXICO

25

B

20

MEXICO
Yucatan

C

15

GUATEMALA
HONDURAS

D

EL SALVADOR
NICARAGUA

10

E

PACIFIC

5

F

OCEAN

Orlando
Tampa
St. Petersburg
Sarasota
L. Okeechobee

UNITED
STATES

C. Canaveral

Grand
Bahama
Fort I.
Lauderdale   Freeport   Gt. Abaco I.

Miami

New Providence I.

C. Sable

Eleuthera I.

Key West

Nassau

Cat I.

Florida Str.

Andros I.

BAHAMAS

S. Sal
or W

Long I.

La Habana
(Havana)
Marianao

Matanzas
Cárdenas
Colón
Batabanó

Sagua la Grande

Sta. Clara
Caibarién

Pinar del Río
San
Antonio

C.
Catoche
El Cuyo

C.
Guane   G. de
Batabanó

I. de Juventud

Cienfuegos

Morón

Camagüey
Nuevitas

Holguin

Ack

Gt. Ina

Antilla

Sancti Spíritus

Jucaro

Ciego de Ávila

Martí

C
U
B
A

Guantán

Bara

Progreso
Temax
El Díaz
Valladolid
Puerto
Morelos
I. de
Cozumel

Mérida

Trinidad

GREATER

Manzanillo
Campechuela ○2000
Bayá

Santiago
de Cuba

Jérémie

Paso de los

Sar

Vigía Chico

Grand Cayman
(U.K.)

Montego Bay

St. Ann's Bay
Savanna la Mar

JAMAICA

Spanish Town
Kingston

P. Antonio

Les Cayes

A

Po

Ciudad Chetumal

Corozal
Belize
Belmopan
Middlesex

Ambergris Cay
Turneffe Is.

BELIZE

Pto. Barrios
Pto. Cortés
Tela

Golfo de Honduras

Trujillo
Iriona

La Ceiba

L. Caratasca

C A R I B

Guatemala
Rosa

Pedro Sula

Comayagua   Wanks or Coco

C. Gracias á Dios

Sta. Ana
San
Salvador
S. Miguel
Chinandega

Tegucigalpa
Juticalpa
Jinotega
Matagalpa
El Gallo

Puerto Cabezas

Providencia
(Col.)

San Andrés
(Col.)

Santa Mart

BARRANQUILLA

G. de Fonseca

León
Managua
Masaya   Granada

Bluefields

L. Nicaragua
Juan

Soledad

Cartagena

Pen. de Nicoya

COSTA RICA
Vol. Irazú
3432

Limón

Colón

PANAMA

Panama

Sincelejo

G. del
Darién

Corozal

Puntarenas
Alajuela
San José
Cartago
3387

3374

David

Chitré

La
Palma   Turbo

El Real

Pto. Wilch

Coiba

Pen. de
Azuero

Arch. de
las Perlas

G. de
Panama

Quibdó

Medellín

Be

G. de Cupica

Manizales
Pereira
Cartago

CO

Pta. Charambirá
Buenaventura

Buga
Arm
enia

Popayán
▲ 4646

Cali
Neiva

5215

5750
Girá

Palmira

ft   m

12 000   4000

9000   3000

6000   2000

4500   1500

3000   1000

1200   400

600   200

0   0

200   600

2000   6000

m   ft

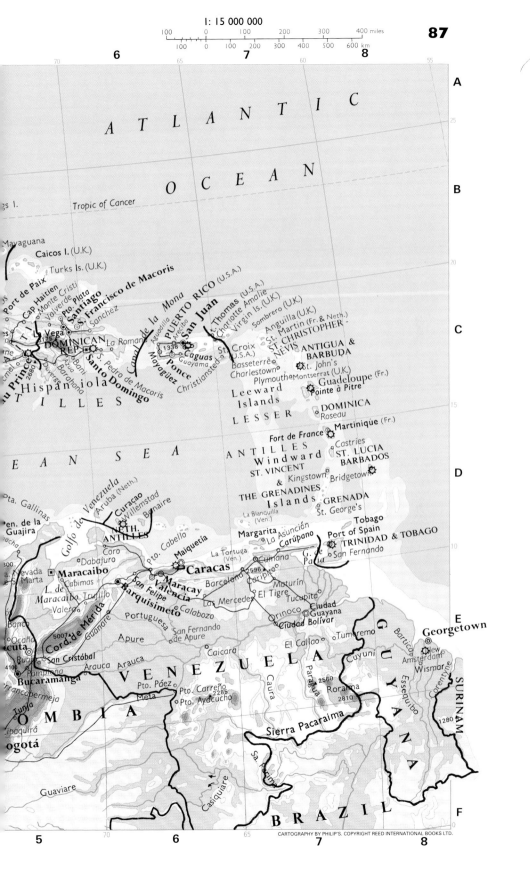

1 : 15 000 000

100   0   100   200   300   400 miles
100   0   100   200   300   400   500   600 km

6   7   8

A T L A N T I C

O C E A N

Tropic of Cancer

A

25

B

20

es I.

Mayaguana

Caicos I. (U.K.)

Turks Is. (U.K.)

Port de Paix
Cap Haitien
Monte Cristi
Valverde   Pto. Plata
Santiago
S. Francisco de Macoris
Sanchez

San Francisco de Macoris

PUERTO RICO (U.S.A.)

St. Thomas (U.S.A.)
Charlotte Amalie
Virgin Is. (U.K.)

Sombrero (U.K.)

Anguilla (U.K.)

St. Martin (Fr. & Neth.)

C

Canal de la Mona

La Romana
Aguadilla   San Juan
Arecibo
1338   Caguas
Guayama   St. Croix
(U.S.A.)
Ponce
Mayagüez   Christiansted

St. CHRISTOPHER -
NEVIS
Basseterre   ANTIGUA &
Charlestown   BARBUDA
St. John's
Plymouth   Montserrat (U.K.)

Guadeloupe (Fr.)
Pointe à Pitre

La Vega
1176
DOMINICAN
REP.
S. Pedro de Macoris
Santo Domingo

u Prince
2680
H i s p a n i o l a

T I L L E S

Leeward
Islands

L E S S E R

DOMINICA
Roseau

15

Fort de France   Martinique (Fr.)

E A N   S E A

A N T I L L E S

Castries

ST. LUCIA

Windward

ST. VINCENT

BARBADOS

D

& Kingstown   Bridgetown

THE GRENADINES

Islands   GRENADA

Pta. Gallinas   Venezuela (Neth.)
Aruba (Neth.)
Curacao
Willemstad   Bonaire
en. de la
Guajira
NETH.
ANTILLES

La Blanquilla
(Ven.)

St. George's

Margarita
La Asunción   Tobago

Carúpano   Port of Spain

Golfo
de Venezuela

Pto. Cabello

Maiquetía
Caracas

La Tortuga
(Ven.)   Cumaná

San Fernando
TRINIDAD & TOBAGO
G. de
Paria

10

Coro
Dabajuro
Maracay

Barcelona
Carúpano

Georgetown

E

Nevada
Marta   Maracaibo
Cabimas   Maracay
Valencia
San Felipe
Barquisimeto
Trujillo
Valera   Calabozo
Portuguesa
San Fernando
de Apure

Caripito
2596
Maturín
El Tigre
Tucupita
Las Mercedes
Orinoco   Ciudad
Guayana
Ciudad Bolívar

Amsterdam
Wismare

L. de
Maracaibo

Banco
Ocaña
cuta   San Cristóbal
Rubio
Pamplona
Bucaramanga

El Callao   Tumeremo
Caicara
Cord de Mérida
Apure
5007

Cuyuni

Barrico
Essequibo

1280

SURINAM

4100
rancabermeja

Arauca   Arauca

V E N E Z U E L A

G
U
Y
A
N
A

Tunja
ipaquirá
ogotá

OMBIA

Pto. Páez
Pto. Carreño
2285
Pto. Ayacucho
Meta

Caura   Paragua
2560
Roraima
2810

Sierra Pacaraima

F

Guaviare

Sa

Casiquiare

B R A Z I L

5   70   6   65   7   8

CARTOGRAPHY BY PHILIP'S. COPYRIGHT REED INTERNATIONAL BOOKS LTD.

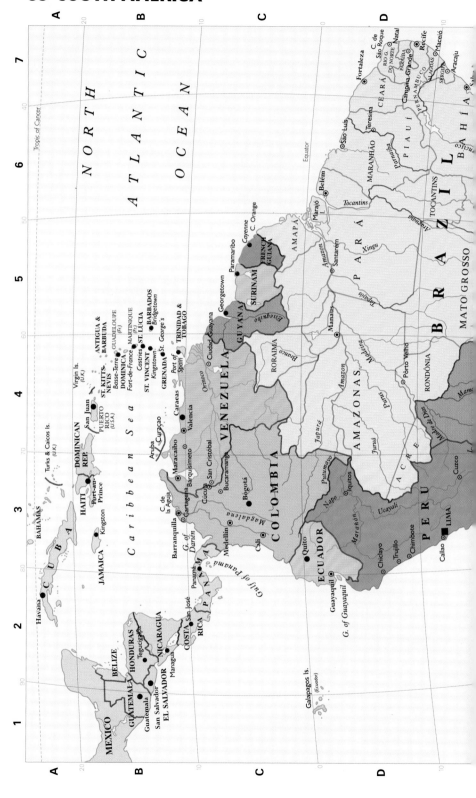

A B C D

7 6 5 4 3 2 1

Tropic of Cancer

N O R T H

A T L A N T I C

O C E A N

Equator

BAHAMAS

Havana

C U B A

JAMAICA Kingston

HAITI DOMINICAN
REP.
Port-au-
Prince

Turks & Caicos Is.
(U.K.)

Caribbean Sea

San Juan
PUERTO
RICO
(U.S.A.)

Virgin Is.
(U.K.)

ANTIGUA &
BARBUDA
ST. KITTS-
NEVIS GUADELOUPE
(Fr.)
Basse-Terre DOMINICA
Fort-de-France MARTINIQUE
Castries ST. LUCIA *(Fr.)*
ST. VINCENT BARBADOS
Kingstown Bridgetown
GRENADA St. George's
Port of TRINIDAD &
Spain TOBAGO

Aruba
Curaçao

Caracas
Valencia
Maracaibo
Barquisimeto
San Cristóbal VENEZUELA
Ciudad Guayana
Orinoco

Georgetown
Paramaribo
Cayenne C. Orange
GUYANA
SURINAM
FRENCH
GUIANA
Essequibo
Demerara

C. de
la Aguja
Barranquilla
Cartagena
Cúcuta
Bucaramanga
Medellín Bogotá COLOMBIA
Cali
Quito
ECUADOR
Guayaquil
G. of Guayaquil

G. of
Darién

PANAMA
Panamá
Gulf of Panama

San José
RICA COSTA
NICARAGUA
Managua
HONDURAS
Tegucigalpa
San Salvador
EL SALVADOR
Guatemala
GUATEMALA
BELIZE
MEXICO

Galápagos Is.
(Ecuador)

RORAIMA
Branco
AMAPA
Marajó
I.
Belém
Santarém
Amazon
Xingu
Tapajós
PARÁ
AMAZONAS
Manaus
Negro
Madeira
Purus
Juruá
Japurá
Putumayo
Napo
Iquitos
Marañón
PERU
Cuzco
LIMA
Callao
Chimbote
Trujillo
Chiclayo
Ucayali
Madre de Dios
ACRE
RONDÔNIA
Mamoré
Guaporé

São Luís
MARANHÃO
Parnaíba
Teresina
PIAUÍ
Fortaleza
CEARÁ
C. de
São Roque
Natal
RIO GRANDE
DO NORTE
PARAÍBA
Campina Grande
PERNAMBUCO
Recife
ALAGOAS
Maceió
SERGIPE
Aracaju
BAHÍA
B R A Z I L
MATO GROSSO
TOCANTINS
Tocantins
Araguaia

20 10 0

40 50 60 70 80 90

20 10 0 10

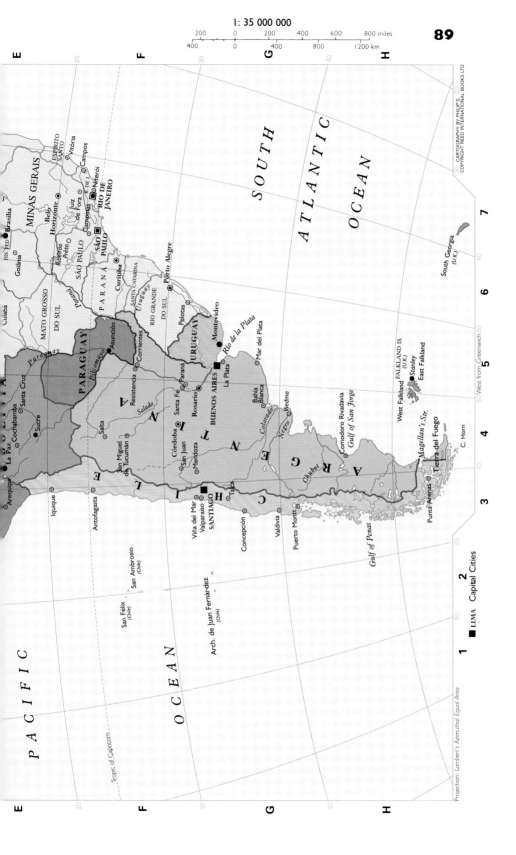

1: 35 000 000

200   0   200   400   600   800 miles
400   0   400   800   1200 km

Projection: Lambert's Azimuthal Equal Area

PACIFIC

OCEAN

SOUTH

ATLANTIC

OCEAN

MINAS GERAIS

ESPIRITO SANTO

Vitória

Campos

Niterói

RIO DE JANEIRO

R. DE J.

DIS. FED. Brasília

Goiânia

Belo Horizonte

Juiz de Fora

Cuiabá

MATO GROSSO DO SUL

Ribeirão Prêto

SÃO PAULO

SÃO PAULO

PARANÁ

Curitiba

SANTA CATARINA

Paranapanema

RIO GRANDE DO SUL

Pôrto Alegre

Pelotas

BOLIVIA

La Paz

Cochabamba

Santa Cruz

Sucre

Arequipa

Iquique

Antofagasta

PARAGUAY

Paraguay

Pilcomayo

Asunción

Corrientes

Resistencia

URUGUAY

Montevideo

Rio de la Plata

Mar del Plata

Salta

San Miguel de Tucumán

Santa Fe

Paraná

Rosario

Córdoba

San Juan

Mendoza

A R G E N T I N A

Salado

BUENOS AIRES

La Plata

Bahía Blanca

Colorado

Vedma

Negro

Chubut

Comodoro Rivadavia

Gulf of San Jorge

C H I L E

Viña del Mar

Valparaíso

SANTIAGO

Talca

Concepción

Valdivia

Puerto Montt

Gulf of Penas

Punta Arenas

Magellan's Str.

Tierra del Fuego

C. Horn

West Falkland

FALKLAND IS. (U.K.)

Stanley

East Falkland

South Georgia (U.K.)

San Ambrosio (Chile)

San Félix (Chile)

Arch. de Juan Fernández (Chile)

Tropic of Capricorn

60° West from Greenwich

LIMA   Capital Cities

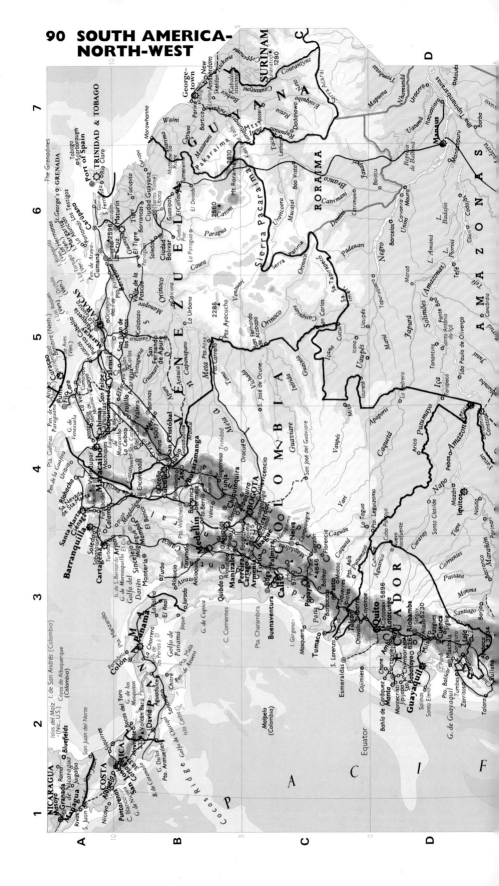

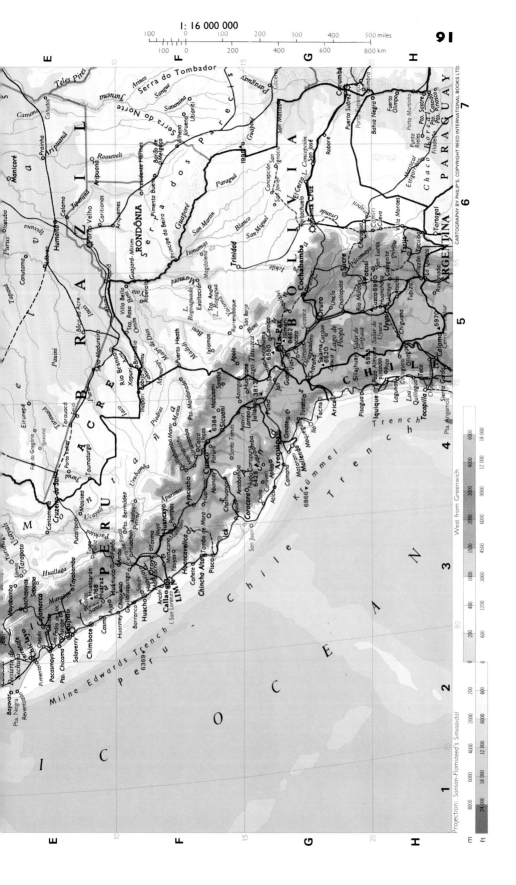

1: 16 000 000

100   0   100   200   300   400   500 miles

100 0   200   400   600   800 km

Projection: Sanson-Flamsteed's Sinusoidal

West from Greenwich

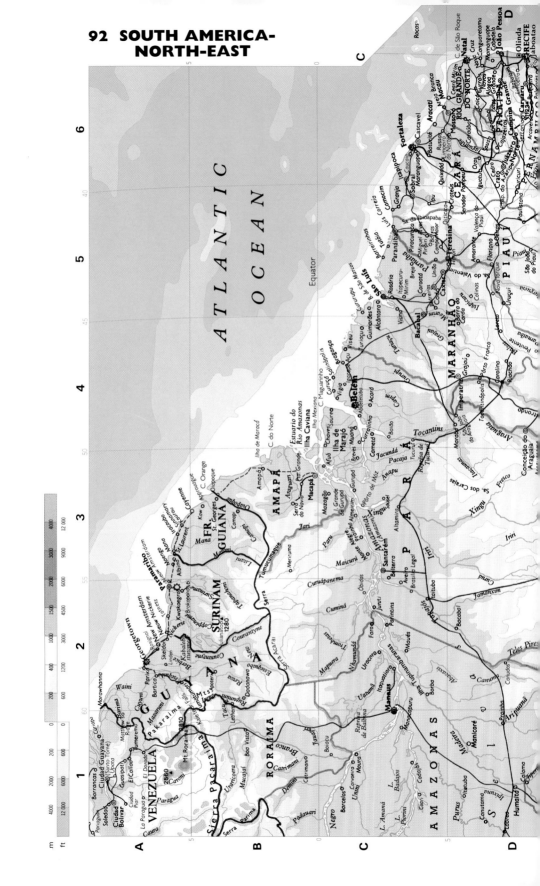

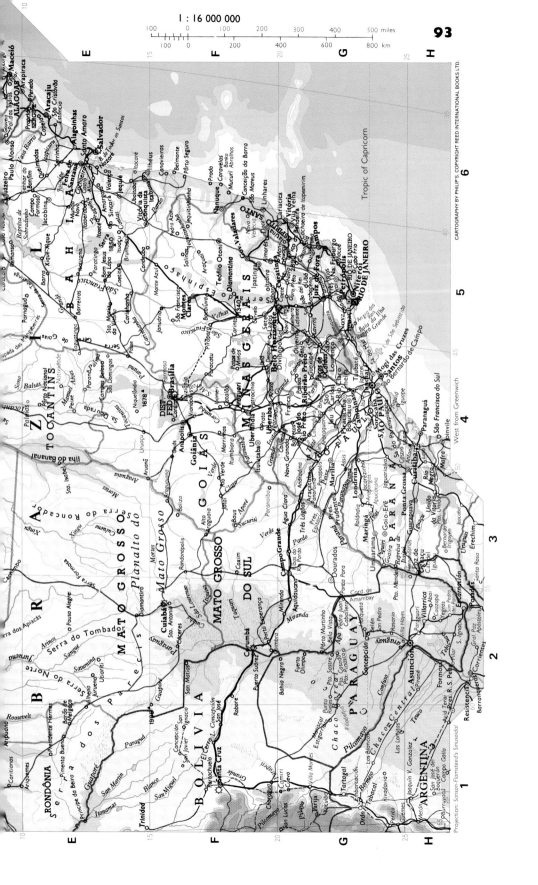

1 : 16 000 000

100   0   100   200   300   400   500 miles

100   0   200   400   600   800 km

Tropic of Capricorn

West from Greenwich

CARTOGRAPHY BY PHILIP'S COPYRIGHT REED INTERNATIONAL BOOKS LTD.

Projection: Sanson-Flamsteed's Sinusoidal

B   R   A   Z   I   L

BAHIA

TOCANTINS

MATO GROSSO

Planalto do Mato Grosso

Serra do Roncador

MATO GROSSO DO SUL

GOIÁS

DIST. FED. Brasília

MINAS GERAIS

ESPÍRITO SANTO

PARANÁ

SÃO PAULO

RIO DE JANEIRO

BOLIVIA

PARAGUAY

ARGENTINA

RONDÔNIA

Serra dos Apiacás

Serra do Tombador

Serra do Norte

Ilha do Bananal

Belo Horizonte

Rio de Janeiro

Niterói

São Paulo

Santos

Salvador

Cuiabá

Goiânia

Goiás

Anápolis

Asunción

Resistencia

Corrientes

Curitiba

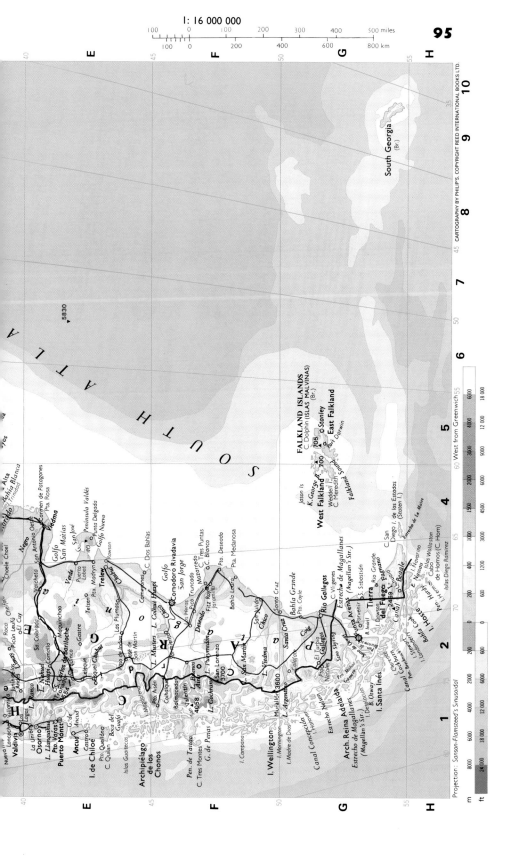

1 : 35 000 000

CARTOGRAPHY BY PHILIP'S.
COPYRIGHT REED INTERNATIONAL BOOKS LTD.
Projection: Zenithal Equidistant

200   0   200   400   600   800 miles
400   0   400   800   1200 km

SOUTHERN OCEAN

East Antarctica

West Antarctica

Bases on King George Island:
Jubany (Argentina)
Com. Ferraz (Brazil)
Ten. Rodolfo Marsh (Chile)
Great Wall (China)
King Sejong (Korea)
Arctowski (Poland)
Artigas (Uruguay)

Ice cap

Permanent ice shelf

Maximum extent of
sea ice

March (Summer) extent
of sea ice

▲3488 Surface elevation and
3700 depth of ice (in metres)

Stanley Permanent bases
(U.K.)

m
4000
12 000
6000
2000

ft
12 000
6000

m
5000
4000
3000
2000
1000
500
0

ft
15 000
12 000
9000
6000
3000
1500
0

Index to Map Pages

The index contains the names of all principal places and features shown on the maps. Physical features composed of a proper name (Erie) and a description (Lake) are positioned alphabetically by the proper name. The description is positioned after the proper name and is usually abbreviated:

Erie, L. **72** **C5**

Where a description forms part of a settlement or administrative name however, it is always written in full and put in its true alphabetical position:

Lake Charles **79** **D7**

Names beginning St. are alphabetized under Saint, but Sankt, Sint, Sant, Santa and San are all spelt in full and are alphabetized accordingly.

The number in bold type which follows each name in the index refers to the number of the map page where that feature or place will be found. This is usually the largest scale at which the place or feature appears.

The letter and figure which are in bold type immediately after the page number give the grid square on the map page, within which the feature is situated.

Rivers carry the symbol ⇢ after their names. A solid square ■ follows the name of a country while an open square ☐ refers to a first order administrative area.

A

A Coruña = La
  Coruña **18** **A1**
Aachen **14** **C4**
Aalborg =
  Ålborg **9** **G9**
Aalen **14** **D6**
Aarau **13** **C8**
Aare ⇢ **13** **C8**
Aarhus = Århus **9** **G10**
Aba **55** **G6**
Ābādān **47** **E7**
Ābādeh **44** **D3**
Abadla **54** **B4**
Abaetetuba . . . **92** **C4**
Abakan **30** **D7**
Abancay **91** **F4**
Abarqū **44** **D3**
Abashiri **32** **F12**
Abay **29** **E8**
Abaya, L. **53** **G6**
Abbay = Nil el
  Azraq ⇢ . . **53** **E5**
Abbeville **12** **A4**
Abbieglassie . . **63** **A4**
Abbot Ice Shelf **96** **B2**
Abbottabad . . **42** **B8**
Abd al Kūrī . . . **49** **E5**
Abéché **53** **F3**
Abeokuta **55** **G5**
Abercorn =
  Mbala **57** **F6**
Abercorn **63** **A5**
Aberdeen,
  Australia **63** **B5**
Aberdeen, *U.K.* **10** **C5**
Aberdeen,
  S. Dak., U.S.A. **76** **C5**

Aberdeen,
  Wash., U.S.A. **80** **B2**
Aberystwyth . . **11** **E4**
Abidjan **55** **G4**
Abilene **78** **C4**
Abitibi L. **69** **D3**
Abkhaz
  Republic ☐ =
  Abkhazia ☐ . **25** **E5**
Abkhazia ☐ . . **25** **E5**
Abminga **62** **A1**
Åbo = Turku . . **9** **F12**
Abohar **42** **D9**
Abomey **55** **G5**
Abong-Mbang . **56** **D2**
Abou-Deïa **53** **F2**
Abri **52** **D5**
Abrolhos, Banka **93** **F6**
Abrud **17** **E6**
Absaroka Range **81** **C9**
Abū al Khaşīb . **47** **E6**
Abū 'Alī **47** **F7**
Abu 'Arīsh . . . **49** **D3**
Abu Dhabi =
  Abū Ẓāby . . . **44** **E3**
Abû Dis **53** **E5**
Abu Hamed . . **53** **E5**
Abū Kamāl . . . **46** **D5**
Abū Madd, Ra's **47** **F4**
Abu Matariq . . **53** **F4**
Abu Rudeis . . . **47** **E3**
Abu Tig **52** **C5**
Abû Zabad . . . **53** **F4**
Abū Ẓāby . . . **44** **E3**
Abuja **55** **G6**
Abukuma-
  Gawa ⇢ . . . **33** **G12**
Abunã **91** **E5**
Abunã ⇢ **91** **E5**

Abut Hd. **65** **E4**
Abwong **53** **G5**
Acaponeta **84** **C3**
Acapulco **84** **D5**
Acarigua **90** **B5**
Accra **55** **G4**
Aceh ☐ **36** **D1**
Achalpur **43** **J10**
Achill I. **11** **E1**
Achinsk **30** **D7**
Acireale **21** **F5**
Acklins I. **86** **B5**
Aconcagua,
  Cerro **94** **C3**
Aconquija, Mt. . **94** **B3**
Açores, Is. dos
  = Azores . . . **50** **C1**
Acraman, L. . . . **62** **B2**
Acre = 'Akko . . **46** **D3**
Acre ☐ **91** **E4**
Acre ⇢ **91** **E5**
Ad Dahnā **47** **F7**
Ad Dammām . . **47** **F7**
Ad Dawhah . . . **44** **E2**
Ad Dilam **47** **G6**
Ad Dīwānīyah . **47** **E6**
Ada **79** **B5**
Adaja ⇢ **18** **B3**
Adam **44** **F4**
Adamaoua,
  Massif de l' . **53** **G1**
Adamawa
  Highlands =
  Adamaoua,
  Massif de l' . **53** **G1**
Adamello, Mte. . **20** **A3**
Adaminaby . . . **63** **C4**
Adam's Bridge **43** **Q11**
Adams Mt. **80** **B3**

Adana **46** **C3**
Adapazarı **46** **B2**
Adarama **53** **E5**
Adare, C. **96** **B15**
Adaut **39** **F4**
Adavale **63** **A3**
Adda ⇢ **20** **B2**
Addis Ababa =
  Addis Abeba **53** **G6**
Addis Abeba . . **53** **G6**
Addis Alem . . . **53** **G6**
Adelaide **62** **B2**
Adelaide I. **96** **A3**
Adelaide Pen. . **70** **B10**
Adélie, Terre . . **96** **A14**
Adélie Land =
  Adélie, Terre **96** **A14**
Aden = Al
  'Adan **49** **E4**
Aden, G. of . . **49** **E4**
Adi **39** **E4**
Adi Ugri **53** **F6**
Adige ⇢ **20** **B4**
Adilabad **43** **K11**
Adirondack Mts. **73** **C8**
Admer **54** **D6**
Admiralty I. . . . **71** **C6**
Admiralty Is. . . **64** **K9**
Ado-Ekiti **55** **G6**
Adonara **39** **F2**
Adoni **43** **M10**
Adour ⇢ **12** **E3**
Adra **18** **D4**
Adrano **21** **F5**
Adrar **54** **C4**
Adré **53** **F3**
Adri **52** **C1**
Adriatic Sea . . . **20** **C5**
Adua **39** **E3**

Adwa 53 F6
Adzhar
  Republic □ =
  Ajaria □ 25 E5
Ægean Sea ... 23 E5
Aerhtai Shan .. 34 B4
Afghanistan ■ . 45 C6
Afgoi 49 G3
'Afif 47 G5
Afognak I. 71 C4
Afuá 92 C3
Afyonkarahisar 46 C2
Agadès =
  Agadez 55 E6
Agadez 55 E6
Agadir 54 B3
Agapa 30 B6
Agartala 41 F8
Agats 39 F5
Agboville 55 G4
Agde 13 E5
Agen 12 D4
Agra 42 F10
Agri → 21 D6
Ağrı Dağı 46 C6
Ağrı Karakose . 46 C5
Agrigento 21 F4
Agrinion 23 E3
Água Clara 93 G3
Agua Prieta .. 84 A3
Aguadas 90 B3
Aguadilla 87 C6
Aguarico → .. 90 D3
Aguas Blancas 94 A3
Aguascalientes 84 C4
Aguilas 19 D5
Agulhas, C. .. 58 E4
Agung 37 F5
Agusan → ... 38 C3
Ahaggar 54 D6
Ahar 46 C6
Ahipara B. 64 A5
Ahiri 43 K12
Ahmadabad ... 43 H8
Ahmadnagar .. 43 K9
Ahmadpur 42 E7
Ahmedabad =
  Ahmadabad . 43 H8
Ahmednagar =
  Ahmadnagar 43 K9
Ahvāz 47 E7
Ahvenanmaa =
  Åland 9 F11
Aḥwar 49 E4
Aigues-Mortes . 13 E6
Aihui 35 A7
Aija 91 E3
Aiken 75 C7
Aimere 39 F2
Aimorés 93 F5
Ain Banaiyan .. 44 F2
Aïn Beïda 54 A6
Aïn Ben Tili .. 54 C3
Aïn-Sefra 54 B4
Ainabo 49 F4
Aïr 55 E6
Aisne → 13 B5
Aiud 17 E6
Aix-en-Provence 13 E6
Aix-la-Chapelle
  = Aachen ... 14 C4
Aix-les-Bains .. 13 D6
Aiyion 23 E4
Aizawl 41 F9
Ajaccio 13 F8
Ajari Rep. =
  Ajaria □ 25 E5
Ajaria □ 25 E5

Ajdâbiyah 52 B3
Ajka 16 E3
'Ajmān 44 E3
Ajmer 42 F9
Ajo, C. de 18 A4
Akaroa 65 E5
Akashi 32 B4
Akelamo 39 D3
Aketi 56 D4
Akharnaí 23 E4
Akhelóös → .. 23 E3
Akhisar 23 E6
Akhmîm 52 C5
Akimiski I. 69 C2
Akita 33 G12
Akjoujt 55 E2
'Akko 46 D3
Akkol 29 E8
Aklavik 70 B6
Akmolinsk =
  Aqmola 29 D8
Akobo → 53 G5
Akola 43 J10
Akordat 53 E6
Akpatok I. 68 B4
Akreïjit 55 E3
Akron 72 D5
Aksai Chin ... 42 B11
Aksaray 46 C3
Aksarka 28 C7
Akşehir 46 C2
Aksu 34 B3
Aksum 53 F6
Aktogay 29 E8
Aktsyabrski ... 17 B9
Aktyubinsk =
  Aqtöbe 29 D6
Aku 55 G6
Akure 55 G6
Akyab = Sittwe 41 G9
Al 'Adan 49 E4
Al Aḥsā 47 F7
Al Amādīyah .. 46 C5
Al Amārah 47 E6
Al 'Aqabah ... 47 E3
Al 'Aramah ... 47 F6
Al Ashkhara ... 45 F4
Al Badi' 47 G6
Al Başrah 47 E6
Al Bayḍā 52 B3
Al Bu'ayrāt al
  Ḥasūn 52 B2
Al Fallūjah 46 D5
Al Fāw 47 E7
Al Fujayrah ... 44 E4
Al Hābah 47 F6
Al Haddār 47 G6
Al Hāmad 47 E4
Al Ḥamar 47 G6
Al Ḥamrā' 47 F4
Al Ḥariq 47 G6
Al Ḥasakah ... 46 C5
Al Ḥawrah 49 E4
Al Ḥayy 47 D6
Al Ḥijāz 47 F4
Al Ḥillah, Iraq . 47 D6
Al Ḥillah,
  Si. Arabia .. 47 G6
Al Hindīyah ... 47 D6
Al Hoceïma ... 54 A4
Al Ḥudaydah .. 49 E3
Al Hufūf 47 F7
Al Ḥulwah 47 G6
Al Ittihad =
  Madīnat ash
  Sha'b 49 E3
Al Jafūrah 47 F7
Al Jaghbūb ... 52 C3
Al Jahrah 47 E6

Al Jalāmīd 47 E4
Al Jawf, Libya . 52 D3
Al Jawf,
  Si. Arabia ... 47 E4
Al Jazirah, Iraq 46 D5
Al Jazirah, Libya 52 C3
Al Jubayl 47 F7
Al Jubaylah ... 47 F6
Al Junaynah .. 53 F3
Al Khābūra .. 44 F4
Al Khalīl 47 E3
Al Kharfah 47 G6
Al Kharj 47 F6
Al Khasab 44 E4
Al Kufrah 52 D3
Al Kūt 47 D6
Al Kuwayt 47 E7
Al Lādhiqīyah . 46 D3
Al Luḥayyah .. 49 D3
Al Madīnah ... 47 F4
Al Majma'ah .. 47 F6
Al Manāmah .. 44 E2
Al Marj 52 B3
Al Mawṣil 46 C5
Al Midhnab ... 47 F6
Al Mish'āb ... 47 E7
Al Mubarraz .. 47 F7
Al Muḥarraq .. 44 E2
Al Mukallā ... 49 E4
Al Mukhā 49 E3
Al Musayyib .. 47 D6
Al Muwaylih .. 47 F3
Al Qadīmah .. 47 G4
Al Qā'iyah ... 47 F5
Al Qāmishli ... 46 C5
Al Qaṣabāt ... 52 B1
Al Qaṣim 47 F5
Al Qaṭif 47 F7
Al Qaṭrūn ... 52 D2
Al Qayṣūmah . 47 E6
Al Quaisūmah . 47 E6
Al Quds =
  Jerusalem ... 47 E3
Al Qurayyāt .. 45 F4
Al Qurnah 47 E6
Al 'Ulā 47 F4
Al Uqaylah ash
  Sharqīgah .. 52 B2
Al Uqayr 47 F7
Al 'Uthmānīyah 47 F7
Al 'Uwaynid ... 47 F6
Al 'Uwayqīlah . 47 E5
Al 'Uyūn 47 F5
Al Wakrah 44 E2
Al Wari'āh 47 F6
Al Yamāmah .. 47 F6
Alabama □ ... 74 C4
Alabama → .. 74 D4
Alaçam Dağları 23 E7
Alagoa Grande 92 D6
Alagoas □ ... 93 D6
Alagoinhas .. 93 E6
Alajuela 86 D3
Alamosa 83 B10
Åland 9 F11
Alania = North
  Ossetia □ ... 25 E5
Alanya 46 C2
Alapayevsk ... 29 D7
Alaşehir 23 E7
Alaska □ 71 B4
Alaska, G. of .. 71 C5
Alaska Peninsula 71 C4
Alaska Range .. 71 B4
Alataw Shankou 34 B3
Alatyr 24 C6
Alausi 90 D3
Alawoona 62 B3
Alba 20 B2

Alba-Iulia 17 E6
Albacete 19 C5
Albacutya, L. .. 62 C3
Albania ■ 23 D3
Albany,
  Australia 60 H2
Albany, Ga.,
  U.S.A. 75 D5
Albany, N.Y.,
  U.S.A. 73 C9
Albany, Oreg.,
  U.S.A. 80 C2
Albany → 69 C2
Albardón 94 C3
Albemarle 75 B7
Alberche → .. 18 C3
Albert, L. 62 C2
Albert L. 57 D6
Albert Nile → . 57 D6
Albertville 13 D7
Albi 12 E5
Albina 92 A3
Alborán 18 E4
Ålborg 9 G9
Alborz, Reshteh-
  ye Kūhhā-ye . 44 B3
Albuquerque . 83 C9
Albury 63 C4
Alcalá de
  Henares 18 B4
Alcalá la Real . 18 D4
Alcamo 21 F4
Alcaníz 19 B5
Alcântara 92 C5
Alcántara,
  Embalse de . 18 C2
Alcantarilla ... 19 D5
Alcaraz, Sierra
  de 19 C4
Alcaudete 18 D3
Alcázar de San
  Juan 18 C4
Alchevsk 25 D4
Alcira 19 C5
Alcoy 19 C5
Aldabra Is. ... 51 G8
Aldan → 31 C10
Alderney 11 G5
Aleg 55 E2
Alegrete 94 B5
Aleisk 29 D9
Aleksandriya =
  Oleksandriya 17 C8
Aleksandrovskoye
  .... 29 C8
Alemania 94 B3
Alençon 12 B4
Aleppo = Ḥalab 46 C4
Alès 13 D6
Alessándria .. 20 B2
Ålesund 8 F9
Aleutian Is. .. 71 C6
Alexander Arch. 71 C6
Alexander I. ... 96 A3
Alexandra,
  Australia 63 C4
Alexandra, N.Z. 65 F3
Alexandria = El
  Iskandarîya .. 52 B4
Alexandria,
  Canada 71 C7
Alexandria,
  Romania ... 22 C5
Alexandria,
  S. Africa ... 59 E5
Alexandria, La.,
  U.S.A. 79 D7
Alexandria, Va.,
  U.S.A. 73 E7

Alexandrina, L.	62 C2	Alvear	94 B5	Amudarya →	29 E6	Angamos, Punta 94 A2
Alexandroúpolis	22 D5	Alvie	62 C3	Amundsen Gulf	70 A7	Ang'angxi 35 B7
Alga	29 E6	Alwar	42 F10	Amundsen Sea	96 B1	Angara → 30 D7
Algarve	18 D1	Alxa Zuoqi 35 C5	Amuntai	37 E5	Angarsk 30 D8
Algeciras	18 D3	Am Dam 53 F3	Amur →	31 D12	Angaston 62 B2
Algemesí	19 C5	Am-Timan ... 53 F3	Amurang	39 D2	Ånge 8 F11
Alger	54 A5	Amâdi, Sudan . 53 G5	Amuri Pass ...	65 E5	Angeles 38 A2
Algeria ■	54 C5	Amadi, Zaïre .. 57 D5	Amyderya =		Angellala 63 A4
Alghero	21 D2	Amadjuak 68 B3	Amudarya →	29 E6	Ångermanälven →
Algiers = Alger	54 A5	Amadjuak L. .. 68 B3	An Nafūd	47 E5	.... 8 F11
Algoa B.	59 E5	Amagasaki 32 B4	An Najaf	47 D6	Angers 12 C3
Alhucemas = Al		Amakusa-Shotō 32 C2	An Nhon	36 B3	Anglesey 11 E4
Hoceïma ..	54 A4	Amaliás 23 F3	An Nu'ayriyah .	47 F7	Ango 57 D5
'Alī al Gharbī ..	47 D6	Amalner 43 J9	Anabar →	30 B9	Angola ■ 58 A3
'Alī Khēl	42 C6	Amambay,		Anaconda	81 B7	Angoulême ... 12 D4
Aliağa	23 E6	Cordillera de 94 A5	Anadolu	46 C2	Angoumois ... 12 D4
Alibo	53 G6	Amangeldy ... 29 D7	'Ānah	46 D5	Angra dos Reis 94 A8
Alicante	19 C5	Amapá 92 B3	Anaheim	82 D4	Angren 29 E8
Alice Springs ..	60 E5	Amapá □ 92 B3	Anakapalle ...	40 J4	Anguilla ■ 87 C7
Aligarh	42 F11	Amarante 92 D5	Analalava	59 G9	Anhui □ 35 C6
Aligūdarz	46 D7	Amargosa 93 E6	Anambas,		Anhwei □ =
Alipur	42 E7	Amarillo 78 B3	Kepulauan ..	37 D3	Anhui □ 35 C6
Alipur Duar ...	41 D7	Amaro, Mte. .. 20 C5	Anambas Is. =		Anjidiv I. 43 M9
Aliwal North ..	59 E5	Amasra 46 B3	Anambas,		Anjou 12 C3
Aljustrel	18 D1	Amasya 46 B3	Kepulauan ..	37 D3	Anjozorobe ... 59 H9
Alkmaar	14 B3	Amazon =		Anamur	46 C3	Anju 35 C7
All American		Amazonas → 92 C3	Anan	32 C4	Anka 55 F6
Canal	82 D5	Amazonas □ .. 90 D6	Anantnag	42 C9	Ankang 35 C5
Allahabad	40 E3	Amazonas → .. 92 C3	Ananyiv	17 E9	Ankara 46 C3
Allanmyo	41 H10	Ambala 42 D10	Anápolis	93 F4	Ankazoabo ... 59 J8
Allegheny → .	73 D5	Ambalavao ... 59 J9	Anār	44 D3	Ankazobe 59 H9
Allegheny		Ambam 56 D2	Anārak	44 C3	Ankoro 57 F5
Plateau	73 F5	Ambanja 59 G9	Anatolia =		Ann Arbor ... 72 C4
Allentown	73 D8	Ambartsevo ... 29 D9	Anadolu	46 C2	Annaba 54 A6
Alleppey	43 Q10	Ambato 90 D3	Añatuya	94 B4	Annan 11 D5
Aller →	14 B5	Ambatolampy . 59 H9	Anaye	53 E1	Annapolis 73 E7
Allier →	13 C5	Ambatondrazaka 59 H9	Anchorage ...	71 B5	Annecy 13 D7
Allora	63 A5	Amberg 15 D6	Ancohuma,		Anning 34 D5
Alma Ata =		Ambergris Cay 85 D7	Nevada	91 G5	Anniston 74 C5
Almaty	29 E8	Amberley 65 E5	Ancón	91 F3	Annobón 51 G4
Almada	18 C1	Ambikapur 40 F4	Ancona	20 C4	Annuello 62 B3
Almadén	18 C3	Ambilobé 59 G9	Ancud	95 E2	Anqing 35 C6
Almansa	19 C5	Ambo 91 F3	Ancud, G. de .	95 E2	Ansbach 14 D6
Almanzor, Pico		Amboise 12 C4	Andalgalá ...	94 B3	Anshan 35 B7
del Moro ..	18 B3	Ambon 39 E3	Andalucía □ ..	18 D3	Anshun 35 D5
Almanzora → .	19 D5	Ambositra 59 J9	Andalusia	74 D4	Ansirabe 59 H9
Almaty	29 E8	Amboyna I. ... 36 C4	Andalusia □ =		Ansongo 55 E5
Almazán	19 B4	Ambriz 56 F2	Andalucía □ ..	18 D3	Ansudu 39 E5
Almeirim	92 C3	Amby 63 A4	Andaman Is. ..	27 H13	Antabamba ... 91 F4
Almelo	14 B4	Amchitka I. ... 71 C1	Andaman Sea .	36 B1	Antakya 46 C4
Almendralejo ..	18 C2	Amderma 28 C7	Andara	58 B4	Antalaha 59 G10
Almería	19 D4	Ameca 84 C4	Anderson	72 D3	Antalya 46 C2
Almora	42 E11	Ameland 14 B3	Anderson → ..	70 B7	Antalya Körfezi 46 C2
Alnwick	10 D6	American		Andes = Andes,		Antananarivo .. 59 H9
Alon	41 F10	Highland 96 B10	Cord. de los .	91 G4	Antarctic Pen. . 96 A4
Alor	39 F2	American		Andes, Cord. de		Antarctica .... 96 C
Alor Setar	36 C2	Samoa ■ ... 65 L13	los	91 G4	Antequera 18 D3
Alps	14 E5	Americus 75 C5	Andhra		Anti Atlas 54 C3
Alsace	13 B7	Amery 71 C10	Pradesh □ ..	43 L11	Antibes 13 E7
Alsask	71 C9	Amery Ice Shelf 96 A10	Andijon	29 E8	Anticosti, I. d' . 69 D4
Alsásua	19 A4	Ames 77 D8	Andikíthira ...	23 G4	Antigua &
Alta Gracia ..	94 C4	Amherst, Burma 41 J11	Andizhan =		Barbuda ■ .. 87 C7
Altagracia ...	90 A4	Amherst,		Andijon	29 E8	Antilla 86 B4
Altai = Aerhtai		Canada 69 D4	Andkhvoy	45 B6	Antioquia 90 B3
Shan	34 B4	Amiata, Mte. .. 20 C3	Andoany	59 G9	Antofagasta ... 94 A2
Altamira	92 C3	Amiens 12 B5	Andong	35 C7	Antofagasta de
Altamura	21 D6	Amirante Is. ... 26 K9	Andorra ■	19 A6	la Sierra ... 94 B3
Altanbulag ...	35 A5	'Ammān 47 E3	Andorra La Vella 19 A6		Antongila,
Altay	34 B3	Amorgós 23 F5	Andreanof Is. .	71 C2	Helodrano .. 59 H9
Altea	19 C5	Amoy = Xiamen 35 D6	Andrewilla ...	62 A2	Antonina 94 B7
Alto Araguaia .	93 F3	Ampanihy 59 J8	Ándria	20 D6	Antrim 11 D3
Alto Molocue ..	59 B7	Ampenan 37 F5	Andriba	59 H9	Antsalova 59 H8
Alton	77 F9	Amper → 15 D6	Andropov =		Antsiranana ... 59 G9
Alton Downs ..	62 A2	Amravati 43 J10	Rybinsk	24 B4	Antsohihy 59 G9
Altoona	73 D6	Amreli 43 J7	Ándros	23 F5	Antwerp =
Altun Shan ...	34 C3	Amritsar 42 D9	Andros I.	86 B4	Antwerpen .. 14 C3
Alūla	49 E5	Amroha 42 E11	Andújar	18 C3	Antwerpen 14 C3
Alusi	39 F4	Amsterdam ... 14 B3	Andulo	56 G3	Anupgarh 42 E8
Alvarado	85 D5	Amstetten ... 15 D8	Aného	55 G5	Anvers =
				Aneto, Pico de .	19 A6	Antwerpen .. 14 C3

Anvers I.

Anvers I. 96 A3
Anxi 34 B4
Anxious B. 62 B1
Anyang 35 C6
Anzhero-
Sudzhensk .. 29 D9
Ánzio 20 D4
Aomori 33 F12
Aosta 20 B1
Aoudéras 55 E6
Aoulef el Arab . 54 C5
Apalachicola ⟶ 74 E5
Apaporis ⟶ .. 90 D5
Aparri 38 A2
Apeldoorn 14 B3
Apennines =
Appennini . . 20 B3
Apia 65 L13
Apiacás, Serra
dos 91 E7
Aplao 91 G4
Apo, Mt. 38 C3
Apollonia =
Marsá Susah 52 B3
Apolo 91 F5
Apoteri 90 C7
Appalachian
Mts. 73 F5
Appennini 20 B3
Appleton 72 B1
Approuague . . 92 B3
Aprília 20 D4
Apucarana 94 A6
Apure ⟶ 90 B5
Apurimac ⟶ . 91 F4
Aqabah = Al
'Aqabah 47 E3
'Aqabah, Khalīj
al 47 E3
Āqcheh 45 B6
Aqīq 53 E6
Aqmola 29 D8
Aqrah 46 C5
Aqtöbe 29 D6
Aquidauana .. 93 G2
Aquitain, Bassin 12 D3
Ar Rachidiya .. 54 B4
Ar Ramādī 46 D5
Ar Raqqah 46 D4
Ar Rass 47 F5
Ar Rifā'ī 47 E6
Ar Riyāḍ 47 F6
Ar Ruṭbah 46 D5
Ar Ruwayḍah . 47 G6
Ara 40 E5
'Arab, Bahr
el ⟶ 53 G4
Arab, Shatt
al ⟶ 47 E7
Arabian Desert
= Es Sahrâ'
Esh Sharqīya 52 C5
Arabian Gulf =
Gulf, The . . 44 E2
Arabian Sea . . . 26 H10
Araç 46 B3
Aracaju 93 E6
Aracataca 90 A4
Aracati 92 C6
Araçatuba 93 G3
Aracena 18 D2
Araçuaí 93 F5
Arad 16 E5
Arada 53 F3
Arafura Sea .. 39 F4
Aragón □ 19 B5
Aragón ⟶ 19 A5
Araguacema .. 92 D4
Araguaia ⟶ .. 92 D4

Araguari 93 F4
Araguari ⟶ .. 92 B4
Arak 54 C5
Arakan Coast .. 41 H9
Arakan Yoma . 41 H10
Araks = Aras,
Rūd-e ⟶ ... 46 B7
Aral 29 E7
Aral Sea 29 E6
Aral Tengizi =
Aral Sea 29 E6
Aralsk = Aral .. 29 E7
Aralskoye More
= Aral Sea .. 29 E6
Aran I. 10 D2
Aran Is. 11 E2
Aranda de
Duero 18 B4
Aranjuez 18 B4
Aranos 58 C3
Araouane 55 E4
Arapgir 46 C4
Arapiraca 93 D6
Arapongas ... 94 A6
Araranguá 94 B7
Araraquara ... 93 G4
Ararat 62 C3
Ararat, Mt. =
Ağrı Dağı ... 46 C6
Araripe,
Chapada do . 92 D6
Aras, Rūd-e ⟶ 46 B7
Arauca 90 B4
Arauca ⟶ 90 B5
Arauco 94 D2
Araxá 93 F4
Araya, Pen. de . 90 A6
Arbatax 21 E2
Arbīl 46 C6
Arbroath 10 C5
Arcachon 12 D3
Archangel =
Arkhangelsk . 28 C5
Arcila = Asilah 54 A3
Arckaringa ... 62 A1
Arckaringa
Cr. ⟶ 62 A2
Arcos de la
Frontera ... 18 D3
Arcot 43 N11
Arcoverde ... 92 D6
Arctic Bay ... 68 A2
Arctic Red River 70 B6
Arda ⟶ 22 D6
Ardabīl 46 C7
Ardahan 46 B5
Ardakān =
Sepīdān ... 44 D3
Ardenne 14 D3
Ardennes =
Ardenne ... 14 D3
Ardestān 44 C3
Ardlethan 63 B4
Ardmore 78 B5
Ardrossan ... 62 B2
Arecibo 87 C6
Areia Branca . 92 C6
Arendal 9 G9
Arequipa 91 G4
Arero 53 H6
Arévalo 18 B3
Arezzo 20 C3
Argamakmur .. 37 E2
Arganda 18 B4
Argentan 12 B3
Argentário, Mte. 20 C3
Argentina ■ .. 94 D3
Argentina Is. . . 96 A3
Argentino, L. .. 95 G2

Argeş ⟶ 22 B6
Arghandab ⟶ 42 D4
Argo 53 E5
Argolikós
Kólpos 23 F4
Árgos 23 F4
Argostólion ... 23 E3
Argungu 55 F5
Århus 9 G10
Arica, Chile .. 91 G4
Arica, Colombia 90 D4
Aridh 47 F6
Arinos ⟶ 91 F7
Aripuanã 91 E6
Aripuanã ⟶ . 90 E6
Ariquemes ... 91 E6
Arizona 94 D3
Arizona □ 83 C7
Arjona 90 A3
Arkalyk =
Arqalyk 29 D7
Arkansas □ ... 79 B7
Arkansas ⟶ . 79 C8
Arkansas City . 78 A5
Arkhangelsk ... 28 C5
Arklow 11 E3
Arkticheskiy,
Mys 30 A7
Arlanzón ⟶ . 18 A3
Arlberg P. 14 E6
Arles 13 E6
Arlington, Va.,
U.S.A. 73 E7
Arlington,
Wash., U.S.A. 80 A2
Arlon 14 D3
Armagh 11 D3
Armavir 25 D5
Armenia 90 C3
Armenia ■ ... 25 E5
Armidale 63 B5
Armstrong ... 69 C2
Arnaud ⟶ 68 B3
Arnauti, C. ... 46 D3
Arnhem 14 C3
Arnhem, C. ... 60 C6
Arnhem Land . 60 C5
Arno ⟶ 20 C3
Arno Bay 62 B2
Arnsberg 14 C5
Aroab 58 D3
Arqalyk 29 D7
Arrabury 62 A3
Arrah = Ara ... 40 E5
Arran 10 D4
Arras 13 A5
Arrecife 54 C2
Arrée, Mts. d' . 12 B2
Arrowtown ... 65 F3
Árta 23 E3
Arthur ⟶ 62 D3
Arthur's Pass . 65 E4
Artigas 94 C5
Artois 12 A5
Artsyz 17 E9
Artvin ⟶ 46 B5
Aru, Kepulauan 39 F4
Aru Is. = Aru,
Kepulauan . 39 F4
Arua 57 D6
Aruanã 93 E3
Aruba ■ 87 D6
Arumpo 62 B3
Arunachal
Pradesh □ .. 41 C10
Arusha 57 E7
Aruwimi ⟶ .. 56 D4
Arxan 35 B6
Arys 29 E7
Arzamas 24 B5

Arzew 54 A4
'As Saffānīyah . 47 E7
As Samāwah .. 47 E6
As
Sulaymānīyah,
Iraq 46 D6
As
Sulaymānīyah,
Si. Arabia ... 47 F6
As Summān .. 47 F6
As Sūq 47 G5
As Suwaydā' .. 46 D4
As Suwayh ... 45 F4
As Şuwayrah .. 47 D6
Asahigawa 32 F12
Asansol 40 F6
Ascension I. ... 51 G2
Aschaffenburg . 14 D5
Aschersleben .. 15 C6
Áscoli Piceno .. 20 C4
Ascope 91 E3
Aseb 49 E3
Asela 53 G6
Asenovgrad ... 22 C5
Ash Shām,
Bādiyat ... 47 D4
Ash Shāmīyah . 47 E6
Ash Shārīqah . 44 E3
Ash Shaṭrah .. 47 E6
Ash Shaykh, J. 46 D3
Ash Shu'aybah 47 F6
Ash Shu'bah .. 47 E6
Ashburton 65 E4
Ashcroft 71 C7
Asheville 75 B6
Ashford,
Australia ... 63 A5
Ashford, U.K. .. 11 F7
Ashgabat = 29 F6
Ashikaga 32 A6
Ashizuri-Zaki . 32 C3
Ashkhabad =
Ashgabat .. 29 F6
Ashmyany ... 17 A7
Ashqelon 47 E3
Ashtabula ... 72 D5
Ashuanipi, L. .. 69 C4
Asia, Kepulauan 39 D4
Asifabad 43 K11
Asike 39 F6
Asilah 54 A3
Asinara 20 D2
Asinara, G. dell' 20 D2
Asino 29 D9
Asipovichy ... 17 B9
'Asīr □ 48 D3
Asir, Ras 49 E5
Asmara =
Asmera ... 53 E5
Asmera 53 E6
Aspiring, Mt. .. 65 F3
Assam □ 41 D9
Assen 14 B4
Assini 55 G4
Assiniboia ... 71 D9
Assiniboine ⟶ 71 D10
Assis 94 A6
Assisi 20 C4
Asti 20 B2
Astipálaia ... 23 F6
Astorga 18 A2
Astoria 80 B2
Astrakhan ... 25 D6
Astrakhan-Bazàr 25 F6
Asturias □ ... 18 A3
Asunción 94 B5
Aswân 52 D5
Aswân High
Dam = Sadd
el Aali 52 D5

Asyût 52 C5
At Ţafilah 47 E3
At Ţā'if 48 C3
Atacama,
  Desierto de . 94 A3
Atacama, Salar
  de 94 A3
Atakpamé 55 G5
Atalaya 91 F4
Atami 32 B6
Atapupu 39 F2
Atår 54 D2
Atasu 29 E8
Atauro 39 F3
Atbara 53 E5
'Atbara → 53 E5
Atbasar 29 D7
Athabasca ... 71 C4
Athabasca → . 70 C8
Athabasca, L. .. 70 C8
Athens =
  Athínai 23 F4
Athens, Ala.,
  U.S.A. 74 B4
Athens, Ga.,
  U.S.A. 75 C6
Athínai 23 F4
Athlone 11 E3
Áthos 23 D5
Athy 11 E3
Ati 53 F2
Atico 91 G4
Atlanta 75 C5
Atlantic City .. 73 E8
Atlas Mts. =
  Haut Atlas .. 54 B3
Atrak =
  Atrek → 44 B3
Atrek → 44 B3
Attawapiskat → 69 C2
Attock 42 C8
Attopeu 36 B3
Attur 43 P11
Atuel → 94 D3
Atyraū 29 E6
Aubagne 13 E6
Aube → 13 B5
Auburn 74 C5
Auburn Ra. .. 63 A5
Aubusson 12 D5
Auch 12 E4
Auckland 64 B6
Aude → 13 E5
Augathella ... 63 A4
Augsburg 14 D6
Augusta, Italy . 21 F5
Augusta, Ga.,
  U.S.A. 75 C7
Augusta, Maine,
  U.S.A. 73 B11
Aunis 12 C3
Auponhia 39 E3
Aurangabad,
  Bihar, India . 40 E5
Aurangabad,
  Maharashtra,
  India 43 K9
Aurich 14 B4
Aurillac 12 D5
Aurora, Colo.,
  U.S.A. 76 F2
Aurora, Ill.,
  U.S.A. 72 D1
Aus 58 D3
Auschwitz =
  Oświęcim .. 16 C4
Austin, Minn.,
  U.S.A. 77 D8
Austin, Tex.,
  U.S.A. 78 D5

Austral Is. =
  Tubuai Is. ... 65 M16
Austral
  Seamount
  Chain 65 M16
Australia ■ ... 60 E6
Australian Alps 63 C4
Australian
  Capital
  Territory □ .. 63 C4
Austria ■ 15 E8
Autun 13 C6
Auvergne 13 D5
Auvergne, Mts.
  d' 13 D5
Auxerre 13 C5
Avallon 13 C5
Aveiro, Brazil .. 92 C2
Aveiro, Portugal 18 B1
Åvej 46 D7
Avellaneda 94 C5
Avellino 21 D5
Aversa 21 D5
Avezzano 20 C4
Aviá Terai 94 B4
Aviemore 10 C5
Avignon 13 E6
Ávila 18 B3
Avilés 18 A3
Avoca → 62 C3
Avola 21 F5
Avranches 12 B3
'Awālī 44 E2
Awash 49 F3
Awatere → ... 65 D6
Awbārī 52 C1
Awjilah 52 C3
Axim 55 H4
Axíós → 23 D4
Ayabaca 90 D3
Ayabe 32 B4
Ayacucho,
  Argentina ... 94 D5
Ayacucho, Peru 91 F4
Ayaguz 29 E9
Ayamonte 18 D2
Ayaviri 91 F4
Âybak 45 B7
Aydın 23 F6
Ayers Rock ... 60 F5
Áyios Evstrátios 23 E5
Aylmer, L. ... 70 B8
Ayn Dār 47 F7
Ayn Zālah 46 C5
Ayr 10 D4
Aytos 22 C6
Ayu, Kepulauan 39 D4
Ayutla 84 D5
Ayvacık 23 E6
Ayvalık 23 E6
Az Zahrān 47 F7
Az Zarqā 46 D4
Az-Zilfī 47 F6
Az Zubayr 47 E6
Az Zuwaytīnah 52 B3
Azamgarh ... 40 D4
Azärbayjan =
  Azerbaijan ■ . 25 E6
Āzarbāyjān-e
  Gharbī □ ... 46 C6
Āzarbāyjān-e
  Sharqī □ 46 C6
Azare 55 F7
Azbine = Aïr .. 55 E6
Azerbaijan ■ .. 25 E6
Azerbaijchan =
  Azerbaijan ■ . 25 E6
Azogues 90 D3
Azores 50 C1

Azov 25 D4
Azov, Sea of .. 25 D4
Azovskoye More
  = Azov, Sea
  of 25 D4
Azovy 28 C7
Azúa 87 C5
Azuaga 18 C3
Azuero, Pen. de 86 E3
Azul 94 D5

B

Ba Don 36 A3
Ba Ria 36 B3
Bab el Mandeb 49 E3
Baba Burnu ... 23 E6
Babadag 22 B7
Babadayhan .. 29 F7
Babaeski 22 D6
Babahoyo 90 D3
Babana 55 F5
Babar 39 F3
Babo 39 E4
Bābol 44 B3
Bābol Sar 44 B3
Baboua 56 C2
Babruysk 17 B9
Babura 55 F6
Babuyan Chan. 38 A2
Babylon 47 D6
Bacabal 92 C5
Bacan,
  Kepulauan .. 39 E3
Bacan, Pulau .. 39 E3
Bacarra 38 A2
Bacău 17 E8
Bachelina ... 29 D7
Back → 70 B9
Backstairs
  Passage ... 62 C2
Bacolod 38 B2
Bad Ischl 15 E7
Bad Kissingen . 14 C6
Bad Lands ... 76 D3
Badagara 43 P9
Badajoz 18 C2
Badalona 19 B7
Badalzai 42 E4
Badampahar .. 40 F6
Badanah 47 E5
Badarinath .. 42 D11
Badas 36 D4
Badas,
  Kepulauan .. 37 D3
Baddo → 42 E4
Bade 39 F5
Baden 15 D9
Baden-Baden . 14 D5
Baden-
  Württemberg □ .. 14 D5
Badgastein 15 E7
Bādghīsāt □ .. 42 B3
Badin 43 G6
Baduen 49 F4
Baena 18 D3
Baeza 18 D4
Bafatá 55 F2
Baffin B. 68 A4
Baffin I. 68 B3
Bafia 56 D2
Bafing → 55 F2
Bafoulabé ... 55 F2
Bāfq 44 D3
Bafra 46 B3

Bāft 44 D4
Bafwasende ... 57 D5
Bagamoyo 57 F7
Baganga 38 C3
Bagansiapiapi . 37 D2
Bagdarin 30 D9
Bagé 94 C6
Baghdād 46 D6
Bagheria 21 E4
Baghlān 42 A6
Baghlān □ ... 45 C7
Baguio 38 A2
Bahamas ■ .. 86 A4
Baharampur .. 40 E7
Bahawalpur .. 42 E7
Bahía =
  Salvador ... 93 E6
Bahía □ 93 E5
Bahía Blanca .. 94 D4
Bahía de
  Caráquez ... 90 D2
Bahía Laura .. 95 F3
Bahía Negra .. 91 H7
Bahr Aouk → . 56 C3
Bahr
  Salamat → . 53 G2
Bahraich 40 D3
Bahrain ■ 44 E2
Baia Mare ... 17 E6
Baïbokoum ... 53 G2
Baidoa 49 G3
Baie Comeau . 69 D4
Ba'ījī 46 D5
Baikal, L. =
  Baykal, Oz. .. 30 D8
Baile Atha Cliath
  = Dublin ... 11 E3
Băileşti 22 B4
Bailundo 58 A3
Bā'ir 47 E4
Baird Mts. ... 71 B3
Bairnsdale ... 63 C4
Baitadi 40 C3
Baiyin 35 C5
Baja 16 E4
Baja, Pta. 84 B1
Baja California . 84 A1
Bajimba, Mt. .. 63 A5
Bakala 56 C4
Bakchar 29 D9
Bakel 55 F2
Baker, L. 70 B10
Baker I. 64 J13
Baker Lake .. 70 B10
Baker Mt. ... 80 A3
Baker's Dozen
  Is. 69 C3
Bakersfield ... 82 C3
Bākhtarān ... 46 D6
Bakı 25 E6
Bakony Forest =
  Bakony
  Hegyseg 16 E3
Bakony Hegyseg 16 E3
Bakouma 56 C4
Baku = Bakı ... 25 E6
Bakutis Coast . 96 B1
Baky = Bakı ... 25 E6
Balabac I. 36 C5
Balabac Str. .. 36 C5
Balabalangan,
  Kepulauan .. 37 E5
Balaghat Ra. .. 43 K10
Balaguer 19 B6
Balaklava,
  Australia 62 B2
Balaklava,
  Ukraine 25 E3
Balakovo 24 C6

Balashov

Balashov 24 C5
Balasore =
Baleshwar .. 40 G6
Balaton 16 E3
Balbina, Reprêsa
de 90 D7
Balchik 22 C7
Balclutha 65 G3
Baleares, Is. .. 19 C7
Balearic Is. =
Baleares, Is. . 19 C7
Baler 38 A2
Baleshwar 40 G6
Bali, Cameroon 55 G7
Bali, Indonesia 37 F5
Bali □ 37 F5
Balıkeşir 23 E6
Balikpapan 37 E5
Balimbing 38 C1
Balipara 41 D9
Baliza 93 F3
Balkan Mts. =
Stara Planina 22 C4
Balkh 45 B6
Balkhash =
Balqash 29 E8
Balkhash, Ozero
= Balqash Köl 29 E8
Balla 41 E8
Ballarat 62 C3
Ballater 10 C5
Ballina, Australia 63 A5
Ballina, Ireland 11 D2
Ballinasloe 11 E2
Ballymena 11 D3
Balmaceda 95 F2
Balmoral 62 C3
Balonne → ... 63 A4
Balqash 29 E8
Balqash Köl ... 29 E8
Balrampur 40 D4
Balranald 62 B3
Balsas → ... 84 D4
Balta 17 D9
Bălţi 17 E8
Baltic Sea 9 G11
Baltimore 73 E7
Baluchistan □ . 42 F4
Balya 23 E6
Bam 44 D4
Bama 53 F1
Bamako 55 F3
Bamba 55 E4
Bambari 56 C4
Bamberg 14 D6
Bambili 57 D5
Bamenda 55 G7
Bāmīān □ 42 B5
Bampūr 45 E5
Ban Don =
Surat Thani . 36 C1
Ban Mê Thuột =
Buon Me
Thuot 36 B3
Banaba 64 K11
Banalia 57 D5
Banam 36 B3
Banamba 55 F3
Bananal, I. do . 93 E3
Banaras =
Varanasi 40 E4
Band-e
Torkestān .. 45 C6
Banda,
Kepulauan . 39 E3
Banda Aceh .. 36 C1
Banda Banda,
Mt. 63 B5
Banda Elat 39 F4

Banda Is. =
Banda,
Kepulauan .. 39 E3
Banda Sea 39 F3
Bandanaira ... 39 E3
Bandar =
Machilipatnam 40 J3
Bandār 'Abbās . 44 E4
Bandar-e Anzalī 46 C7
Bandar-e
Bushehr =
Būshehr 44 D2
Bandar-e Chārak 44 E3
Bandar-e
Deylam 47 E7
Bandar-e
Khomeynı ... 47 E7
Bandar-e
Lengeh 44 E3
Bandar-e
Ma'shur 47 E7
Bandar-e
Nakhīlū 44 E3
Bandar-e Rīg .. 44 D2
Bandar-e
Torkeman ... 44 B3
Bandar
Maharani =
Muar 37 D2
Bandar
Penggaram =
Batu Pahat .. 37 D2
Bandar Seri
Begawan ... 36 D5
Bandawe 59 A6
Bandeira, Pico
da 93 G5
Bandera 94 B4
Bandiagara ... 55 F4
Bandırma 23 D7
Bandon 11 F2
Bandundu 56 E3
Bandung 37 F3
Banff, Canada . 71 C8
Banff, U.K. 10 C5
Banfora 55 F4
Bang Saphan . 36 B1
Bangalore 43 N10
Bangassou ... 56 D4
Banggai,
Kepulauan .. 39 E2
Banggi, P. 36 C5
Banghāzī 52 B3
Bangka, P.,
Sulawesi,
Indonesia ... 39 D3
Bangka, P.,
Sumatera,
Indonesia ... 37 E3
Bangka, Selat . 37 E3
Bangkinang .. 37 D2
Bangko 37 E2
Bangkok 36 B2
Bangladesh ■ . 41 E7
Bangong Co .. 42 B11
Bangor, Down,
U.K. 11 D4
Bangor,
Caerns. & Mers.,
U.K. 11 E4
Bangor, U.S.A. 73 B11
Bangued 38 A2
Bangui 56 D3
Bangweulu, L. . 57 G6
Bani 87 C5
Banī Walīd ... 52 B1
Banīnah 52 B3
Bāniyās 46 D3
Banja Luka 20 B6

Banjarmasin .. 37 E4
Banjul 55 F1
Bankipore 40 E5
Banks I. 70 A7
Banks Pen. 65 E5
Banks Str. 62 D4
Bankura 40 F6
Banningville =
Bandundu .. 56 E3
Bannu 42 C7
Banská Bystrica 16 D4
Banswara 43 H9
Bantry 11 F2
Banu 42 B6
Banyak,
Kepulauan .. 37 D1
Banyo 55 G7
Banzare Coast . 96 A13
Banzyville =
Mobayi 56 D4
Baoding 35 C6
Baoji 35 C5
Baoshan 34 D4
Baotou 35 B6
Bapatla 40 K3
Ba'qūbah 46 D6
Bar,
Montenegro, Yug.
.... 22 C2
Bar, Ukraine .. 17 D8
Bar-le-Duc 13 B6
Barabai 37 E5
Barabinsk 29 D8
Baracaldo 19 A4
Baracoa 86 B5
Barahona 87 C5
Barail Range .. 41 E9
Barakhola 41 E9
Barakpur 41 F7
Barakula 63 A5
Baramula 42 B9
Baran 43 G10
Baranavichy ... 17 B8
Baranof I. 71 C6
Barão de
Melgaço 91 F6
Barapasi 39 E5
Barat Daya,
Kepulauan .. 39 F3
Barbacena ... 93 G5
Barbacoas 90 C3
Barbados ■ .. 87 D8
Barbastro 19 A6
Barberton 59 D6
Barcellona Pozzo
di Gotto ... 21 E5
Barcelona, Spain 19 B7
Barcelona,
Venezuela ... 90 A6
Barcelos 90 D6
Bardai 52 D2
Barddhaman .. 40 F6
Bardejov 16 D5
Bardera 49 G3
Bardi, Ra's 47 F4
Bardīyah 52 B4
Bareilly 42 E11
Barentu 53 E6
Barfleur, Pte. de 12 B3
Barga 34 C3
Bargal 49 E5
Barhi 40 E5
Bari 20 D6
Bari Doab 42 D8
Barīm 49 E3
Barinas 90 B4
Baring, C. 70 B8
Bârîs 52 D5
Barisal 41 F8
Barisan, Bukit . 37 E2

Barito → 37 E4
Barkly Tableland 60 D6
Barlee, L. 60 F2
Barletta 20 D6
Barmedman ... 63 B4
Barmer 42 G7
Barmera 62 B3
Barnato 63 B4
Barnaul 29 D9
Barnsley 11 E6
Barnstaple 11 F4
Baro 55 G6
Baroda =
Vadodara ... 43 H8
Barpeta 41 D8
Barquísimeto .. 90 A5
Barra, Brazil ... 93 E5
Barra, U.K. 10 C3
Barra do Corda 92 D4
Barra do Piraí . 93 G5
Barra Falsa, Pta.
da 59 C7
Barraba 63 B5
Barrackpur =
Barakpur ... 41 F7
Barraigh =
Barra 10 C3
Barranca, Lima,
Peru 91 F3
Barranca,
Loreto, Peru . 90 D3
Barrancabermeja
.... 90 B4
Barrancas 90 B6
Barrancos 18 C2
Barranqueras .. 94 B5
Barranquilla ... 90 A4
Barras 92 C5
Barreiras 93 E5
Barreirinhas .. 92 C5
Barreiro 18 C1
Barreiros 92 D6
Barretos 93 G4
Barrier Ra. ... 62 B3
Barrington Tops 63 B5
Barringun 63 A4
Barrow 71 A4
Barrow-in-
Furness 11 D5
Barry 11 F5
Barsi 43 K9
Barsoi 40 E6
Barstow 82 C4
Bartica 90 B7
Bartın 46 B3
Bartle Frere .. 61 D8
Bartlesville ... 79 A6
Barysaw 17 A9
Bāsa'idū 44 E3
Basankusa ... 56 D3
Basarabeasca . 17 E9
Basel 13 C7
Bashi Channel . 35 D7
Bashkir Republic
=
Bashkortostan □
.... 29 D6
Bashkortostan □ 29 D6
Basilan 38 C2
Basilan Str. ... 38 C2
Basildon 11 F7
Basim =
Washim 43 J10
Basingstoke .. 11 F6
Basle = Basel . 13 C7
Basoka 56 D4
Basongo 56 E4
Basque
Provinces =
País Vasco □ 19 A4

Basra = Al
  Başrah 47 E6
Bass Str. 61 H8
Bassano 71 C8
Bassano del
  Grappa 20 B3
Bassas da India 59 C7
Bassein 41 J10
Basseterre 87 C7
Bassikounou .. 55 E3
Bastak 44 E3
Bastar 40 H3
Basti 40 D4
Bastia 13 E8
Bastogne 14 C3
Bata 56 D1
Bataan 38 B2
Batabanó 86 B3
Batabanó, G. de 86 B3
Batac 38 A2
Batagoy 31 C11
Batamay 31 C10
Batang 34 C4
Batangafo 56 C3
Batangas 38 B2
Batanta 39 E4
Bateman's B. .. 63 C5
Batemans Bay . 63 C5
Bath 11 F5
Bathurst,
  Australia 63 B4
Bathurst,
  Canada 69 D4
Bathurst, C. ... 70 A7
Bathurst Harb. . 62 D4
Bathurst Inlet . 70 B9
Batinah 44 F4
Batlow 63 C4
Batman 46 C7
Batna 54 A6
Baton Rouge .. 79 D8
Batouri 56 D2
Battambang ... 36 B2
Battipáglia ... 21 D5
Battle Creek .. 72 C3
Battle Harbour . 69 C5
Battleford 71 C9
Batu 49 F2
Batu, Kepulauan 37 E1
Batu Is. = Batu,
  Kepulauan .. 37 E1
Batu Pahat ... 37 D2
Batuata 39 F2
Batumi 25 E5
Baturaja 37 E2
Baturité 92 C6
Bau 37 D4
Baubau 39 F2
Bauchi 55 F6
Bauer, C. 62 B1
Baukau 39 F3
Bauru 93 G4
Baús 93 F3
Bautzen 15 C8
Bavaria =
  Bayern □ ... 15 D6
Bawdwin 41 F11
Bawean 37 F4
Bawku 55 F4
Bawlake 41 H11
Bay, L. de 38 B2
Bay City 72 C4
Bay View 64 C7
Bayamo 86 B4
Bayan Har Shan 34 C4
Bayan Hot =
  Alxa Zuoqi .. 35 C5
Bayanaüyl 29 D8
Bayanhongor .. 35 B5

Bayázeh 44 C3
Baybay 38 B2
Bayburt 46 B5
Bayern □ 15 D6
Bayeux 12 B3
Bayındır 23 E6
Baykal, Oz. ... 30 D8
Baykit 30 C7
Baykonur =
  Bayqongyr .. 29 E7
Bayombong ... 38 A2
Bayonne 12 E3
Bayovar 91 E2
Bayqongyr 29 E7
Bayram-Ali =
  Bayramaly .. 29 F7
Bayramaly 29 F7
Bayramiç 23 E6
Bayreuth 15 D6
Bayrūt 46 D3
Baytown 79 E6
Baza 19 D4
Bazaruto, I. do . 59 C7
Beachport 62 C2
Beagle, Canal . 95 G3
Beardmore
  Glacier 96 C14
Béarn 12 E3
Beauce, Plaine
  de la 12 B4
Beaudesert ... 63 A5
Beaufort 36 C5
Beaufort Sea .. 66 B6
Beaufort West . 58 E4
Beaumont 79 D6
Beaune 13 C6
Beauvais 12 B5
Beauval 71 C9
Beaver 71 B5
Beaver → 71 C9
Beawar 42 F9
Bečej 22 B3
Béchar 54 B4
Bedford 11 E6
Beenleigh 63 A5
Be'er Sheva ... 47 E3
Beersheba =
  Be'er Sheva . 47 E3
Befale 56 D4
Bega 63 C4
Behbehān 47 E7
Behshahr 44 B3
Bei Jiang → .. 35 D6
Bei'an 35 B7
Beibei 35 D5
Beihai 35 D5
Beijing 35 C6
Beilpajah 63 B3
Beinn na
  Faoghla =
  Benbecula .. 10 C3
Beira 59 B6
Beirut = Bayrūt 46 D3
Beitbridge 59 C6
Beja, *Portugal* . 18 C2
Béja, *Tunisia* .. 54 A6
Bejaia 54 A6
Béjar 18 B3
Bejestān 44 C4
Békéscsaba ... 16 E5
Bela, *India* 40 E4
Bela, *Pakistan* . 42 F5
Bela Crkva ... 22 B3
Bela Vista 94 A5
Belarus ■ 24 C2
Belau = Palau ■ 38 C4
Belawan 37 D1
Belaya Tserkov
  = Bila Tserkva 17 D10

Belcher Is. 69 C3
Belém 92 C4
Belén 94 A5
Belet Uen 49 G4
Belev 24 C4
Belfast 11 D4
Belfort 13 C7
Belgaum 43 M9
Belgium ■ 14 C3
Belgorod 24 C4
Belgorod-
  Dnestrovskiy
  = Bilhorod-
  Dnistrovskyy 25 D3
Belgrade =
  Beograd 22 B3
Beli Drim → .. 22 C3
Belinga 56 D2
Belinyu 37 E3
Beliton Is. =
  Belitung 37 E3
Belitung 37 E3
Belize ■ 85 D7
Belize City 85 D7
Belkovskiy,
  Ostrov 30 B11
Bell Bay 62 D4
Bell Peninsula . 68 B2
Bell Ville 94 C4
Bella Unión ... 94 C5
Bella Vista ... 94 B5
Bellary 43 M10
Bellata 63 A4
Belle-Ile 12 C2
Belle Isle 69 C5
Belle Isle, Str. of 69 C5
Belleville,
  Canada 69 D3
Belleville, *U.S.A.* 77 F10
Bellin =
  Kangirsuk .. 68 C4
Bellingen 63 B5
Bellingham ... 80 A2
Bellingshausen
  Sea 96 A2
Bellinzona 13 C8
Belluno 20 A4
Belmont 63 B5
Belmonte 93 F6
Belmopan 85 D7
Belo Horizonte 93 F5
Belo-Tsiribihina 59 H8
Beloit 77 D10
Belokorovichi . 17 C9
Belomorsk ... 28 C4
Belonia 41 F8
Belorussia ■ =
  Belarus ■ ... 24 C2
Belovo 29 D9
Beloye, Ozero . 24 A4
Beloye More .. 28 C4
Belozersk 24 A4
Beltana 62 B2
Belterra 92 C3
Beltsy = Bălţi . 17 E8
Belukha 29 E9
Beluran 36 C5
Belyy, Ostrov . 28 B8
Belyy Yar 29 D9
Bembéréke ... 55 F5
Bemidji 77 B7
Ben Gardane .. 52 B1
Ben Lomond,
  N.S.W.,
  Australia 63 B5
Ben Lomond,
  Tas., Australia 62 D4
Ben Nevis 10 C4
Bena 55 F6

Bena Dibele ... 56 E4
Benagerie 62 B3
Benalla 63 C4
Benambra, Mt. . 63 C4
Benares =
  Varanasi 40 E4
Benavente 18 A3
Benbecula 10 C3
Benbonyathe .. 62 B2
Bend 80 C3
Bender Beila .. 49 F5
Bendery =
  Tighina 17 E9
Bendigo 63 C3
Benevento 20 D5
Bengal, Bay of . 41 J7
Bengbu 35 C6
Benghazi =
  Banghāzī ... 52 B3
Bengkalis 37 D2
Bengkulu 37 E2
Bengkulu □ ... 37 E2
Benguela 58 A2
Beni 57 D5
Beni → 91 F5
Beni Abbès ... 54 B4
Beni Mazâr ... 52 C5
Beni Mellal ... 54 B3
Beni Ounif 54 B4
Beni Suef 52 C5
Benidorm 19 C5
Benin ■ 55 G5
Benin, Bight of 55 H5
Benin City 55 G6
Benjamin
  Constant 90 D4
Benoni 59 D5
Bent 44 E4
Benteng 39 F2
Benton Harbor 72 C2
Benue → 55 G6
Benxi 35 B7
Beo 38 D3
Beograd 22 B3
Beppu 32 C2
Berati 23 D2
Berau, Teluk .. 39 E4
Berber 53 E5
Berbera 49 E4
Berbérati 56 D3
Berbice → ... 90 B7
Berdichev =
  Berdychiv ... 17 D9
Berdsk 29 D9
Berdyansk ... 25 D4
Berdychiv ... 17 D9
Berebere 39 D3
Bereda 49 E5
Berehove 17 D6
Berekum 55 G4
Berestechko .. 17 C7
Bereza 17 B7
Berezhany ... 17 D7
Berezina =
  Byarezina → 17 B10
Berezniki 28 D6
Berezovo 28 C7
Berga 19 A6
Bergama 23 E6
Bérgamo 20 B2
Bergen 8 F9
Bergerac 12 D4
Berhala, Selat . 37 E2
Berhampore =
  Baharampur . 40 E7
Berhampur ... 40 H5
Bering Sea ... 71 C1
Bering Strait .. 71 B3
Berja 19 D4

Berkeley 80 G2
Berkner I. 96 B4
Berlin 15 B7
Bermejo →, Formosa, Argentina ... 94 B5
Bermejo →, San Juan, Argentina ... 94 C3
Bermuda ■ ... 67 F13
Bern 13 C7
Bernardo de Irigoyen 94 B6
Bernburg 15 C6
Berne = Bern . 13 C7
Bernina, Piz ... 13 C8
Beroroha 59 J9
Beroun 16 D2
Berrechid 54 B3
Berri 62 B3
Berry, Australia 63 B5
Berry, France .. 12 C5
Bershad 17 D9
Bertoua 56 D2
Berwick-upon-Tweed 10 D6
Besalampy 59 H8
Besançon 13 C7
Besar 37 E5
Besni 46 C4
Bessarabiya ... 17 E9
Bessarabka = Basarabeasca 17 E9
Bessemer 74 C4
Betanzos 18 A1
Bétaré Oya ... 56 C2
Bethanien 58 D3
Bethel 71 B3
Bethlehem, S. Africa 59 D5
Bethlehem, U.S.A. 73 D8
Béthune 13 A5
Bethungra 63 B4
Betioky 59 J8
Betroka 59 J9
Betsiamites .. 69 D4
Bettiah 40 D5
Betul 43 J10
Betung 37 D4
Beverley 11 E6
Beverly Hills .. 82 C3
Beyla 55 G3
Beyneu 29 E6
Beypazarı 46 B2
Beyşehir Gölü . 46 C2
Bezhitsa 24 C3
Béziers 13 E5
Bezwada = Vijayawada . 40 J3
Bhachau 43 H7
Bhadrakh 40 G6
Bhadravati ... 43 N9
Bhagalpur ... 40 E6
Bhakra Dam ... 42 D10
Bhamo 41 E11
Bhandara 43 J11
Bhanrer Ra. ... 43 H11
Bharat = India ■ 43 J10
Bharatpur 42 F10
Bhatpara 41 F7
Bhaunagar = Bhavnagar . 43 J8
Bhavnagar ... 43 J8
Bhawanipatna . 40 H3
Bhilsa = Vidisha 43 H10
Bhilwara 43 G9
Bhima → 43 L10

Bhimavaram .. 40 J3
Bhind 42 F11
Bhiwandi 43 K8
Bhiwani 42 E10
Bhola 41 F8
Bhopal 43 H10
Bhubaneshwar 40 G5
Bhuj 43 H6
Bhumiphol Dam = Phumiphon, Khuan 36 A1
Bhusaval 43 J9
Bhutan ■ 41 D8
Biafra, B. of = Bonny, Bight of 56 D1
Biak 39 E5
Biała Podlaska . 17 B6
Białogard 16 A2
Białystok 17 B6
Biaro 39 D3
Biarritz 12 E3
Bibala 58 A2
Biberach 14 D5
Bibiani 55 G4
Bida 55 G6
Bidar 43 L10
Biddeford 73 C10
Bidon 5 = Poste Maurice Cortier 54 D5
Bié, Planalto de 58 A3
Biel 13 C7
Bielefeld 14 B5
Biella 20 B2
Bielsk Podlaski 17 B6
Bielsko-Biała .. 16 D4
Bien Hoa 36 B3
Bienne = Biel . 13 C7
Bienville, L. ... 69 C3
Big Horn Mts. = Bighorn Mts. 81 C10
Big Spring 78 C3
Big Trout L. ... 69 C2
Biğa 23 D6
Bigadiç 23 E7
Biggar 71 C9
Biggenden 63 A5
Bighorn → ... 81 B10
Bighorn Mts. .. 81 C10
Bihać 20 B5
Bihar 40 E5
Bihar □ 40 E5
Bihor, Munţii .. 17 E6
Bijagós, Arquipélago dos 55 F1
Bijapur, Karnataka, India 43 L9
Bijapur, Mad. P., India 40 H3
Bijār 46 D6
Bijeljina 20 B7
Bijnor 42 E11
Bikaner 42 E8
Bikini Atoll 64 H11
Bila Tserkva ... 17 D10
Bilara 42 F8
Bilaspur 40 F4
Bilauk Taungdan 36 B2
Bilbao 19 A4
Bilbo = Bilbao . 19 A4
Bilecik 46 B2
Bilhorod-Dnistrovskyy 25 D3
Bilir 31 C11
Billings 81 C9
Billiton Is. = Belitung 37 E3

Bilma 53 E1
Biloxi 79 D9
Biltine 53 F3
Bima 39 F1
Bimbo 56 D3
Bina-Etawah .. 43 G11
Binalbagan ... 38 B2
Binalong 63 B4
Bīnālūd, Kūh-e 44 B4
Binatang 37 D4
Binda 63 A4
Bindle 63 A4
Bindura 59 B6
Bingara, N.S.W., Australia 63 A5
Bingara, Queens., Australia 63 A3
Bingham Canyon 81 E7
Binghamton .. 73 C8
Bingöl 46 C5
Binh Dinh = An Nhon 36 B3
Binh Son 36 A3
Binjai 37 D1
Binnaway 63 B4
Binongko 39 F2
Bintan 37 D2
Bintulu 37 D4
Bintuni 39 E4
Binzert = Bizerte 54 A6
Bioko 56 D1
Bir 43 K9
Bir Autrun 53 E4
Bir Mogrein ... 54 C2
Bîr Ungât 52 D5
Bira 39 E4
Birao 53 F3
Birchip 62 C3
Birecik 46 C4
Bireuen 36 C1
Bîrlad 17 E8
Bîrlad → 17 F8
Birmingham, U.K. 11 E6
Birmingham, U.S.A. 74 C4
Birmitrapur ... 40 F5
Birni Nkonni .. 55 F6
Birnin Kebbi .. 55 F5
Birr 11 E3
Birrie → 63 A4
Birsk 29 D6
Birur 43 N9
Bisa 39 E3
Biscay, B. of .. 12 D2
Biscoe Bay 96 B17
Biscoe Is. 96 A3
Bishkek 29 E8
Biskra 54 B6
Bislig 38 C3
Bismarck 76 B4
Bismarck Arch. 64 K9
Bissagos = Bijagós, Arquipélago dos 55 F1
Bissau 55 F1
Bistriţa 17 E7
Bistriţa → ... 17 E8
Bitam 56 D2
Bitkine 53 F2
Bitlis 46 C5
Bitola 22 D3
Bitolj = Bitola . 22 D3
Bitter L. = Buheirat-Murrat-el-Kubra 52 B5

Bitterfontein .. 58 E3
Bitterroot Range 81 B6
Biu 53 F1
Biwa-Ko 32 B5
Biysk 29 D9
Bizerte 54 A6
Bjelovar 20 B6
Black Forest = Schwarzwald 14 D5
Black Hills ... 76 D3
Black Sea 25 E3
Black Volta → 55 G4
Blackball 65 E4
Blackburn 11 E5
Blackpool 11 E5
Blackwater → . 11 E2
Blagodarnoye = Blagodarnyy 25 D5
Blagodarnyy .. 25 D5
Blagoevgrad .. 22 C4
Blagoveshchensk 31 D10
Blairmore 71 D8
Blanc, Mont .. 13 D7
Blanca, B. 95 D4
Blanca Peak ... 83 B10
Blanche, C. ... 62 B1
Blanche, L. ... 62 A2
Blanes 19 B7
Blantyre 59 B6
Blayney 63 B4
Blednaya, Gora 28 B7
Blenheim 65 D5
Blida 54 A5
Bligh Sound .. 65 F2
Blitta 55 G5
Blodgett Iceberg Tongue 96 A13
Bloemfontein . 59 D5
Bloemhof 59 D5
Blois 12 C4
Bloomington, Ill., U.S.A. .. 77 E10
Bloomington, Ind., U.S.A. .. 72 E2
Blue Mts., Oreg., U.S.A. 80 C4
Blue Mts., Pa., U.S.A. 73 D7
Blue Nile = Nîl el Azraq → . 53 E5
Blue Ridge Mts. 75 A7
Bluefields 86 D3
Bluff 65 G3
Blumenau 94 B7
Bo 55 G2
Bo Duc 36 B3
Bo Hai 35 C6
Boa Vista 90 C6
Boatman 63 A4
Bobadah 63 B4
Bobbili 40 H4
Bobo-Dioulasso 55 F4
Bóbr → 16 B2
Bobraomby, Tanjon' i ... 59 G9
Bobruysk = Babruysk ... 17 B9
Bôca do Acre .. 91 E5
Boca Raton ... 75 F7
Bocaiúva 93 F5
Bocanda 55 G4
Bocaranga ... 56 C3
Bochnia 16 D5
Bochum 14 C4
Boda 56 D3
Bodaybo 30 D9
Boden 8 E12
Bodensee 13 C8
Bodhan 43 K10

Name	Page	Ref
Bodø	8	E10
Bodrog →	16	D5
Bodrum	23	F6
Boende	56	E4
Boffa	55	F2
Bogalusa	79	D9
Bogan Gate	63	B4
Boggabilla	63	A5
Boggabri	63	B5
Bogo	38	B2
Bogong, Mt.	63	C4
Bogotá	90	C4
Bogotol	29	D9
Bogra	41	E7
Bogué	55	E2
Bohemian Forest = Böhmerwald	15	D7
Bohena Cr. →	63	B4
Böhmerwald	15	D7
Bohol	38	C2
Bohol Sea	38	C2
Bohotleh	49	F4
Boise	80	D5
Boise City	78	A2
Bojador C.	54	C2
Bojana →	22	D2
Bojnūrd	44	B4
Boké	55	F2
Bokhara →	63	A4
Bokoro	53	F2
Bokote	56	E4
Bokungu	56	E4
Bol	53	F1
Bolama	55	F1
Bolan Pass	42	E5
Bolbec	12	B4
Bole	34	B3
Bolekhiv	17	D6
Bolekhov = Bolekhiv	17	D6
Bolesławiec	16	C2
Bolgrad = Bolhrad	17	F9
Bolhrad	17	F9
Bolinao C.	38	A1
Bolívar, Argentina	94	D4
Bolívar, Colombia	90	C3
Bolivia ■	91	G6
Bollon	63	A4
Bolobo	56	E3
Bologna	20	B3
Bologoye	24	B3
Bolomba	56	D3
Bolong	38	C2
Bolsena, L. di	20	C3
Bolshereche	29	D8
Bolshevik, Ostrov	30	B8
Bolshoi Kavkas = Caucasus Mountains	25	E5
Bolshoy Atlym	28	C7
Bolshoy Begichev, Ostrov	30	B9
Bolshoy Lyakhovskiy, Ostrov	31	B12
Bolton	11	E5
Bolu	46	B2
Bolvadin	46	C2
Bolzano	20	A3
Bom Despacho	93	F4
Bom Jesus da Lapa	93	E5
Boma	56	F2
Bomaderry	63	B5
Bombala	63	C4
Bombay	43	K8
Bomboma	56	D3
Bomili	57	D5
Bomongo	56	D3
Bomu →	56	D4
Bon, C.	52	A1
Bonaire	87	D6
Bonang	63	C4
Bonavista	69	D5
Bondo	56	D4
Bondoukou	55	G4
Bone, Teluk	39	E2
Bonerate	39	F2
Bonerate, Kepulauan	39	F2
Bong Son = Hoai Nhon	36	B3
Bongandanga	56	D4
Bongor	53	F2
Bonifacio	13	F8
Bonn	14	C4
Bonney, L.	62	C3
Bonny, Bight of	56	D1
Bonoi	39	E5
Bontang	37	D5
Bonthain	39	F1
Bonthe	55	G2
Bontoc	38	A2
Boolaboolka L.	62	B3
Booligal	63	B3
Boonah	63	A5
Boorindal	63	B4
Boorowa	63	B4
Boothia, Gulf of	68	A2
Boothia Pen.	70	A10
Booué	56	E2
Bor, Serbia, Yug.	22	B4
Bôr, Sudan	53	G5
Borama	49	F3
Borås	9	G10
Borāzjān	44	D2
Borba	90	D7
Borda, C.	62	C2
Bordeaux	12	D3
Bordertown	62	C3
Bordj Fly Ste. Marie	54	C4
Bordj-in-Eker	54	D6
Bordj Omar Driss	54	C6
Bordj-Tarat	54	C6
Borger	78	B3
Borisoglebsk	24	C5
Borisov = Barysaw	17	A9
Borja	90	D3
Borkou	53	E2
Borkum	14	B4
Borley, C.	96	A9
Borneo	37	D4
Bornholm	9	G11
Boromo	55	F4
Borongan	38	B3
Borovichi	24	B3
Borşa	17	E7
Borūjerd	46	D7
Boryslav	17	D6
Bosa	21	D2
Bosanska Gradiška	20	B6
Bosaso	49	E4
Boshan	35	C6
Boshrūyeh	44	C4
Bosna →	20	B7
Bosna i Hercegovina = Bosnia-Herzegovina ■	20	B6
Bosnia-Herzegovina ■	20	B6
Bosnik	39	E5
Bosobolo	56	D3
Bosporus = Karadeniz Boğazı	22	D7
Bossangoa	56	C3
Bossembélé	53	G2
Bosso	53	F1
Bosten Hu	34	B3
Boston, U.K.	11	E6
Boston, U.S.A.	73	C10
Botany B.	63	B5
Bothnia, G. of	8	F12
Bothwell	62	D4
Botletle →	58	C4
Botoşani	17	E8
Botswana ■	58	C4
Botucatu	94	A7
Bou Djébéha	55	E4
Bou Izakarn	54	C3
Bouaké	55	G3
Bouar	56	C3
Bouârfa	54	B4
Bouca	56	C3
Bougie = Bejaia	54	A6
Bougouni	55	F3
Boulder	76	E2
Boulder Dam = Hoover Dam	82	B5
Boulogne-sur-Mer	12	A4
Boultoum	55	F7
Bouna	55	G4
Boundiali	55	G3
Bourbonnais	13	C5
Bourem	55	E4
Bourg-en-Bresse	13	C6
Bourg-St.-Maurice	13	D7
Bourges	12	C5
Bourgogne	13	C6
Bourke	63	B4
Bournemouth	11	F6
Bousso	53	F2
Boutilimit	55	E2
Bowen Mts.	63	C4
Bowman I.	96	A12
Bowmans	62	B2
Bowral	63	B5
Bowraville	63	B5
Boyne →	11	E3
Boyni Qara	45	B6
Boz Dağları	23	E7
Bozburun	23	F7
Bozcaada	23	E6
Bozdoğan	23	F7
Bozeman	81	C8
Bozen = Bolzano	20	A3
Bozoum	56	C3
Bra	20	B1
Brač	20	C6
Bracciano, L. di	20	C4
Brach	52	C1
Bräcke	8	F11
Brad	17	E6
Bradenton	75	F6
Bradford	11	E6
Braemar	62	B2
Braga	18	B1
Bragança, Brazil	92	C4
Bragança, Portugal	18	B2
Brahmanbaria	41	F8
Brahmani →	40	G6
Brahmaputra →	41	F7
Braidwood	63	C4
Brăila	17	F8
Branco →	90	D6
Brandenburg = Neubrandenburg	15	B7
Brandenburg	15	B7
Brandenburg □	15	B7
Brandon	71	D10
Brandvlei	58	E4
Braniewo	16	A4
Bransfield Str.	96	A4
Brantford	69	D2
Branxholme	62	C3
Brasiléia	91	F5
Brasília	93	F4
Braşov	17	F7
Brassey, Banjaran	36	D5
Bratislava	16	D3
Bratsk	30	D8
Braunau	15	D7
Braunschweig	14	B6
Brava	49	G3
Bravo del Norte, R. → = Grande, Rio →	79	F5
Brawley	82	D5
Bray	11	E3
Bray, Pays de	12	B4
Brazil ■	93	E4
Brazos →	79	E6
Brazzaville	56	E3
Brčko	20	B7
Breaksea Sd.	65	F2
Bream B.	64	A6
Bream Hd.	64	A6
Brecon	11	F5
Breda	14	C3
Bredasdorp	58	E4
Bredbo	63	C4
Bregenz	14	E5
Brejo	92	C5
Bremen	14	B5
Bremerhaven	14	B5
Bremerton	80	B2
Brenner P.	15	E6
Bréscia	20	B3
Breslau = Wrocław	16	C3
Bressanone	20	A3
Brest, Belarus	17	B6
Brest, France	12	B1
Brest-Litovsk = Brest	17	B6
Bretagne	12	B2
Brett, C.	64	A6
Brewarrina	63	A4
Brezhnev = Naberezhnyye Chelny	29	D6
Bria	56	C4
Briançon	13	D7
Bribie I.	63	A5
Bridgeport	73	D9
Bridgetown	87	D8
Bridgewater	69	D4
Bridgewater, C.	62	C3
Bridlington	11	D6
Bridport	62	D4
Brig	13	C7
Brigham City	81	E7
Bright	63	C4
Brighton, Australia	62	C2
Brighton, U.K.	11	F6
Bríndisi	21	D6
Brinkworth	62	B2
Brisbane	63	A5
Brisbane →	63	A5

Bristol, U.K. ... 11 F5
Bristol, U.S.A. . 75 A6
Bristol B. 71 C3
Bristol Channel 11 F4
Britstown 58 E4
Brittany =
Bretagne 12 B2
Brive-la-
Gaillarde 12 D4
Brixen =
Bressanone . 20 A3
Brlik 29 E8
Brno 16 D3
Brocken 14 C6
Brodeur Pen. .. 68 A2
Brodnica 16 B4
Brody 17 C7
Broken Hill =
Kabwe 59 A5
Broken Hill 62 B3
Bronte Park ... 62 D4
Brooks Ra. 71 B5
Brooloo 63 A5
Broome 60 D3
Broughton
Island 68 B4
Brown, Pt. 62 B1
Brownsville .. 78 B5
Brownwood ... 78 D4
Bruay-en-Artois 13 A5
Bruce, Mt. ... 60 E2
Bruck an der
Mur 15 E8
Bruges =
Brugge 14 C2
Brugge 14 C2
Brumado ... 93 E5
Brunei = Bandar
Seri Begawan 36 D5
Brunei ■ 36 D4
Brunner, L. ... 65 E4
Brunswick =
Braunschweig 14 B6
Brunswick ... 75 D7
Brunswick, Pen.
de 95 G2
Brusque 94 B7
Brussel 14 C3
Brussels =
Brussel 14 C3
Bruthen 63 C4
Bruxelles =
Brussel 14 C3
Bryan 79 D5
Bryan, Mt. 62 B2
Bryansk 24 C3
Bu Craa 54 C2
Buapinang ... 39 E2
Buayan 38 C3
Bucak 46 C2
Bucaramanga . 90 B4
Buchach 17 D7
Buchanan ... 55 G2
Buchans ... 69 D5
Bucharest =
Bucureşti .. 22 B6
Buckie 10 C5
Buckleboo ... 62 B2
Bucureşti ... 22 B6
Budalin ... 41 F10
Budapest ... 16 E4
Budd Coast ... 96 A12
Bude 11 F4
Budennovsk ... 25 E5
Budgewoi 63 B5
Budjala 56 D3
Buenaventura . 90 C3
Buenos Aires .. 94 C5
Buenos Aires, L. 95 F2
Buffalo 73 C6

Bug → =
Buh → 25 D3
Bug → 16 B5
Buga 90 C3
Bugel, Tanjung 37 F4
Bugsuk 36 C5
Bugun Shara .. 35 B5
Buguruslan ... 29 D6
Buh → 25 D3
Buheirat-Murrat-
el-Kubra 52 B5
Buir Nur 35 B6
Bujumbura ... 57 E5
Bukama 57 F5
Bukavu 57 E5
Bukene 57 E6
Bukhara =
Bukhoro 29 F7
Bukhoro 29 F7
Bukittinggi ... 37 E2
Bukoba 57 E6
Bula 39 E4
Bulahdelah ... 63 B5
Bulan 38 B2
Bulandshahr .. 42 E10
Bulawayo ... 59 C5
Buldan 23 E7
Bulgan 35 B5
Bulgaria ■ ... 22 C5
Bulgroo 63 A3
Bulgunnia ... 62 B1
Bulhar 49 E3
Buli, Teluk ... 39 D3
Buliluyan, C. .. 36 C5
Bulli 63 B5
Bulloo → 63 A3
Bulloo Downs . 62 A3
Bulloo L. 62 A3
Bulls 64 D6
Bulo Burti ... 49 G4
Bulsar = Valsad 43 J8
Bulukumba ... 39 F2
Bulun 30 B10
Bumba 56 D4
Bumhpa Bum . 41 D11
Buna 57 D7
Bunbah, Khalīj . 52 B3
Bunbury 60 G2
Buncrana ... 11 D3
Bundaberg ... 63 A5
Bundi 43 G9
Bundoran ... 11 D2
Bungo-Suidō .. 32 C3
Bunia 57 D6
Bunji 42 B9
Buntok 37 E4
Bunyu 37 D5
Buol 39 D2
Buon Me Thuot 36 B3
Buorkhaya, Mys 31 B11
Buqayq 47 F7
Bur Acaba ... 49 G3
Bûr Safâga .. 52 C5
Bûr Sa'îd ... 52 B5
Bûr Sûdân ... 53 E6
Burao 49 F4
Buraydah 47 F6
Buraymī, Al
Wāhāt al 44 E3
Burcher 63 B4
Burdur 46 C2
Burdwan =
Barddhaman 40 F6
Burgas 22 C6
Burgersdorp . 59 E5
Burgos 18 A4
Burgundy =
Bourgogne . 13 C6
Burhaniye ... 23 E6
Burhanpur 43 J10

Burias 38 B2
Buriram 36 A2
Burji 53 G6
Burkina Faso ■ 55 F4
Burlington,
Iowa, U.S.A. . 77 E9
Burlington, N.C.,
U.S.A. 75 A8
Burlington, Vt.,
U.S.A. 73 B9
Burlyu-Tyube .. 29 E8
Burma ■ 41 G11
Burnie 62 D4
Burnley 11 E5
Burnside → ... 70 B9
Burqān 47 E6
Burqin 34 B3
Burra 62 B2
Burren Junction 63 B4
Burrendong
Dam 63 B4
Burrinjuck Res. . 63 C4
Burruyacú ... 94 B4
Bursa 23 D7
Burtundy ... 62 B3
Buru 39 E3
Burundi ■ ... 57 E6
Burutu 55 G6
Bury St.
Edmunds ... 11 E7
Buşayyah ... 47 E6
Büshehr 44 D2
Büshehr □ 44 D2
Bushire =
Büshehr 44 D2
Businga 56 D4
Busra ash Shām 46 D4
Busto Arsízio . 20 B2
Busu-Djanoa .. 56 D4
Busuanga ... 38 B1
Buta 57 D4
Butare 57 E5
Butaritari ... 64 J12
Butembo ... 57 D5
Butha Qi 35 B7
Buton 39 E2
Butte 81 B7
Butterworth ... 36 C2
Butuan 38 C3
Butung = Buton 39 E2
Buturlinovka .. 24 C5
Buxtehude ... 14 B5
Buy 24 B5
Büyük
Menderes → 23 F6
Büyükçekmece 22 D7
Buzău 17 F8
Buzău → 17 F8
Buzen 32 C2
Buzi → 59 B6
Buzuluk 29 D6
Byarezina → .. 17 B10
Bydgoszcz ... 16 B3
Byelarus =
Belarus ■ ... 24 C2
Byelorussia =
Belarus ■ ... 24 C2
Bykhaw 17 B10
Bykhov =
Bykhaw 17 B10
Bylot I. 68 A3
Byrd, C. 96 A3
Byrock 63 B4
Byron Bay ... 63 A5
Byrranga, Gory 30 B8
Byrranga Mts. =
Byrranga,
Gory 30 B8
Bytom 16 C4
Bytów 16 A3

C

Ca Mau = Quan
Long 36 C3
Caála 58 A3
Cabanatuan ... 38 A2
Cabedelo 92 D7
Cabimas 90 A4
Cabinda 56 F2
Cabinda □ 56 F2
Cabinet Mts. .. 81 B6
Cabo Blanco .. 95 F3
Cabo Frio 93 G5
Cabo Pantoja . 90 D3
Cabonga,
Réservoir ... 69 D3
Caboolture ... 63 A5
Cabora Bassa
Dam =
Cahora Bassa
Dam 59 B6
Cabot Str. ... 69 D5
Cabra 18 D3
Cabrera 19 C7
Cabriel → 19 C5
Čačak 22 C3
Cáceres, Brazil . 91 G7
Cáceres, Spain 18 C2
Cachimbo, Serra
do 93 D2
Cachoeira ... 93 E6
Cachoeira de
Itapemirim . 93 G5
Cachoeira do
Sul 94 C6
Cacólo 56 G3
Caconda 58 A3
Cacongo 56 F2
Cadibarrawirracanna,
L. 62 A2
Cadiz, Phil. ... 38 B2
Cádiz, Spain .. 18 D2
Cádiz, G. de .. 18 D2
Cadney Park .. 62 A1
Caen 12 B3
Caeté 93 F5
Caetité 93 E5
Cagayan → .. 38 A2
Cagayan de Oro 38 C2
Cágliari 21 E2
Cágliari, G. di . 21 E2
Caguas 87 C6
Cahora Bassa
Dam 59 B6
Cahors 12 D4
Cahuapanas .. 91 E3
Cahul 17 F9
Caia 59 B7
Caibarién ... 86 B4
Caicara 90 B5
Caicó 92 D6
Caicos Is. 87 B5
Caird Coast ... 96 B5
Cairns 61 D8
Cairo = El
Qâhira 52 B5
Cairo 79 A9
Caiundo 58 B3
Caiza 91 H5
Cajamarca ... 91 E3
Cajàzeiras ... 92 D6
Calabar 55 H6
Calábria □ 21 E2
Calafate 95 G2
Calahorra 19 A5
Calais 12 A4
Calama, Brazil . 91 E6
Calama, Chile . 94 A3

Calamar,
 Bolívar,
 Colombia ... 90 A4
Calamar,
 Vaupés,
 Colombia ... 90 C4
Calamian Group 38 B1
Calamocha 19 B5
Calang 36 D1
Calapan 38 B2
Călăraşi 22 B6
Calatayud 19 B5
Calauag 38 B2
Calavite, C. ... 38 B2
Calbayog 38 B2
Calca 91 F4
Calcutta 41 F7
Caldas da
 Rainha 18 C1
Caldera 94 B2
Caldwell 80 D5
Caledon 58 E3
Caledon → .. 59 E5
Calgary 71 C8
Cali 90 C3
Calicut 43 P9
California ☐ ... 82 B3
California, G. de 84 B2
Calingasta 94 C3
Callabonna, L. . 62 A3
Callao 91 F3
Caloundra 63 A5
Caltagirone .. 21 F5
Caltanissetta .. 21 F5
Calulo 56 G2
Calunda 58 A4
Calvi 13 E8
Calviá 19 C7
Calvinia 58 E3
Camabatela .. 56 F3
Camacupa 58 A3
Camaguey 86 B4
Camaná 91 G4
Camargo 91 H5
Camargue 13 E6
Camarones ... 95 E3
Cambay =
 Khambhat .. 43 H8
Cambay, G. of =
 Khambat, G.
 of 43 J8
Cambodia ■ .. 36 B3
Cambrai 13 A5
Cambrian Mts. . 11 E5
Cambridge, N.Z. 64 B6
Cambridge, U.K. 11 E7
Cambridge,
 U.S.A. 73 C10
Cambridge Bay 70 B9
Cambridge G. . 60 C4
Cambundi-
Catembo ... 56 G3
Camden, Ark.,
 U.S.A. 79 C7
Camden, N.J.,
 U.S.A. 73 E8
Cameroon ■ .. 56 C2
Cameroun, Mt. . 56 D1
Cametá 92 C4
Caminha 18 B1
Camira Creek .. 63 A5
Camissombo .. 56 F4
Camocim 92 C5
Camopi → .. 92 B3
Campana, I. ... 95 F1
Campánia ☐ .. 21 D5
Campbell Town 62 D4
Campbellton .. 69 D4
Campbelltown . 63 B5

Campbeltown . 10 D4
Campeche 85 D6
Campeche, B. de 85 D6
Camperdown .. 62 C3
Campina Grande 92 D6
Campinas 94 A7
Campo 56 D1
Campo Belo ... 93 G4
Campo Formoso 93 E5
Campo Grande 93 G3
Campo Maíor . 92 C5
Campo Mourão 93 G3
Campoalegre .. 90 C3
Campobasso .. 20 D5
Campos 93 G5
Campos Belos . 93 E4
Campuya → .. 90 D4
Camrose 71 C8
Çan 23 D6
Can Tho 36 B3
Canada ■ 70 C8
Cañada de
 Gómez 94 C4
Çanakkale 23 D6
Çanakkale
 Boğazı 23 D6
Cananea 84 A2
Canarias, Is. ... 54 C1
Canary Is. =
 Canarias, Is. . 54 C1
Canavieiras ... 93 F6
Canbelego ... 63 B4
Canberra 63 C4
Candala 49 E4
Candelo 63 C4
Candia =
 Iráklion 23 G5
Canea = Khaniá 23 G5
Canelones 94 C5
Cañete, Chile .. 94 D2
Cañete, Peru .. 91 F3
Cangas de
 Narcea 18 A2
Canguaretama . 92 D6
Canguçu 94 C6
Canicattì 21 F4
Canigou, Mt. .. 12 E5
Canipaan 36 C5
Çankırı 46 B3
Cann River 63 C4
Cannanore ... 43 P9
Cannes 13 E7
Canoas 94 B6
Canora 71 C9
Canowindra ... 63 B4
Cantabria ☐ ... 18 A4
Cantabrian Mts.
 = Cantábrica,
 Cordillera ... 18 A3
Cantábrica,
 Cordillera ... 18 A3
Cantal, Plomb
 du 13 D5
Canterbury 11 F7
Canterbury ☐ . 65 E4
Canterbury
 Bight 65 F4
Canterbury
 Plains 65 E4
Canton =
 Guangzhou . 35 D6
Canton 72 D5
Canudos 91 E7
Canutama 91 E6
Cap-Haïtien .. 87 C5
Cap St.-Jacques
 = Vung Tau . 36 B3
Capaia 56 F4
Capanaparo → 90 B5
Cape Barren I. . 62 D4

Cape Breton I. . 69 D4
Cape Coast ... 55 G4
Cape Dorset .. 68 B3
Cape Dyer 68 B4
Cape Girardeau 79 A9
Cape Jervis ... 62 C2
Cape Town ... 58 E3
Cape Verde Is. ■ 50 E1
Cape York
 Peninsula ... 61 C7
Capela 93 E6
Capim → ... 92 C4
Capraia 20 C2
Capri 21 D5
Caprivi Strip . 58 B4
Captain's Flat . 63 C4
Caquetá → ... 90 D5
Caracal 22 B5
Caracas 90 A5
Caracol 93 D5
Caradoc 62 B3
Carajás, Serra
 dos 92 D3
Carangola 93 G5
Caransebeş ... 17 F6
Caratasca, L. .. 86 C3
Caratinga 93 F5
Caraúbas 92 D6
Caravaca 19 C5
Caravelas 93 F6
Caraveli 91 G4
Carballo 18 A1
Carbonara, C. . 21 E2
Carbondale ... 79 A9
Carbonear ... 69 D5
Carbonia 21 E2
Carcassonne .. 12 E5
Carcross 70 B6
Cardamon Hills 43 Q10
Cárdenas 86 B3
Cardiff 11 F5
Cardigan B. ... 11 E4
Cardston 71 D8
Carei 17 E6
Careysburg ... 55 G2
Carhué 94 D4
Caria 23 F7
Cariacica 93 G5
Caribbean Sea . 86 D5
Cariboo Mts. .. 71 C7
Carinda 63 B4
Carinhanha ... 93 E5
Carinthia ☐ =
 Kärnten ☐ .. 15 E7
Caripito 90 A6
Caritianas 91 E6
Carlisle 11 D5
Carlow 11 E3
Carlsbad 78 C1
Carmacks 70 B6
Carman 71 D10
Carmarthen ... 11 F4
Carmaux 12 D5
Carmelo 94 C5
Carmen 90 B3
Carmen de
 Patagones .. 95 E4
Carmona 18 D3
Carnac 12 C2
Carnarvon,
 Australia 60 E1
Carnarvon,
 S. Africa ... 58 E4
Carnegie, L. ... 60 F3
Carnic Alps =
 Karnische
 Alpen 20 A4
Carniche Alpi =
 Karnische
 Alpen 20 A4

Carnot 56 D3
Carnot, C. 62 B2
Carolina 92 D4
Carolina 92 D4
Caroline I. 65 K15
Caroline Is. ... 64 J10
Caroní → ... 90 B6
Caroníe =
 Nébrodi,
 Monti 21 F5
Caroona 63 B5
Carpathians ... 16 D5
Carpaţii
 Meridionali .. 17 F7
Carpentaria, G.
 of 60 C6
Carpentras 13 D6
Carpi 20 B3
Carpolac =
 Morea 62 C3
Carrara 20 B3
Carrauntoohill . 11 E2
Carrick-on-Suir 11 E3
Carrieton 62 B2
Carrizal Bajo .. 94 B2
Carşamba ... 46 B4
Carson City ... 80 F4
Cartagena,
 Colombia ... 90 A3
Cartagena,
 Spain 19 D5
Cartago,
 Colombia ... 90 C3
Cartago,
 Costa Rica .. 86 E3
Carterton 65 D6
Carthage 79 A6
Cartwright 68 C5
Caruaru 92 D6
Carúpano 90 A6
Carvoeiro 90 D6
Carvoeiro, C. .. 18 C1
Casablanca ... 54 B3
Cascade Ra. ... 80 B3
Cascais 18 C1
Cascavel 94 A6
Cáscina 20 C3
Caserta 20 D5
Casiguran 38 A2
Casilda 94 C4
Casino 63 A5
Casiquiare → . 90 C5
Casma 91 E3
Caspe 19 B5
Casper 81 D10
Caspian Sea .. 29 E6
Cassinga 58 B3
Castellammare
 di Stábia 21 D5
Castellón de la
 Plana 19 C5
Castelo Branco 18 C2
Castelsarrasin . 12 D4
Castelvetrano . 21 F4
Casterton 62 C3
Castilla La
 Mancha ☐ .. 18 C4
Castilla y
 Leon ☐ 18 A3
Castlebar 11 E2
Castleblaney .. 11 D3
Castlemaine .. 63 C3
Castlereagh → 63 B4
Castres 12 E5
Castries 87 D7
Castro 95 E2
Castro Alves .. 93 E6
Castuera 18 C3
Cat I. 86 B4
Catacáos 91 E2

Catalão

Catalão 93 F4
Çatalca 22 D7
Catalonia =
Cataluña □ .. 19 B6
Cataluña □ ... 19 B6
Catamarca 94 B3
Catanduanes .. 38 B2
Catanduva 93 G4
Catánia 21 F5
Catanzaro 21 E6
Catarman 38 B2
Cateel 38 C3
Catoche, C. .. 85 C7
Catrimani 90 C6
Catskill Mts. ... 73 C8
Cauca → .. 90 B4
Caucaia 92 C6
Caucasus
  Mountains .. 25 E5
Caúngula 56 F3
Cauquenes 94 D2
Caura → ... 90 B6
Cauvery → ... 43 P11
Caux, Pays de . 12 B4
Cavan 11 D3
Cavendish .. 62 C3
Caviana, I. 92 B3
Cavite 38 B2
Cawndilla L. .. 62 B3
Cawnpore =
  Kanpur 40 D3
Caxias 92 C5
Caxias do Sul . 94 B6
Caxito 56 F2
Cayambe 90 C3
Cayenne 92 A3
Cayman Is. ■ .. 67 H11
Cazombo 58 A4
Ceadâr-Lunga . 17 E9
Ceanannus Mor 11 E3
Ceará =
  Fortaleza 92 C6
Ceará □ 92 D6
Ceará Mirim .. 92 D6
Cebollar 94 B3
Cebu 38 B2
Cecil Plains ... 63 A5
Cedar L. 71 C10
Cedar Rapids .. 77 E9
Cedro 92 D6
Ceduna 62 B1
Cefalù 21 E5
Cegléd 16 E4
Celaya 84 C4
Celebes =
  Sulawesi □ . 39 E2
Celebes Sea ... 39 D2
Celje 20 A5
Celle 14 B6
Celtic Sea 11 F2
Central,
  Cordillera ... 90 C3
Central African
  Rep. ■ 56 C4
Central Makran
  Range 42 F4
Cephalonia =
  Kefallinía ... 23 E3
Ceram = Seram 39 E3
Ceram Sea =
  Seram Sea . 39 E3
Ceres, Argentina 94 B4
Ceres, S. Africa 58 E3
Cerignola 20 D5
Cerigo = Kíthira 23 F4
Çerkeş 46 B3
Çerkeşköy 22 D6
Cervera 19 B6
Cesena 20 B4

Česka Republika
  = Czech
  Rep. ■ 16 D2
České
  Budějovice .. 16 D2
Ceskomoravská
  Vrchovina ... 16 D2
Çeşme 23 E6
Cessnock 63 B5
Cetinje 22 C2
Cetraro 21 E5
Ceuta 18 E3
Cévennes 13 D5
Ceyhan 46 C3
Ceylon = Sri
  Lanka ■ 43 R12
Chachapoyas .. 91 E3
Chachran 42 E7
Chad ■ 53 E2
Chad, L. =
  Tchad, L. ... 53 F1
Chadileuvú → .. 94 D3
Chadyr-Lunga =
  Ceadâr-Lunga 17 E9
Chagai Hills ... 42 E3
Chagos Arch. .. 27 K11
Chāh Bahār ... 45 E5
Chāh Gay Hills 45 D6
Chahar Burjak . 42 D3
Chaibasa 40 F5
Chakhānsūr ... 42 D3
Chakradharpur . 40 F5
Chakwal 42 C8
Chala 91 G4
Chalcis =
  Khalkís 23 E4
Chalhuanca ... 91 F4
Chalisgaon ... 43 J9
Chalky Inlet .. 65 G2
Challapata ... 91 G5
Chalon-sur-
  Saône 13 C6
Châlons-en-
  Champagne . 13 B6
Chaman 42 D5
Chamba, India . 42 C10
Chamba,
  Tanzania .. 59 A7
Chambal → .. 42 F11
Chambéry 13 D6
Chamical 94 C3
Chamonix-Mont
  Blanc 13 D7
Champagne ... 13 B6
Champaign ... 72 D1
Champlain, L. . 73 B9
Chañaral 94 B2
Chandigarh ... 42 D10
Chandler 62 A1
Chandmani ... 34 B4
Chandpur 41 F8
Chandrapur ... 43 K11
Chang Jiang → 35 C7
Changanacheri 43 Q10
Changane → . 59 C6
Changchiak'ou
  = Zhangjiakou 35 B6
Ch'angchou =
  Changzhou .. 35 C6
Changchun ... 35 B7
Changde 35 D6
Changhai =
  Shanghai ... 35 C7
Changjiang ... 35 E5
Changsha 35 D6
Changting ... 35 D6
Changzhi 35 C6
Changzhou ... 35 C6
Channapatna .. 43 N10

Channel Is. 11 G5
Chantada 18 A2
Chanthaburi ... 36 B2
Chantrey Inlet . 70 B10
Chanute 79 A6
Chao Phraya → 36 B2
Chao'an 35 D6
Chaoyang 35 D6
Chapala, L. de . 84 C4
Chapayevsk ... 24 C6
Chapra =
  Chhapra 40 E5
Chār 54 D2
Chara 31 D9
Charadai 94 B5
Charagua 91 G6
Charaña 91 G5
Chardara 29 E7
Chardzhou =
  Chärjew 29 F7
Charente → .. 12 D3
Chari → 53 F1
Chārīkār 42 B6
Chärjew 29 F7
Charleroi 14 C3
Charles, C. ... 73 F8
Charleston, S.C.,
  U.S.A. 75 C8
Charleston,
  W. Va., U.S.A. 72 E5
Charlesville ... 56 F4
Charleville 63 A4
Charleville-
  Mézières 13 B6
Charlotte 75 B7
Charlotte Amalie 87 C7
Charlottesville . 73 E6
Charlottetown . 69 D4
Charlton 62 C3
Charolles 13 C6
Charouine 54 C4
Charters Towers 61 E8
Chartres 12 B4
Chascomús ... 94 D5
Châteaubriant . 12 C3
Châteaulin 12 B1
Châteauroux .. 12 C4
Châtellerault .. 12 C4
Chatham, N.B.,
  Canada 69 D4
Chatham, Ont.,
  Canada 69 D2
Chatham, U.K. . 11 F7
Chatrapur 40 H5
Chattahoochee →
  .... 74 D5
Chattanooga .. 74 B5
Chauk 41 G10
Chaukan La ... 41 D11
Chaumont 13 B6
Chaves, Brazil . 92 C4
Chaves, Portugal 18 B2
Chavuma 58 A4
Cheb 16 C1
Cheboksary ... 24 B6
Chech, Erg 54 D4
Chechenia □ .. 25 E6
Checheno-
  Ingush
  Republic =
  Chechenia □ 25 E6
Chechnya =
  Chechenia □ 25 E6
Cheduba I. 41 H9
Cheepie 63 A4
Chegga 54 C3
Chegutu 59 B5
Chehalis 80 B2
Cheju Do 35 C7

Chekiang =
  Zhejiang □ .. 35 D7
Cheleken 29 F6
Chelforó 95 D3
Chelkar =
  Shalqar 29 E6
Chelkar Tengiz,
  Solonchak .. 29 E7
Chełm 17 C6
Chełmno .. 16 B4
Chelmsford ... 11 F7
Cheltenham ... 11 F5
Chelyabinsk .. 29 D7
Chemnitz 15 C7
Chen Xian 35 D6
Chenab → ... 42 D7
Chencha 53 G6
Chenchiang =
  Zhenjiang .. 35 C6
Chengchou =
  Zhengzhou .. 35 C6
Chengde 35 B6
Chengdu 35 C5
Chengjiang ... 34 D5
Ch'engtu =
  Chengdu ... 35 C5
Cheo Reo 36 B3
Cheom Ksan .. 36 B2
Chepén 91 E3
Chepes 94 C3
Cher → 12 C4
Cherbourg ... 12 B3
Cherchell 54 A5
Cherdyn 28 C6
Cherepanovo .. 29 D9
Cherepovets .. 24 B4
Chergui, Chott
  ech 54 B5
Cherikov =
  Cherykaw .. 17 B10
Cherkasy 24 D3
Cherlak 29 D8
Chernaya 30 B6
Chernigov =
  Chernihiv .. 24 C3
Chernihiv 24 C3
Chernivtsi ... 17 D7
Chernobyl =
  Chornobyl .. 17 C10
Chernovtsy =
  Chernivtsi ... 17 D7
Chernyakhovsk 28 D3
Cherquenco ... 94 D2
Cherrapunji ... 41 E8
Cherven 17 B9
Chervonohrad . 17 C7
Cherykaw 17 B10
Chesapeake .. 73 F7
Chesapeake B. . 73 E7
Cheshskaya
  Guba 28 C5
Chester, U.K. .. 11 E5
Chester, U.S.A. 73 E8
Chesterfield ... 11 D6
Chesterfield, Is. 64 L10
Chesterfield Inlet 70 B10
Chesterton Ra. . 63 A4
Cheviot Hills .. 10 D5
Chew Bahir ... 53 H6
Cheyenne 76 E2
Cheyenne → . 76 C4
Chhapra 40 E5
Chhatarpur ... 43 G11
Chhindwara .. 43 H11
Chhlong 36 B3
Chi → 36 A2
Chiamussu =
  Jiamusi 35 B8
Chiange 58 B2

Colima

Chiapa → 85 D6
Chiávari 20 B2
Chiavenna 20 A2
Chiba 32 B7
Chibemba 58 B2
Chibia 58 B2
Chibougamau . 69 D3
Chibuk 53 F1
Chicacole =
Srikakulam .. 40 H4
Chicago 72 D2
Chichagof I. ... 71 C6
Chichibu 32 A6
Ch'ich'ihaerh =
Qiqihar 35 B7
Chickasha 78 B5
Chiclana de la
Frontera 18 D2
Chiclayo 91 E3
Chico 80 F3
Chico →,
Chubut,
Argentina ... 95 E3
Chico →,
Santa Cruz,
Argentina ... 95 F3
Chicoutimi 69 D3
Chidambaram . 43 P11
Chidley, C. 68 B4
Chiemsee 15 E7
Chiengi 57 F5
Chiese → 20 B3
Chieti 20 C5
Chiguana 91 H5
Chihli, G. of =
Bo Hai 35 C6
Chihuahua 84 B3
Chiili 29 E7
Chik Bollapur .. 43 N10
Chikmagalur .. 43 N9
Chilapa 84 D5
Chilas 42 B9
Chilaw 43 R11
Childers 63 A5
Chile ■ 94 D2
Chilete 91 E3
Chililabombwe 59 A5
Chilin = Jilin .. 35 B7
Chilka L. 40 H5
Chillán 94 D2
Chiloé, I. de .. 95 E2
Chilpancingo .. 84 D5
Chiluage 56 F4
Chilwa, L. 59 B7
Chimbay 29 E6
Chimborazo ... 90 D3
Chimbote 91 E3
Chimkent =
Shymkent .. 29 E7
Chimoio 59 B6
Chin □ 41 F9
China ■ 35 C5
Chinan = Jinan 35 C6
Chinandega ... 85 E7
Chincha Alta .. 91 F3
Chinchilla 63 A5
Chinchou =
Jinzhou 35 B7
Chinde 59 B7
Chindwin → .. 41 G10
Chingola 59 A5
Ch'ingtao =
Qingdao 35 C7
Chinguetti 54 D2
Chinhoyi 59 B6
Chiniot 42 D8
Chinju 35 C7
Chinnampo ... 35 C7
Chinon 12 C4

Chinsali 57 G6
Chióggia 20 B4
Chíos = Khíos . 23 E6
Chipata 59 A6
Chiquián 91 F3
Chiquinquira .. 90 B4
Chirchiq 29 E7
Chiromo 59 B7
Chisamba 59 A5
Chisapani Garhi 40 D5
Chişinău 17 E9
Chistopol 24 B7
Chita 30 D9
Chitado 58 B2
Chitembo 58 A3
Chitral 42 B7
Chitré 86 E3
Chittagong ... 41 F8
Chittagong □ .. 41 E8
Chittaurgarh .. 43 G9
Chittoor 43 N11
Chiusi 20 C3
Chivasso 20 B1
Chivilcoy 94 B4
Chkalov =
Orenburg ... 29 D6
Choele Choel . 95 D3
Chojnice 16 B3
Cholet 12 C3
Choluteca 85 E7
Choma 59 B5
Chomutov 16 C1
Chon Buri 36 B2
Chone 90 D3
Chŏngjin 35 B7
Chŏngju 35 C7
Chongqing ... 35 D5
Chŏnju 35 C7
Chonos, Arch.
de los 95 F2
Chop 17 D6
Chornobyl ... 17 C10
Chortkiv 17 D7
Chorzów 16 C4
Chos-Malal .. 94 D2
Chōshi 32 B7
Choszczno ... 16 B2
Chotila 43 H7
Choybalsan ... 35 B6
Christchurch .. 65 E5
Christiana 59 D5
Christmas I. =
Kiritimati ... 65 J15
Chu = Shu ... 29 E8
Ch'uanchou =
Quanzhou ... 35 D6
Chubut → 95 E3
Chudskoye, Oz. 24 B2
Chūgoku-Sanchi 32 B3
Chula Vista .. 82 D4
Chulucanas ... 90 E2
Chulym → ... 29 D9
Chumbicha ... 94 B3
Chumphon ... 36 B1
Chuna → 30 D7
Chunchŏn ... 35 C7
Chungking =
Chongqing .. 35 D5
Chunya 57 F6
Chuquibamba . 91 G4
Chuquicamata . 94 A3
Chur 13 C8
Churachandpur 41 E9
Churchill 70 C10
Churchill →,
Man., Canada 70 C10
Churchill →,
Nfld., Canada 69 C4

Churchill, C. ... 70 C10
Churchill Falls . 69 C4
Churchill L. ... 71 C9
Churu 42 E9
Chushal 42 C11
Chusovoy 28 D6
Chuvash
Republic □ =
Chuvashia □ 24 B6
Chuvashia □ .. 24 B6
Cicero 72 D2
Ciechanów 16 B5
Ciego de Avila . 86 B4
Ciénaga 90 A4
Cienfuegos ... 86 B3
Cieszyn 16 D4
Cieza 19 C5
Cijara, Pantano
de 18 C3
Cilacap 37 F3
Cilician Gates P. 46 C3
Cill Chainnigh =
Kilkenny ... 11 E3
Cimişlia 17 E9
Cimone, Mte. .. 20 B3
Cîmpina 17 F7
Cîmpulung ... 17 F7
Cinca → 19 B6
Cincar 20 C6
Cincinnati 72 E3
Çine 23 F7
Cinto, Mte. 13 E8
Circle 71 B5
Cirebon 37 F3
Citlaltépetl 84 D5
Città di Castello 20 C4
Ciudad Bolívar 90 B6
Ciudad
Chetumal ... 85 D7
Ciudad del
Carmen 85 D6
Ciudad Delicias
= Delicias .. 84 B3
Ciudad Guayana 90 B6
Ciudad Juárez . 84 A3
Ciudad Madero 84 C5
Ciudad Mante . 84 C5
Ciudad Obregón 84 B3
Ciudad Real .. 18 C4
Ciudad Rodrigo 18 B2
Ciudad Trujillo
= Santo
Domingo ... 87 C6
Ciudad Victoria 84 C5
Civitanova
Marche 20 C4
Civitavécchia .. 20 C3
Cizre 46 C5
Clanwilliam ... 58 E3
Clare 62 B2
Clarence →,
Australia ... 63 A5
Clarence →,
N.Z. 65 E5
Clarence, I. ... 95 G2
Clarence I. ... 96 A4
Clarie Coast .. 96 A13
Clarke I. 62 D4
Clarksdale ... 79 B8
Clarksville 74 A4
Claveria 38 A2
Clear, C. 11 F2
Clearwater ... 75 F6
Clearwater Mts. 81 B6
Clermont-
Ferrand 13 D5
Cleveland,
Australia 63 A5
Cleveland,
U.S.A. 72 D5

Clifden 65 G2
Clifton 63 A5
Clifton Hills .. 62 A2
Clinton, Canada 71 C7
Clinton, N.Z. .. 65 G3
Clinton, U.S.A. 77 E9
Clinton Colden
L. 70 B9
Clones 11 D3
Clonmel 11 E3
Clovis 78 B2
Cluj-Napoca ... 17 E6
Clunes 62 C3
Clutha → 65 G3
Clyde 65 F3
Clyde → 10 D4
Clyde, Firth of . 10 D4
Clyde River ... 68 A4
Coari 90 D6
Coast Mts. ... 70 C7
Coast Ranges . 80 F2
Coats I. 68 B2
Coats Land ... 96 B5
Coatzacoalcos . 85 D6
Cobalt 69 D3
Cobar 63 B4
Cóbh 11 F2
Cobham 62 B3
Cobija 91 F5
Cobourg 69 D3
Cobram 63 C4
Cóbué 59 A6
Coburg 15 C6
Cocanada =
Kakinada ... 40 J4
Cochabamba . 91 G5
Cochin 43 Q10
Cochin China =
Nam-Phan .. 36 B3
Cochrane 69 D2
Cochrane, L. .. 95 F2
Cockburn 62 B3
Cockburn, Canal 95 G2
Coco → 86 D3
Cocobeach ... 56 D1
Cod, C. 73 C10
Codajás 90 D6
Codó 92 C5
Cœur d'Alene . 80 B5
Coffeyville 79 A6
Coffin B. 62 B2
Coffin Bay
Peninsula ... 62 B2
Coffs Harbour . 63 B5
Cognac 12 D3
Cohuna 63 C3
Coiba, I. 86 E3
Coig → 95 G3
Coihaique 95 F2
Coimbatore ... 43 P10
Coimbra, Brazil 91 G7
Coimbra,
Portugal 18 B1
Coín 18 D3
Cojimies 90 C3
Colac 62 C3
Colatina 93 F5
Colbeck, C. ... 96 B17
Colbinabbin ... 63 C3
Colchester ... 11 F7
Colebrook ... 62 D4
Coleman 71 D8
Coleraine,
Australia 62 C3
Coleraine, U.K. 11 D3
Coleridge, L. .. 65 E4
Colesberg 59 E5
Colhué Huapi, L. 95 F3
Colima 84 D4

Colinas

Colinas 92 D5
Coll 10 C3
Collaguasi 91 H5
Collarenebri .. 63 A4
Collingwood .. 64 D5
Colmar 13 B7
Colo ➝ 63 B5
Cologne = Köln 14 C4
Colomb-Béchar
= Béchar .. 54 B4
Colômbia 93 G4
Colombia ■ ... 90 C4
Colombo 43 R11
Colón, Cuba ... 86 B3
Colón, Panama 85 F9
Colonia 94 C5
Colonia Dora .. 94 B4
Colonsay 10 C3
Colorado □ ... 83 A10
Colorado ➝,
Argentina ... 95 D4
Colorado ➝,
N. Amer. 82 E5
Colorado
Plateau 83 B7
Colorado
Springs 76 F2
Columbia, Mo.,
U.S.A. 77 F8
Columbia, S.C.,
U.S.A. 75 B7
Columbia,
Tenn., U.S.A. 74 B4
Columbia ➝ .. 80 B1
Columbia,
District of □ . 73 E7
Columbia Basin 80 B4
Columbretes, Is. 19 C6
Columbus, Ga.,
U.S.A. 74 C5
Columbus,
Miss., U.S.A. 74 C3
Columbus, Ohio,
U.S.A. 72 E4
Colville ➝ 71 A4
Colville, C. 64 B6
Colwyn Bay ... 11 E5
Comácchio 20 B4
Comallo 95 E2
Comilla 41 F8
Comino, C. 21 D2
Committee B. . 68 B2
Commonwealth
B. 96 A14
Commoron
Cr. ➝ 63 A5
Communism Pk.
=
Kommunizma,
Pik 29 F8
Como 20 B2
Como, L. di ... 20 B2
Comodoro
Rivadavia ... 95 F3
Comorin, C. ... 43 Q10
Comoro Is. =
Comoros ■ . 51 H8
Comoros ■ ... 51 H8
Comox 71 D7
Compiègne ... 13 B5
Compton Downs 63 B4
Comrat 17 E9
Conakry 55 G2
Conara Junction 62 D4
Concarneau ... 12 C2
Conceição da
Barra 93 F6
Concepción,
Bolivia 91 G6

Concepción,
Chile 94 D2
Concepción,
Paraguay ... 94 A5
Concepción, L. 91 G6
Concepción del
Oro 84 C4
Concepción del
Uruguay 94 C5
Conchos ➝ .. 84 B3
Concord, N.C.,
U.S.A. 75 B7
Concord, N.H.,
U.S.A. 73 C10
Concordia,
Argentina ... 94 C5
Concórdia, Brazil 90 D5
Condamine ... 63 A5
Condeúba 93 E5
Condobolin ... 63 B4
Conegliano ... 20 B4
Confuso ➝ ... 94 B5
Congo =
Zaïre ➝ .. 56 F2
Congo
(Kinshasa) =
Zaïre ■ ... 56 E4
Congo ■ 56 E3
Conjeeveram =
Kanchipuram 43 N11
Conlea 63 B3
Conn, L. 11 D2
Connacht □ .. 11 E2
Connecticut □ . 73 D9
Connecticut ➝ 73 D9
Connemara ... 11 E2
Conoble 63 B3
Cononaco ➝ . 90 D3
Conran, C. 63 C4
Conselheiro
Lafaiete 93 G5
Constance =
Konstanz ... 14 E5
Constance, L. =
Bodensee ... 13 C8
Constanţa 22 B7
Constantine .. 54 A6
Constitución .. 94 D2
Contai 40 G6
Contamana ... 91 E4
Contas ➝ 93 E6
Conway, L. ... 62 A2
Coober Pedy .. 62 A1
Cooch Behar =
Koch Bihar .. 41 D7
Cook, B. 95 H2
Cook, Mt. 65 E4
Cook Inlet ... 71 C4
Cook Is. 65 L14
Cook Strait ... 65 D6
Cooktown 61 D8
Coolabah 63 B4
Cooladdi 63 A4
Coolah 63 B4
Coolamon 63 B4
Coolangatta ... 63 A5
Cooma 63 C4
Coonabarabran 63 B4
Coonamble ... 63 B4
Coondapoor .. 43 N9
Coongie 62 A3
Coongoola ... 63 A4
Cooninie, L. .. 62 A2
Cooper Cr. ➝,
N. Terr.,
Australia 60 C5
Cooper Cr. ➝,
S. Austral.,
Australia 62 A2

Cooroy 63 A5
Cootamundra . 63 B4
Cooyar 63 A5
Copenhagen =
København .. 9 G10
Copiapó 94 B2
Copiapó ➝ ... 94 B2
Copley 62 B2
Copper Center . 71 B5
Copper Cliff ... 69 D2
Coppermine .. 70 B8
Coppermine ➝ 70 B8
Coquimbo 94 C2
Corabia 22 C5
Coracora 91 G4
Coral Harbour . 68 B2
Coral Sea 64 L10
Corato 21 D6
Corby 11 E6
Corcaigh = Cork 11 F2
Corcubión 18 A1
Córdoba,
Argentina ... 94 C4
Córdoba, Spain 18 D3
Córdoba, Sierra
de 94 C4
Cordon 38 A2
Cordova 71 B5
Corfu = Kérkira 23 E2
Coria 18 C2
Corigliano
Cálabro 21 E6
Corinna 62 D4
Corinth =
Kórinthos ... 23 F4
Corinth, G. of =
Korinthiakós
Kólpos 23 E4
Corinto 93 F5
Cork 11 F2
Çorlu 22 D6
Corner Brook .. 69 D5
Corneşti 17 E9
Cornwall 69 D3
Corny Pt. 62 B2
Coro 90 A5
Coroatá 92 C5
Corocoro 91 G5
Coroico 91 G5
Coromandel ... 64 B6
Corona 62 B3
Coronation Gulf 70 B8
Coronation I. .. 96 A4
Coronel 94 D2
Coronel Dorrego 94 D4
Coronel Suárez 94 D4
Corowa 63 C4
Corpus Christi . 78 F5
Corque 91 G5
Corrib, L. 11 E2
Corrientes 94 B5
Corrientes ➝ . 90 D4
Corrientes, C.,
Colombia ... 90 B3
Corrientes, C.,
Mexico 84 C3
Corse 13 E8
Corse, C. 13 E8
Corsica = Corse 13 E8
Corsicana 79 C5
Corte 13 E8
Çorum 46 B3
Corumbá 91 G7
Corumbá de
Goiás 93 F4
Corunna = La
Coruña 18 A1
Corvallis 80 C2
Cosenza 21 E6

Costa Blanca .. 19 C5
Costa Brava ... 19 B7
Costa del Sol .. 18 D3
Costa Dorada . 19 B6
Costa Rica ■ .. 86 E3
Cotabato 38 C2
Cotagaita 91 H5
Côte d'Azur ... 13 E7
Côte-d'Ivoire ■
= Ivory
Coast ■ 55 G3
Cotentin 12 B3
Cotonou 55 G5
Cotopaxi 90 D3
Cotswold Hills . 11 F5
Cottbus 15 C8
Couedic, C. du . 62 C2
Coulman I. 96 B15
Council 71 B3
Council Bluffs .. 77 E7
Courantyne ➝ 90 B7
Courtrai =
Kortrijk 14 C2
Coventry 11 E6
Covilhã 18 B2
Covington 72 E3
Cowal, L. 63 B4
Cowangie 62 C3
Cowarie 62 A2
Cowell 62 B2
Cowra 63 B4
Coxim 93 F3
Cox's Bazar .. 41 G8
Cozumel, I. de . 85 C7
Craboon 63 B4
Cracow =
Kraków 16 C4
Cracow 63 A5
Cradock 59 E5
Craigavon 11 D3
Crailsheim ... 14 D6
Craiova 22 B4
Cranbrook,
Australia 62 D4
Cranbrook,
Canada 71 D8
Crateús 92 D5
Crato 92 D6
Crawley 11 F6
Crazy Mts. ... 81 B8
Crema 20 B2
Cremona 20 B3
Cres 20 B5
Cressy 62 C3
Crete = Kríti .. 23 G5
Créteil 12 B5
Creus, C. de .. 19 A7
Creuse ➝ 12 C4
Crewe 11 E5
Criciúma 94 B7
Crimean Pen. =
Krymskyy
Pivostriv 25 D3
Crişul Alb ➝ . 16 E5
Crişul Negru ➝ 16 E5
Crna Gora =
Montenegro □ 22 C2
Crna Gora 22 C3
Crna Reka ➝ . 22 D3
Croatia ■ 20 B6
Crocker,
Banjaran 36 C5
Croker, C. 60 C5
Cromer 11 E7
Cromwell 65 F3
Cronulla 63 B5
Crookwell 63 B4
Cross Sound .. 71 C6
Crotone 21 E6

Crows Nest ... 63 A5
Crowsnest Pass 71 D8
Cruz Alta 94 B6
Cruz del Eje ... 94 C4
Cruzeiro 93 G5
Cruzeiro do Sul 91 E4
Crystal Brook . 62 B2
Csongrád 16 E5
Cuamba 59 A7
Cuando → .. 58 B4
Cuarto → 94 C4
Cuba ■ 86 B4
Cubango → .. 58 B4
Cuchi 58 A3
Cúcuta 90 B4
Cuddalore ... 43 P11
Cuddapah 43 M11
Cudgewa 63 C4
Cuenca, Ecuador 90 D3
Cuenca, Spain . 19 B4
Cuenca,
Serranía de . 19 C5
Cuernavaca ... 84 D5
Cuevas del
Almanzora .. 19 D5
Cuevo 91 H6
Cuiabá 93 F2
Cuiabá → 93 F2
Cuima 58 A3
Cuito → 58 B4
Cukai 36 D2
Culcairn 63 C4
Culgoa → 63 A4
Culiacán 84 C3
Culion 38 B2
Cullarin Ra. ... 63 B4
Cullera 19 C5
Culuene → ... 93 E3
Culverden 65 E5
Cumaná 90 A6
Cumberland,
Canada 71 D7
Cumberland,
U.S.A. 73 E6
Cumberland
Pen. 68 B4
Cumberland
Plateau 72 F3
Cumberland Sd. 68 B4
Cumborah 63 A4
Cumbum 43 M11
Cummins 62 B2
Cumnock 63 B4
Cunene → ... 58 B2
Cúneo 20 B1
Cunnamulla ... 63 A4
Cupica, G. de .. 90 B3
Curaçao 87 D6
Curaray → ... 90 D4
Curiapo 90 B6
Curicó 94 C2
Curitiba 94 B7
Currabubula .. 63 B5
Currais Novos . 92 D6
Curralinho ... 92 C4
Curraweena ... 63 B4
Currie 62 C3
Curtea de Argeş 17 F7
Curtis Group .. 62 C4
Curuápanema →
.... 92 C2
Curuçá 92 C4
Curup 37 E2
Cururupu 92 C5
Curuzú Cuatiá . 94 B5
Cuttaburra → . 63 A3
Cuttack 40 G5
Cuvier I. 64 B6
Cuxhaven 14 B5

Cuyo 38 B2
Cuzco, Bolivia . 91 H5
Cuzco, Peru .. 91 F4
Cwmbran 11 F5
Cyclades =
Kikládhes ... 23 F5
Cygnet 62 D4
Cyprus ■ 46 D3
Cyrenaica 52 C3
Cyrene =
Shaḥḥāt 52 B3
Czech Rep. ■ .. 16 D2
Częstochowa .. 16 C4

D

Da Hinggan Ling 35 B7
Da Lat 36 B3
Da Nang 36 A3
Da Qaidam ... 34 C4
Da Yunhe → . 35 C7
Daba Shan ... 35 C5
Dabakala 55 G4
Dabo 37 E2
Dabola 55 F2
Daboya 55 G4
Dacca = Dhaka 41 F8
Dacca =
Dhaka □ ... 41 E8
Dachau 15 D6
Dadanawa 90 C7
Dadra and
Nagar
Haveli □ ... 43 J8
Dadu 42 F5
Daet 38 B2
Dagana 55 E1
Dagestan □ ... 25 E6
Daghestan
Republic =
Dagestan □ . 25 E6
Dagö = Hiiumaa 24 B1
Dagupan 38 A2
Dahlak Kebir .. 49 D3
Dahod 43 H9
Dahomey =
Benin ■ ... 55 G5
Dahra 55 E1
Daiō-Misaki ... 32 B5
Dairût 52 C5
Dakar 55 F1
Dakhla 54 D1
Dakhla, El
Wâhât el- ... 52 C4
Dakhovskaya ... 25 E5
Đakovica 22 C3
Dalaman → .. 23 F7
Dalandzadgad . 35 B5
Dālbandīn 42 E4
Dalby 63 A5
Dalhousie 69 D4
Dali 34 D5
Dalian 35 C7
Daliang Shan . 34 D5
Dallarnil 63 A5
Dallas 79 C5
Dalmacija 20 C6
Dalmatia =
Dalmacija ... 20 C6
Daloa 55 G3
Dalton 74 B5
Dalton Iceberg
Tongue 96 A13
Daly Waters ... 60 D5
Daman 43 J8
Damanhûr 52 B5

Damar 39 F3
Damaraland ... 58 C3
Damascus =
Dimashq 46 D4
Damāvand 44 C3
Damāvand,
Qolleh-ye ... 44 C3
Damba 56 F3
Dāmghān 44 B3
Damiel 18 C4
Damietta =
Dumyât 52 B5
Dammam = Ad
Dammām ... 47 F7
Damoh 43 H11
Dampier 60 E2
Dampier, Selat 39 E4
Danbury 73 D9
Dandeldhura .. 40 C3
Dandeli 43 M9
Dandenong ... 63 C4
Dandong 35 B7
Danger Is. =
Pukapuka ... 65 L14
Danger Pt. 58 E3
Dangora 55 F6
Danilov 24 B5
Dankhar Gompa 42 C11
Dannemora ... 9 F11
Dannevirke ... 64 D7
Dante 49 E5
Danube =
Dunărea → . 17 F9
Danville 75 A8
Danzig =
Gdańsk 16 A4
Dao 38 B2
Daoud = Aïn
Beïda 54 A6
Dar el Beida =
Casablanca . 54 B3
Dar es Salaam . 57 F7
Dārāb 44 D3
Daraj 54 B7
Darband 42 B8
Darbhanga ... 40 D5
Dardanelles =
Çanakkale
Boğazı 23 D6
Dârfûr 53 F3
Dargai 42 B7
Dargan Ata ... 29 E7
Dargaville 64 A5
Darıca 22 D7
Darién, G. del . 90 B3
Darjeeling =
Darjiling 40 D7
Darjiling 40 D7
Darling → ... 62 B3
Darling Downs 63 A5
Darling Ra. ... 60 G2
Darlington ... 11 D6
Darłowo 16 A3
Darmstadt ... 14 D5
Darnah 52 B3
Darnley, C. ... 96 A10
Darnley B. ... 70 B7
Dart, C. 96 B18
Dartmoor 11 F5
Dartmouth ... 69 D4
Dartmouth, L. . 63 A4
Darvaza 29 E6
Darvel, Teluk . 37 D5
Darwha 43 J10
Darwin 60 C5
Daryoi Amu =
Amudarya → 29 E6
Dās 44 E3
Dashhowuz ... 29 E6

Dasht → 42 G2
Dasht-e Kavīr .. 44 C3
Dasht-e Mārgow 42 D3
Datça 23 F6
Datia 43 G11
Datong 35 B6
Datu, Tanjung . 37 D3
Datu Piang 38 C2
Daugava → .. 24 B1
Daugavpils ... 24 B2
Daulpur 42 F10
Dauphin 71 C9
Dauphiné 13 D6
Davangere ... 43 M9
Davao 38 C3
Davao, G. of .. 38 C3
Dāvar Panāh .. 45 E5
Davenport 77 E9
David 86 E3
David Gorodok
= Davyd
Haradok 17 B8
Davis Sea ... 96 A11
Davis Str. 68 B5
Davos 13 C8
Davyd Haradok 17 B8
Dawson 70 B6
Dawson, I. 95 G2
Dax 12 E3
Daxian 35 C5
Daxue Shan .. 34 C5
Daylesford ... 62 C3
Dayr az Zawr .. 46 D5
Dayton 72 E3
Daytona Beach 75 E7
Dead Sea 47 E3
Deakin 60 G4
Deal I. 62 C4
Deán Funes .. 94 C4
Death Valley .. 82 B4
Deba Habe ... 55 F7
Debar 22 D3
Dębica 16 C5
Debre Markos . 53 F6
Debre Tabor .. 53 F6
Debrecen 16 E5
Decatur, Ala.,
U.S.A. 74 B4
Decatur, Ill.,
U.S.A. 77 F10
Deccan 43 L11
Děčín 16 C2
Dedéagach =
Alexandroúpolis
.... 22 D5
Dédougou 55 F4
Dee → 10 C5
Deepwater ... 63 A5
Degeh Bur ... 49 F3
Deggendorf ... 15 D7
Deh Bīd 44 D3
Dehi Titan ... 42 C3
Dehibat 54 B7
Dehkareqan ... 46 C6
Dehra Dun ... 42 D11
Dej 17 E6
Dekese 56 E4
Delano 82 C3
Delaware □ ... 73 E8
Delaware → .. 73 E8
Delegate 63 C4
Delgado, C. ... 57 G8
Delgo 53 D5
Delhi 42 E10
Delice → 46 C3
Delicias 84 B3
Delmenhorst .. 14 B5
Delmiro Gouveia 93 D6
Deloraine 62 D4

Delungra

Delungra 63 A5
Delvinë 23 E3
Demanda, Sierra
  de la 18 A4
Demavand =
  Damāvand .. 44 C3
Demba 56 F4
Dembecha 53 F6
Dembidolo 53 G5
Demini → 90 D6
Demirci 23 E7
Demirköy 22 D6
Dempo 37 E2
Den Haag = 's-
  Gravenhage . 14 B3
Den Helder ... 14 B3
Denau 29 F7
Dendang 37 E3
Denia 19 C6
Denial B. 62 B1
Deniliquin 63 C3
Denman Glacier 96 A11
Denmark ■ ... 9 G9
Denmark Str. .. 66 C17
Denpasar 37 F5
Denton 78 C5
D'Entrecasteaux
  Is. 61 B9
Denver 76 F2
Deoghar 40 E6
Deolali 43 K8
Deoria 40 D4
Deosai Mts. ... 42 B9
Dêqên 34 D4
Dera Ghazi Khan 42 D7
Dera Ismail
  Khan 42 D7
Derbent 25 E6
Derby, Australia 60 D3
Derby, U.K. ... 11 E6
Derg, L. 11 E2
Dergaon 41 D9
Derry =
  Londonderry 11 D3
Derudub 53 E6
Des Moines ... 77 E8
Des Moines → 77 E9
Desaguadero → 91 G5
Dese 49 E2
Desna → 17 C10
Desolación, I. .. 95 G2
Despeñaperros,
  Paso 18 C4
Dessau 15 C7
Dessye = Dese 49 E2
D'Estrees B. ... 62 C2
Detmold 14 C5
Detroit 72 C4
Deutsche Bucht 14 A5
Deva 17 F6
Devakottai 43 Q11
Deventer 14 B4
Devonport,
  Australia ... 62 D4
Devonport, N.Z. 64 B6
Dewas 43 H10
Deyhūk 44 C4
Deyyer 44 E2
Dezfūl 47 D7
Dezhneva, Mys 31 C16
Dhafra 44 F3
Dhahira 44 F4
Dhahran = Az̧
  Z̧ahrān 47 F7
Dhaka 41 F8
Dhaka □ 41 E8
Dhamar 49 E3
Dhamtari 40 G3
Dhanbad 40 F6

Dhangarhi 40 C3
Dhankuta 40 D6
Dhar 43 H9
Dharmapuri ... 43 N11
Dharwad 43 M9
Dhaulagiri 40 C4
Dhenkanal ... 40 G5
Dhíkti Óros ... 23 G5
Dhírfis 23 E4
Dhodhekánisos 23 F6
Dhrol 43 H7
Dhubaibah ... 44 F3
Dhuburi 41 D7
Dhule 43 J9
Dhut → 49 E5
Diafarabé 55 F4
Diamantina ... 93 F5
Diamantina → 62 A2
Diamantino .. 93 E2
Diamond
  Harbour 40 F7
Diapaga 55 F5
Dībā 44 E4
Dibaya 56 F4
Dibaya-Lubue . 56 E3
Dibbi 49 G3
Dibrugarh 41 D10
Dickson 28 B9
Didiéni 55 F3
Diébougou 55 F4
Diefenbaker L. . 71 C9
Dieppe 12 B4
Digby 69 D4
Digges Is. 68 B3
Dighinala 41 F9
Digne-les-Bains 13 D7
Digos 38 C3
Digul → 39 F5
Dihang → ... 41 D10
Dijlah, Nahr → 47 E6
Dijon 13 C6
Dikimdya 31 D10
Dikson =
  Dickson 28 B9
Dikwa 53 F1
Dili 39 F3
Dilling 53 F4
Dillingham 71 C4
Dilolo 56 G4
Dilston 62 D4
Dimashq 46 D4
Dimbokro 55 G4
Dimboola 62 C3
Dîmboviţa → . 22 B6
Dimitrovgrad,
  Bulgaria 22 C5
Dimitrovgrad,
  Russia 24 C6
Dimitrovo =
  Pernik 22 C4
Dinagat 38 B3
Dinajpur 41 E7
Dinan 12 B2
Dinant 14 C3
Dinar 46 C2
Dīnār, Kūh-e .. 44 D2
Dinara Planina . 20 B6
Dinard 12 B2
Dinaric Alps =
  Dinara Planina 20 B6
Dindigul 43 P11
Dingle 11 E1
Dinguiraye ... 55 F2
Dingwall 10 C4
Dinosaur
  National
  Monument .. 81 E9
Diourbel 55 F1
Dipolog 38 C2

Dir 42 B7
Diré 55 E4
Dire Dawa 49 F3
Dirico 58 B4
Dirranbandi ... 63 A4
Disa 43 G8
Disappointment,
  L. 60 E3
Disaster B. 63 C4
Discovery B. .. 62 C3
Disteghil Sar .. 42 A9
Distrito
  Federal □ ... 93 F4
Diu 43 J7
Divinópolis ... 93 G5
Divnoye 25 D5
Dīwāl Kol 42 B5
Diyarbakır 46 C5
Diz Chah 44 C3
Djado 52 D1
Djakarta =
  Jakarta 37 F3
Djambala 56 E2
Djanet 54 D6
Djawa = Jawa . 37 F3
Djelfa 54 B5
Djema 57 C5
Djenné 55 F4
Djerba, I. de .. 52 B1
Djerid, Chott .. 54 B6
Djibo 55 F4
Djibouti 49 E3
Djibouti ■ 49 E3
Djolu 56 D4
Djougou 55 G5
Djoum 56 D2
Djourab 53 E2
Djugu 57 D6
Dmitriya
  Lapteva, Proliv 31 B12
Dnepr → =
  Dnipro → .. 25 D3
Dneprodzerzhinsk
  =
  Dniprodzerzhynsk
  .... 25 D3
Dnepropetrovsk
  =
  Dnipropetrovsk
  .... 25 D4
Dnestr → =
  Dnister → .. 17 E10
Dnestrovski =
  Belgorod ... 24 C4
Dnieper =
  Dnipro → .. 25 D3
Dniester =
  Dnister → .. 17 E10
Dnipro → .. 25 D3
Dniprodzerzhynsk
  .... 25 D3
Dnipropetrovsk 25 D4
Dnister → .. 17 E10
Dnistrovskyy
  Lyman 17 E10
Dnyapro =
  Dnipro → .. 25 D3
Doba 53 G2
Doberai, Jazirah 39 E4
Doblas 94 D4
Dobo 39 F4
Doboj 20 B7
Dobreta-Turnu-
  Severin 22 B4
Dobrich 22 C6
Dobruja 22 B7
Dobrush 17 B10
Dodecanese =
  Dhodhekánisos
  .... 23 F6

Dodge City 78 A3
Dodoma 57 F7
Dogi 42 C3
Doha = Ad
  Dawhah 44 E2
Dohazari 41 F9
Doi 39 D3
Dois Irmãos, Sa. 93 D5
Dolbeau 69 D3
Dole 13 C6
Dolomites =
  Dolomiti ... 20 A3
Dolomiti 20 A3
Dolores 94 D5
Dolphin, C. 95 G5
Dolphin and
  Union Str. .. 70 B8
Dombarovskiy . 29 D6
Dominica ■ ... 87 C7
Dominican
  Rep. ■ 87 C5
Domo 49 F4
Domodóssola . 20 A2
Domville, Mt. .. 63 A5
Don →, Russia 25 D4
Don →, U.K. .. 10 C5
Don Benito ... 18 C3
Donald 62 C3
Donau =
  Dunărea → . 17 F9
Donauwörth .. 14 D6
Doncaster ... 11 E6
Dondo, Angola 56 F2
Dondo, Mozam. 59 B6
Dondo, Teluk . 39 D2
Donegal 11 D2
Donets → ... 25 D5
Donetsk 25 D4
Donggala 39 E1
Dongola 53 E5
Dongou 56 D3
Dongting Hu .. 35 D6
Dongxing 35 D5
Donington, C. . 62 B2
Donnelly's
  Crossing 64 A5
Donostia = San
  Sebastián ... 19 A5
Dora Báltea → 20 B2
Dorchester, C. . 68 B3
Dordogne → . 12 D3
Dordrecht ... 14 C3
Dori 55 F4
Dornbirn 14 E5
Dorohoi 17 E8
Döröö Nuur ... 34 B4
Dorrigo 63 B5
Dortmund ... 14 C4
Doruma 57 D5
Dos Bahías, C. . 95 E3
Dos Hermanas 18 D3
Dosso 55 F5
Dothan 74 D5
Douai 13 A5
Douala 56 D1
Douarnenez .. 12 B1
Double Island
  Pt. 63 A5
Doubs → 13 C6
Doubtful Sd. .. 65 F2
Doubtless B. .. 64 A5
Douentza 55 F4
Douglas, U.K. . 11 D4
Douglas, U.S.A. 83 E8
Doumé 56 D2
Dourados 94 A6
Douro → 18 B1
Dover, Australia 62 D4
Dover, U.K. ... 11 F7

112

Dover, Str. of	11	F7	Dumfries	11	D5	Dymer	17	C10	Edward →	62	C3

Dover, Str. of .. 11 F7
Dovrefjell 8 F9
Dowlat Yār 45 C6
Dowlatābād ... 44 D4
Dowshī 45 C7
Draa, Oued → 54 C2
Drăgăşani 22 B5
Dragichyn 17 B7
Dragoman,
  Prokhod 22 C4
Draguignan ... 13 E7
Drake 63 A5
Drakensberg .. 59 E5
Dráma 22 D5
Drammen 9 G10
Drau =
  Drava → ... 20 B7
Drava → 20 B7
Dresden 15 C7
Dreux 12 B4
Drina → 20 B7
Drini → 22 C2
Drochia 17 D8
Drogichin =
  Dragichyn .. 17 B7
Drogobych =
  Drohobych .. 17 D6
Drohobych ... 17 D6
Droichead Atha
  = Drogheda . 11 E3
Dromedary, C. . 63 C5
Drumheller ... 71 C8
Drummond Pt. . 62 B2
Drut → 17 B10
Drygalski I. 96 A11
Dschang 55 G7
Dubā 47 F3
Dubai = Dubayy 44 E3
Dubăsari 17 E9
Dubăsari Vdkhr. 17 E9
Dubawnt → .. 70 B9
Dubawnt, L. .. 70 B9
Dubayy 44 E3
Dubbo 63 B4
Dublin, Ireland 11 E3
Dublin, U.S.A. . 75 C6
Dubno 17 C7
Dubossary =
  Dubăsari 17 E9
Dubossary
  Vdkhr. =
  Dubăsari
  Vdkhr. 17 E9
Dubovka 24 D5
Dubréka 55 G2
Dubrovitsa =
  Dubrovytsya 17 C8
Dubrovnik 20 C7
Dubrovytsya .. 17 C8
Dudhi 40 E4
Dudinka 30 C6
Duero =
  Douro → ... 18 B1
Dugi Otok 20 B5
Duisburg 14 C4
Dukelský
  Průsmyk .. 16 D5
Dukhān 44 E2
Duki 42 D6
Duku 55 F7
Dulce → 94 C4
Dulit, Banjaran 37 D4
Duluth 77 B8
Dum Duma ... 41 D10
Dum Hadjer .. 53 F2
Dumaguete ... 38 C2
Dumai 37 D2
Dumaran 38 B1

Dumfries 11 D5
Dumyât 52 B5
Dún Dealgan =
  Dundalk 11 D3
Dun Laoghaire 11 E3
Duna =
  Dunărea → . 17 F9
Dunaj =
  Dunărea → . 17 F9
Dunakeszi ... 16 E4
Dunărea → .. 17 F9
Dunaújváros .. 16 E4
Dunav =
  Dunărea → . 17 F9
Dunback 65 F4
Dunbar 10 D5
Duncan 78 B5
Dundalk 11 D3
Dundee,
  S. Africa 59 D6
Dundee, U.K. .. 10 C5
Dundoo 63 A3
Dunedin 65 F4
Dunfermline .. 10 C5
Dungarvan ... 11 E3
Dungog 63 B5
Dungu 57 D5
Dunhuang ... 34 B4
Dunkerque .. 12 A5
Dunkirk =
  Dunkerque .. 12 A5
Dunkwa 55 G4
Dúnleary = Dun
  Laoghaire ... 11 E3
Dunolly 62 C3
Dunqul 52 D5
Dunstan Mts. .. 65 F3
Durance → .. 13 E6
Durango =
  Victoria de
  Durango 84 C4
Durango 19 A4
Durazno 94 C5
Durazzo =
  Durrësi 22 D2
Durban 59 D6
Düren 14 C4
Durg 40 G3
Durgapur 40 F6
Durham, U.K. . 11 D6
Durham, U.S.A. 75 B8
Durham Downs 63 A4
Durmitor 22 C2
Durrësi 22 D2
Dursunbey ... 23 E7
D'Urville,
  Tanjung ... 39 E5
D'Urville I. 65 D5
Dusa Mareb .. 49 F4
Dushak 29 F7
Dushanbe ... 29 F7
Dusky Sd. ... 65 F2
Düsseldorf ... 14 C4
Dutch Harbor .. 71 C3
Duwādimi ... 47 F6
Duyun 35 D5
Düzce 46 B2
Duzdab =
  Zāhedān 45 D5
Dvina,
  Severnaya → 28 C5
Dvinsk =
  Daugavpils .. 24 B2
Dwarka 43 H6
Dyatlovo =
  Dzyatlava ... 17 B7
Dyer, C. 68 B4
Dyer Plateau .. 96 B3
Dyersburg 79 A9

Dymer 17 C10
Dynevor Downs 63 A3
Dzamin Üüd .. 35 B6
Dzerzhinsk ... 24 B5
Dzhambul =
  Zhambyl 29 E8
Dzhankoy ... 25 D3
Dzhardzhan ... 30 C10
Dzhezkazgan =
  Zhezqazghan 29 E7
Dzhizak =
  Jizzakh 29 E7
Działdowa ... 16 B5
Dzierzoniów ... 16 C3
Dzungaria =
  Junggar Pendi 34 B3
Dzungarian Gate
  = Alataw
  Shankou ... 34 B3
Dzuumod ... 35 B5
Dzyarzhynsk .. 17 B8
Dzyatlava 17 B7

E

Earnslaw, Mt. . 65 F3
East Bengal ... 41 E7
East Beskids =
  Vychodné
  Beskydy 17 D5
East C. 64 B8
East China Sea 35 C7
East Falkland . 95 G5
East Kilbride .. 10 D4
East London . 59 E5
East Main =
  Eastmain ... 69 C3
East Orange .. 73 D8
East St. Louis . 77 F9
East Toorale .. 63 B4
Eastbourne, N.Z. 65 D6
Eastbourne, U.K. 11 F7
Eastern Ghats . 43 N11
Eastmain 69 C3
Eastmain → .. 69 C3
Eau Claire 77 C9
Eberswalde-
  Finow 15 B7
Ebolowa 56 D2
Ebro → 19 B6
Eceabat 23 D6
Ech Cheliff .. 54 A5
Echo Bay 70 B8
Echuca 63 C3
Ecija 18 D3
Ecuador ■ 90 D3
Ed Dâmer ... 53 E5
Ed Debba 53 E5
Ed Dueim ... 53 F5
Edd 49 E3
Eddystone Pt. . 62 D4
Édea 56 D2
Eden → 14 C5
Eder → 14 C5
Édhessa 23 D4
Edievale ... 65 F3
Edinburg 78 F4
Edinburgh .. 10 D5
Edinița 17 D8
Edirne 22 D6
Edithburgh .. 62 C2
Edmonton 71 C8
Edmundston .. 69 D4
Edremit 23 E6
Edremit Körfezi 23 E6
Edson 71 C8

Edward → ... 62 C3
Edward, L. 57 E5
Edward VII Land 96 C17
Edwards Plateau 78 D3
Égadi, Ísole .. 21 F4
Eger = Cheb .. 16 C1
Eger 16 E5
Egersund 9 G9
Egmont, C. ... 64 C5
Egmont, Mt. .. 64 C6
Eğridir 46 C2
Eğridir Gölü .. 46 C2
Egypt ■ 52 C5
Eibar 19 A4
Eidsvold 63 A5
Eifel 14 C4
Eigg 10 C3
Eil 49 F4
Eildon, L. .. 63 C4
Eindhoven .. 14 C3
Eire = Ireland ■ 11 E3
Eirunepé 91 E5
Eisenach 14 C6
Eisenerz 15 E8
Eivissa = Ibiza . 19 C6
Eketahuna ... 64 D6
Ekibastuz 29 D8
El Aaiún 54 C2
El Alamein .. 52 B4
El Aricha 54 B4
El 'Arîsh 52 B5
El Asnam = Ech
  Cheliff 54 A5
El Bawiti 52 C4
El Bayadh ... 54 B5
El Callao 90 B6
El Centro .. 82 D5
El Cerro 91 G6
El Cuy 95 D3
El Cuyo 85 C7
El Dere 49 G4
El Diviso 90 C3
El Djouf 55 D3
El Dorado, Ark.,
  U.S.A. 79 C7
El Dorado,
  Kans., U.S.A. 78 A5
El Dorado,
  Venezuela .. 90 B6
El Escorial 18 B3
El Faiyûm ... 52 C5
El Fâsher 53 F4
El Ferrol 18 A1
El Fuerte 84 B3
El Gal 49 E5
El Geneina = Al
  Junaynah ... 53 F3
El Geteina ... 53 F5
El Gîza 52 C5
El Iskandarîya . 52 B4
El Jadida 54 B3
El Jebelein .. 53 F5
El Kab 53 E5
El Kala 54 A6
El Kamlin .. 53 E5
El Kef 54 A6
El Khandaq .. 53 E5
El Khârga 52 C5
El Khartûm .. 53 E5
El Khartûm
  Bahrî 53 E5
El Laqâwa 53 F4
El Mafâza ... 53 F5
El Mahalla el
  Kubra 52 B5
El Mansûra .. 52 B5
El Minyâ 52 C5
El Obeid 53 F5
El Odaiya 53 F4

El Oued

El Oued ... 54 B6
El Paso ... 83 E9
El Prat de Llobregat ... 19 B7
El Puerto de Santa María . 18 D2
El Qâhira ... 52 B5
El Qantara ... 52 B5
El Qasr ... 52 C4
El Reno ... 78 B5
El Salvador ■ . 85 E7
El Shallal ... 52 D5
El Suweis ... 52 C5
El Tigre ... 90 B6
El Tocuyo ... 90 B5
El Turbio ... 95 G2
El Uqsur ... 52 C5
El Vigía ... 90 B4
El Wak ... 57 D8
El Wuz ... 53 E5
Elat ... 47 E3
Elâzığ ... 46 C4
Elba ... 20 C3
Elbasani ... 22 D3
Elbe → ... 14 B5
Elbert, Mt. ... 83 A9
Elbeuf ... 12 B4
Elbidtan ... 46 C4
Elbing = Elbląg 16 A4
Elbląg ... 16 A4
Elbrus ... 25 E5
Elburz Mts. = Alborz, Reshteh-ye Kūhhā-ye .. 44 B3
Elche ... 19 C5
Elda ... 19 C5
Elde → ... 15 B6
Eldoret ... 57 D7
Elektrostal ... 24 B4
Elephant I. ... 96 A4
Eleuthera ... 86 A4
Elgin, U.K. ... 10 C5
Elgin, U.S.A. ... 72 C1
Elgon, Mt. ... 57 D6
Eliase ... 39 F4
Elisabethville = Lubumbashi . 59 A5
Elista ... 25 D5
Elizabeth, Australia ... 62 B2
Elizabeth, U.S.A. 73 D8
Elizabeth City . 75 A9
Elizabethton ... 75 A6
Ełk ... 16 B6
Elkhart ... 72 D3
Elkhovo ... 22 C6
Elko ... 80 E6
Ellery, Mt. ... 63 C4
Ellesmere, L. .. 65 G5
Ellice Is. = Tuvalu ■ .. 64 K12
Elliston ... 62 B1
Ellore = Eluru . 40 J3
Ellsworth Land 96 B2
Ellsworth Mts. . 96 B2
Elmore ... 63 C3
Elmshorn ... 14 B5
Eltham ... 64 C6
Eluru ... 40 J3
Elvas ... 18 C2
Elx = Elche ... 19 C5
Ely ... 11 E7
Emämrüd 44 B3
Emba = Embi . 29 E6
Emba → =
Embi → 29 E6
Embarcación .. 94 A4
Embi ... 29 E6

Embi → ... 29 E6
Embrun ... 13 D7
Embu ... 57 E7
Emden ... 14 B4
Emerson ... 71 D10
Emet ... 23 E7
Emi Koussi ... 53 E2
Emine, Nos ... 22 C6
Emmen ... 14 B4
Empalme ... 84 B2
Empangeni ... 59 D6
Empedrado . 94 B5
Empty Quarter = Rub' al Khali ... 48 D4
Ems → ... 14 B4
En Nahud ... 53 F4
Enaratoli ... 39 E5
Enare = Inarijärvi ... 8 E13
Encanto, C. ... 38 A2
Encarnación .. 94 B5
Encounter B. .. 62 C2
Ende ... 39 F2
Enderby Land . 96 A9
Enewetak Atoll 64 H11
Enez ... 23 D6
Engadin ... 13 C9
Engaño, C. ... 38 A2
Engels ... 24 C6
Enggano ... 37 F2
Engkilili ... 37 D4
England □ ... 11 E6
Englewood ... 76 F2
English Bazar = Ingraj Bazar . 40 E7
English Channel 11 F5
Enid ... 78 A5
Enna ... 21 F5
Ennedi ... 53 E3
Enngonia ... 63 A4
Ennis ... 11 E2
Enniskillen ... 11 D3
Enns → ... 15 D8
Enschede ... 14 B4
Ensenada ... 84 A1
Entebbe ... 57 D6
Entroncamento 18 C1
Enugu ... 55 G6
Enugu Ezike ... 55 G6
Eólie, Ís. ... 21 E5
Épernay ... 13 B5
Ephesus ... 23 F6
Épinal ... 13 B7
Epukiro ... 58 C3
Equatorial Guinea ■ ... 56 D1
Er Rahad ... 53 F5
Er Rif ... 54 A4
Er Roseires ... 53 F5
Eräwadī Myit = Irrawaddy → 41 K10
Erbil = Arbil . 46 C6
Erciyaş Dağı ... 46 C3
Érd ... 16 E4
Erdek ... 23 D6
Erebus, Mt. ... 96 B15
Erechim ... 94 B6
Ereğli, Konya, Turkey ... 46 C3
Ereğli, Zonguldak, Turkey ... 46 B2
Eresma → ... 18 B3
Erewadi Myitwanya .. 41 K10
Erfurt ... 15 C6
Ergani ... 46 C4
Érice ... 21 E4

Erie ... 73 C5
Erie, L. ... 72 C5
Erigavo ... 49 E4
Erímanthos ... 23 F3
Erimo-misaki .. 32 F12
Eritrea ■ ... 49 E2
Erlangen ... 15 D6
Ermenak ... 46 C3
Ermoúpolis = Síros ... 23 F5
Ernakulam = Cochin ... 43 Q10
Erne, Lower L. . 11 D3
Erode ... 43 P10
Eromanga ... 63 A3
Erramala Hills . 43 M11
Ertis → = Irtysh → ... 28 C7
Erzgebirge 15 C7
Erzincan ... 46 C4
Erzurum ... 46 C5
Es Sahrâ' Esh Sharqîya ... 52 C5
Es Sînâ' ... 52 C5
Esbjerg ... 9 G9
Esch-sur-Alzette 13 B6
Escuinapa ... 84 C3
Eşfahān ... 44 C2
Esh Sham = Dimashq ... 46 D4
Esil → = Ishim → ... 29 D8
Eskilstuna ... 9 G11
Eskimo Pt. ... 70 B10
Eskişehir ... 46 C2
Esla → ... 18 B2
Eslāmābād-e Gharb ... 46 D6
Eşme ... 23 E7
Esmeraldas ... 90 C3
Esperance ... 60 G3
Esperanza ... 94 C4
Espichel, C. ... 18 C1
Espinal ... 90 C4
Espinazo, Sierra del = Espinhaço, Serra do 93 F5
Espinhaço, Serra do ... 93 F5
Espírito Santo □ 93 G5
Esquel ... 95 E2
Esquina ... 94 C5
Essaouira ... 54 B3
Essen ... 14 C4
Essequibo → . 90 B7
Esslingen ... 14 D5
Estados, I. de Los ... 95 G4
Estância ... 93 E6
Estevan ... 71 D9
Estonia ■ ... 24 B2
Estrêla, Serra da 18 B2
Estremoz ... 18 C2
Estrondo, Serra do ... 92 D4
Esztergom ... 16 E4
Etadunna ... 62 A2
Étampes ... 12 B5
Etawah ... 42 F11
Ethiopia ■ ... 49 F2
Etna ... 21 F5
Etosha Pan ... 58 B3
Euboea = Évvoia ... 23 E5
Eucumbene, L. 63 C4
Eufaula ... 74 D5
Eugene ... 80 C2
Eugowra ... 63 B4

Eulo ... 63 A4
Euphrates = Furāt, Nahr al → ... 47 E6
Eureka ... 80 E1
Euroa ... 63 C4
Europa, I. ... 59 J8
Europa, Picos de 18 A3
Europa, Pta. de 18 D3
Europa Pt. = Europa, Pta. de ... 18 D3
Evans Head ... 63 A5
Evanston ... 72 C2
Evansville ... 72 F2
Everard, L. ... 62 B1
Everest, Mt. ... 40 C6
Everett ... 80 B2
Everglades National Park 75 G7
Evinayong ... 56 D2
Évora ... 18 C2
Évreux ... 12 B4
Évros → ... 22 D6
Evry ... 12 B5
Évvoia ... 23 E5
Ewo ... 56 E2
Exaltación ... 91 F5
Exeter ... 11 F5
Exmoor ... 11 F5
Exmouth ... 11 F5
Extremadura □ 18 C2
Eyasi, L. ... 57 E6
Eyre (North), L. 62 A2
Eyre (South), L. 62 A2
Eyre, L. ... 60 F6
Eyre Cr. → ... 62 A2
Eyre Mts. ... 65 F3
Eyre Pen. ... 62 C2
Ezine ... 23 E6

F

F.Y.R.O.M. = Macedonia ■ 22 D3
Fabriano ... 20 C4
Facatativá ... 90 C4
Fachi ... 55 E7
Fada ... 53 E3
Fada-n-Gourma 55 F5
Faddeyevskiy, Ostrov ... 31 B12
Fādilī ... 47 F7
Faenza ... 20 B3
Færoe Is. = Føroyar ... 6 C4
Fǎgǎras ... 17 F7
Fagnano, L. ... 95 G3
Fahraj ... 45 D4
Fahūd ... 44 F4
Fair Isle ... 10 B6
Fairbanks ... 71 B5
Fairfield, Ala., U.S.A. ... 74 C4
Fairfield, Calif., U.S.A. ... 80 F2
Fairlie ... 65 F4
Fairweather, Mt. 71 C6
Faisalabad ... 42 D8
Faizabad ... 40 D4
Fakfak ... 39 E4
Falaise ... 12 B3
Falam ... 41 F9
Falconara Maríttima ... 20 C4
Falcone, C. ... 20 D2

Faleshty =
Fălești 17 E8
Fălești 17 E8
Falkland Is. □ . 95 G5
Falkland Sd. ... 95 G5
Fall River 73 D10
Falmouth 11 F4
False B. 58 E3
Fălticeni 17 E8
Falun 9 F11
Famagusta ... 46 D3
Fandriana 59 J9
Fano 20 C4
Fao = Al Fāw . 47 E7
Faradje 57 D5
Farafangana .. 59 J9
Farāh 42 C3
Farāh □ 42 C3
Faranah 55 F2
Farasān, Jazā'ir 49 D3
Farasan Is. =
Farasān,
Jazā'ir ... 49 D3
Fareham 11 F6
Farewell, C. ... 64 D5
Farewell C. =
Farvel, Kap . 68 C6
Farghona 29 E8
Fargo 76 B6
Farim 55 F1
Farīmān 45 C4
Farina 62 B2
Faro, Brazil .. 92 C2
Faro, Portugal . 18 D2
Farrāshband .. 44 D3
Farrell Flat .. 62 B2
Farrukhabad-
cum-
Fatehgarh .. 42 F11
Fārs □ 44 D3
Fársala 23 E4
Fartak, Râs ... 47 E3
Farvel, Kap .. 68 C6
Fāryāb □ 45 B6
Fasā 44 D3
Fasano 21 D6
Fastiv 17 C9
Fastov = Fastiv 17 C9
Fatagar, Tanjung 39 E4
Fatehgarh 42 F11
Fatehpur, Raj.,
India 42 F9
Fatehpur, Ut. P.,
India 40 E3
Favara 21 F4
Favignana 21 F4
Faya-Largeau . 53 E2
Fayd 47 F5
Fayetteville,
Ark., U.S.A. . 79 A6
Fayetteville,
N.C., U.S.A. . 75 B8
Fazilka 42 D9
Fdérik 54 D2
Feather → ... 80 F3
Featherston ... 65 D6
Fécamp 12 B4
Fehmarn 15 A6
Feilding 64 D6
Feira de Santana 93 E6
Feldkirch 14 E5
Felipe Carrillo
Puerto 85 D7
Felixstowe 11 F7
Fengjie 35 C5
Fenyang 35 C6
Feodosiya 25 D4
Ferdows 44 C4
Ferfer 49 F4

Fergana =
Farghona ... 29 E8
Fermo 20 C4
Fernando de
Noronha 92 C7
Fernando Póo =
Bioko 56 D1
Ferozepore =
Firozpur 42 D9
Ferrara 20 B3
Ferreñafe ... 91 E3
Ferret, C. 12 D3
Ferrol = El
Ferrol 18 A1
Fès 54 B4
Feshi 56 F3
Feteşti 22 B6
Fetlar 10 A6
Feuilles → ... 68 C3
Fezzan 52 C2
Fianarantsoa .. 59 J9
Fianga 53 G2
Ficksburg 59 D5
Fieri 23 D2
Figeac 12 D5
Figueira da Foz 18 B1
Figueras 19 A7
Figuig 54 B4
Fiji ■ 64 L12
Filiatrá 23 F3
Filyos → 46 B3
Fingöe 59 A6
Finike 46 C2
Finisterre, C. .. 18 A1
Finke → 62 A2
Finland ■ 8 F13
Finland, G. of . 9 G13
Finlay → 70 C7
Finley 63 C4
Finniss, C. ... 62 B1
Fiora → 20 C3
Firat = Furāt,
Nahr al → .. 47 E6
Firenze 20 C3
Firozabad 42 F11
Firozpur 42 D9
Firūzābād 44 D3
Firūzkūh 44 C3
Fish → 58 D3
Fishguard 11 E4
Fitri, L. 53 F2
Fitz Roy 95 F3
Fiume = Rijeka 20 B5
Fizi 57 E5
Flagstaff 83 C7
Flaming Gorge
Reservoir ... 81 E9
Flamingo, Teluk 39 F5
Flandre 14 C2
Flathead L. 81 B6
Flattery, C. 80 A1
Flensburg 14 A5
Flers 12 B3
Flesko, Tanjung 39 D2
Flin Flon 71 C9
Flinders → ... 61 D7
Flinders I. 62 C4
Flinders Ras. .. 62 B2
Flint 72 C4
Flint I. 65 L15
Flinton 63 A4
Florence =
Firenze 20 C3
Florence, Ala.,
U.S.A. 74 B4
Florence, S.C.,
U.S.A. 75 B8
Florence, L. ... 62 A2
Flores 39 F2

Flores Sea 39 F2
Florești 17 E9
Floriano 92 D5
Florianópolis . 94 B7
Florida 94 C5
Florida □ 75 F6
Florida, Straits
of 86 B3
Flórina 23 D3
Florø 8 F9
Fluk 39 E3
Flushing =
Vlissingen . 14 C2
Fly → 61 B7
Flying Fish, C. . 96 B1
Foça 23 E6
Focşani 17 F8
Fóggia 20 D5
Föhr 14 A5
Foix 12 E4
Foligno 20 C4
Folkestone ... 11 F7
Fond du Lac .. 77 D10
Fondi 20 D4
Fonsagrada ... 18 A2
Fonseca, G. de 85 E7
Fontainebleau . 13 B5
Fonte Boa 90 D5
Fontenay-le-
Comte 12 C3
Foochow =
Fuzhou 35 D6
Forbes 63 B4
Ford's Bridge . 63 A4
Forécariah ... 55 G2
Forestier Pen. . 62 D4
Forfar 10 C5
Forlì 20 B4
Formentera ... 19 C6
Formentor, C. de 19 C7
Former
Yugoslav
Republic of
Macedonia =
Macedonia ■ 22 D3
Fórmia 20 D4
Formosa =
Taiwan ■ ... 35 D7
Formosa 94 B5
Formosa, Serra 93 E3
Formosa Bay . 57 E8
Føroyar .....
Føroyar 6 C4
Føroyar 6 C1
Forrest 62 C3
Forsayth 61 D7
Forst 15 C8
Forster 63 B5
Fort Albany .. 69 C2
Fort Collins ... 76 E2
Fort-de-France . 87 D7
Fort de Possel =
Possel 56 C3
Fort Dodge ... 77 D7
Fort Franklin .. 70 B7
Fort Good-Hope 70 B7
Fort Hertz =
Putao 41 D11
Fort Lauderdale 75 F7
Fort Liard 70 B7
Fort Macleod . 71 D8
Fort MacMahon 54 C5
Fort McPherson 70 B6
Fort Miribel .. 54 C5
Fort Myers ... 75 F7
Fort Norman . 70 B7
Fort Peck L. .. 81 B10
Fort Pierce ... 75 F7
Fort Portal ... 57 D6

Fort Providence 70 B8
Fort Resolution 70 B8
Fort Rosebery =
Mansa 57 G5
Fort Rupert =
Waskaganish 69 C3
Fort Sandeman 42 D6
Fort Scott 79 A6
Fort Severn ... 69 C2
Fort Shevchenko 29 E6
Fort Simpson . 70 B7
Fort Smith,
Canada 70 B8
Fort Smith,
U.S.A. 79 B6
Fort Trinquet =
Bir Mogrein . 54 C2
Fort Walton
Beach 74 D4
Fort Wayne ... 72 D3
Fort William ... 10 C4
Fort Worth ... 78 C5
Fort Yukon ... 71 B5
Fortaleza 92 C6
Forūr 44 E3
Foshan 35 D6
Fossano 20 B1
Fougamou ... 56 E2
Fougères 12 B3
Foula 10 A5
Foumban 55 G7
Foúrnoi 23 F6
Fouta Djalon . 55 F2
Foveaux Str. .. 65 G3
Foxe Basin .. 68 B3
Foxe Chan. .. 68 B2
Foxe Pen. 68 B3
Foxton 64 D6
Foz do Gregório 91 E4
Foz do Iguaçu . 94 B6
Franca 93 G4
Francavilla
Fontana 21 D6
France ■ 13 C5
Frances 62 C3
Franceville 56 E2
Franche-Comté 13 C6
Francistown .. 59 C5
Frankfurt,
Brandenburg,
Germany ... 15 B8
Frankfurt,
Hessen,
Germany ... 14 C5
Fränkische Alb 15 D6
Franklin B. 70 B7
Franklin I. 96 B15
Franklin Mts. .. 70 B7
Franklin Str. .. 70 A10
Frankston 63 C4
Frantsa Iosifa,
Zemlya 28 A6
Franz 69 D2
Franz Josef
Land =
Frantsa Iosifa,
Zemlya 28 A6
Fraser → 71 D7
Fraser I. 63 A5
Fraserburgh .. 10 C5
Fray Bentos ... 94 C5
Fredericksburg 73 E7
Fredericton .. 69 D4
Fredrikstad ... 9 G10
Freeport,
Bahamas ... 86 A4
Freeport, U.S.A. 77 D10
Freetown 55 G2
Fregenal de la
Sierra 18 C2

Freibourg =
Fribourg 13 C7
Freiburg 14 E4
Freire 95 D2
Freising 15 D6
Freistadt 15 D8
Fréjus 13 E7
Fremont 82 B2
French
Guiana ■ ... 92 B3
French
Polynesia ■ . 65 M16
Fresco ➤ 92 D3
Freshfield, C. . . 96 A14
Fresnillo 84 C4
Fresno 82 B3
Freycinet Pen. . 62 D4
Fria, C. 58 B2
Frías 94 B3
Fribourg 13 C7
Friedrichshafen 14 E5
Friendly Is. =
Tonga ■ 65 L13
Frobisher B. .. 68 B4
Frobisher Bay =
Iqaluit 68 B4
Frome, L. 62 B2
Frome Downs . 62 B2
Front Range ... 81 E11
Frosinone 20 D4
Frunze =
Bishkek 29 E8
Frutal 93 F4
Frýdek-Místek . 16 D4
Fuchou =
Fuzhou 35 D6
Fuengirola 18 D3
Fuentes de
Oñoro 18 B2
Fuerte ➤ 84 B3
Fuerte Olimpo . 91 H7
Fuerteventura . 54 C2
Fuhai 34 B3
Fuji-San 32 B6
Fujian □ 35 D6
Fujinomiya .. 32 B6
Fujisawa 32 B6
Fukien =
Fujian □ 35 D6
Fukuchiyama .. 32 B4
Fukui 32 A5
Fukuoka 32 C2
Fukushima 33 G12
Fukuyama 32 B3
Fulda 14 C5
Fulda ➤ 14 C5
Fullerton 82 D4
Funabashi 32 B7
Funchal 54 B1
Fundación 90 A4
Fundão 18 B2
Fundy, B. of ... 69 D4
Funtua 55 F6
Furāt, Nahr
al ➤ 47 E6
Furneaux Group 62 D4
Fürstenwalde .. 15 B8
Fürth 14 D6
Fury and Hecla
Str. 68 B2
Fusagasuga ... 90 C4
Fushun 35 B7
Futuna 64 L12
Fuxin 35 B7
Fuyuan 35 B8
Fuzhou 35 D6
Fyn 9 G10

G

Gabela 56 G2
Gabès 54 B7
Gabès, G. de . . 52 B1
Gabon ■ 56 E2
Gaborone 59 C5
Gabrovo 22 C5
Gachsārān 44 D2
Gadag 43 M9
Gadarwara ... 43 H11
Gadhada 43 J7
Gadsden 74 B4
Gadwal 43 L10
Gafsa 54 B6
Gagnoa 55 G3
Gagnon 69 C4
Gaillimh =
Galway 11 E2
Gainesville, Fla.,
U.S.A. 75 E6
Gainesville, Ga.,
U.S.A. 75 B6
Gainesville,
Tex., U.S.A. . 78 C5
Gairdner, L. ... 62 B2
Galangue 58 A3
Galashiels 10 D5
Galați 17 F9
Galatina 21 D7
Galcaio 49 F4
Galdhøpiggen . 8 F9
Galela 39 D3
Galesburg 77 E9
Galich 24 B5
Galicia □ 18 A2
Galilee, Sea of
= Yam
Kinneret .. 46 D3
Gallabat 53 F6
Gállego ➤ ... 19 B5
Gallegos ➤ .. 95 G3
Gallinas, Pta. .. 90 A4
Gallipoli =
Gelibolu ... 23 D6
Gallípoli 21 D6
Gällivare 8 E12
Galloway, Mull
of 11 D4
Galong 63 B4
Galveston 79 E6
Gálvez 94 C4
Galway 11 E2
Galway B. 11 E2
Gambaga 55 F4
Gambela 53 G5
Gambia ■ 55 F1
Gambia ➤ 55 F1
Gambier Is. ... 62 C2
Gamboma 56 E3
Gan Jiang ➤ .. 35 D6
Ganāveh 44 D2
Gäncä 25 E6
Gand = Gent .. 14 C2
Ganda 58 A2
Gandak ➤ 40 E5
Gandava 42 E5
Gander 69 D5
Gandhi Sagar . 43 G9
Gandi 55 F6
Gandía 19 C5
Ganedidalem =
Gani 39 E3
Ganga ➤ 41 F8
Ganganagar .. 42 E8
Gangara 55 F6
Gangaw 41 F10
Gangdisê Shan 40 B3

Ganges =
Ganga ➤ .. 41 F8
Gangtok 41 D7
Gani 39 E3
Gansu □ 34 C5
Ganta 55 G3
Gantheaume, C. 62 C2
Ganyem 39 E6
Ganzhou 35 D6
Gaoua 55 F4
Gaoual 55 F2
Gaoxiong 35 D7
Gap 13 D7
Gar 34 C2
Garabogazköl
Aylagy 29 E6
Garanhuns ... 92 D6
Garawe 55 H3
Gard 49 F4
Garda, L. di ... 20 B3
Garden City ... 78 A3
Garden Grove . 82 D4
Gardēz 42 C6
Gargano, Mte. . 20 D5
Garigliano ➤ . 20 D4
Garland 81 E7
Garm 29 F8
Garmisch-
Partenkirchen 14 E6
Garmsär 44 C3
Garoe 49 F4
Garonne ➤ .. 12 D3
Garoua 53 G1
Garrison Res. =
Sakakawea, L. 76 B4
Garry, L. 70 B9
Garsen 57 E8
Garvie Mts. ... 65 F3
Garwa =
Garoua ... 53 G1
Gary 72 D2
Garzê 34 C5
Garzón 90 C3
Gasan Kuli =
Hasan Kuli . 29 F6
Gascogne 12 E4
Gascogne, G. de 12 D2
Gascony =
Gascogne ... 12 E4
Gashaka 55 G7
Gaspé 69 D4
Gaspé, C. de .. 69 D4
Gaspé, Pén. de 69 D4
Gasteiz = Vitoria 19 A4
Gastonia 75 B7
Gastre 95 E3
Gata, C. de ... 19 D4
Gata, Sierra de 18 B2
Gateshead 11 D6
Gauhati 41 D8
Gavāter 45 E5
Gávdhos 23 G5
Gävle 8 F11
Gawilgarh Hills 43 J10
Gawler 62 B2
Gaxun Nur ... 34 B5
Gaya, India ... 40 E5
Gaya, Niger .. 55 F5
Gayndah 63 A5
Gaysin =
Haysyn 17 D9
Gayvoron =
Hayvoron .. 17 D9
Gaza 47 E3
Gaziantep ... 46 C4
Gazli 29 E7
Gdańsk 16 A4

Gdańska, Zatoka 16 A4
Gdov 24 B2
Gdynia 16 A4
Gebe 39 D3
Gebeit Mine ... 52 D6
Gebze 22 D7
Gedaref 53 F6
Gede, Tanjung 37 F3
Gediz ➤ 23 E6
Gedser 9 H10
Geelong 63 C3
Geesthacht ... 14 B6
Geidam 53 F1
Geili 53 E5
Geita 57 E6
Gejiu 34 D5
Gela 21 F5
Geladi 49 F4
Gelehun 55 G2
Gelibolu 23 D6
Gelsenkirchen . 14 C4
Gemas 37 D2
Gemena 56 D3
Gemerek 46 C4
Gemlik 23 D7
General Acha . 94 D4
General Alvear,
Buenos Aires,
Argentina .. 94 D5
General Alvear,
Mendoza,
Argentina .. 94 D3
General Artigas 94 B5
General
Belgrano .. 94 D5
General Guido . 94 D5
General Juan
Madariaga .. 94 D5
General La
Madrid 94 D4
General
MacArthur .. 38 B3
General Martin
Miguel de
Güemes ... 94 A4
General Paz .. 94 B5
General Pico .. 94 D4
General Pinedo 94 B4
General Santos 38 C3
General Villegas 94 D4
Geneva =
Genève 13 C7
Geneva, L. =
Léman, L. ... 13 C7
Genève 13 C7
Genil ➤ 18 D3
Gennargentu,
Mti. del 21 D2
Genoa =
Génova ... 20 B2
Genoa 63 C4
Génova 20 B2
Génova, G. di . 20 C2
Gent 14 C2
Georga, Zemlya 28 A5
George 58 E4
George ➤ ... 68 C4
George, L.,
N.S.W.,
Australia 63 C4
George, L.,
S. Austral.,
Australia 62 C3
George, L.,
Uganda ... 57 D6
George River =
Kangiqsualujjuaq
.... 68 C4
George Sound . 65 F2

George Town .	36	C2
George V Land	96	A14
George VI Sound	96	B3
Georgetown, Gambia	55	F2
Georgetown, Guyana	90	B7
Georgia □	75	C6
Georgia ■	25	E5
Georgian B.	69	D2
Georgiu-Dezh = Liski	24	C4
Georgiyevsk	25	E5
Gera	15	C7
Geral de Goiás, Serra	93	E4
Geraldton	60	F1
Gerede	46	B3
Gereshk	42	D4
Gerlogubi	49	F4
Germany ■	14	C6
Germiston	59	D5
Gerona	19	B7
Geser	39	E4
Getafe	18	B4
Gettysburg	73	E7
Getz Ice Shelf	96	B18
Ghaghara →	40	E5
Ghana ■	55	G4
Ghanzi	58	C4
Ghardaïa	54	B5
Gharyān	52	B1
Ghat	54	D7
Ghawdex = Gozo	21	F5
Ghayl	47	G6
Ghazal, Bahr el →, Chad	53	F2
Ghazâl, Bahr el →, Sudan	57	C6
Ghazaouet	54	A4
Ghaziabad	42	E10
Ghazipur	40	E4
Ghaznī	42	C6
Ghaznī □	42	C6
Ghèlinsor	49	F4
Ghent = Gent	14	C2
Ghizao	42	C4
Ghowr □	42	C4
Ghugus	43	K11
Ghūrīān	42	B2
Gia Lai = Pleiku	36	B3
Gian	38	C3
Giarabub = Al Jaghbūb	52	C3
Giarre	21	F5
Gibraltar ■	18	D3
Gibraltar, Str. of	18	E3
Gibson Desert	60	E4
Gidole	53	G6
Giessen	14	C5
Gifu	32	B5
Giglio	20	C3
Gijón	18	A3
Gila →	82	D5
Gīlān □	46	C7
Gilbert Is.	64	J12
Gilgandra	63	B4
Gilgit	42	B9
Gilles, L.	62	B2
Gilmore	63	C4
Gimbi	53	G6
Gin Gin	63	A5
Ginir	49	F3
Giohar	49	G4
Gióna, Óros	23	E4
Girardot	90	C4
Giresun	46	B4
Girga	52	C5
Giridih	40	E6
Girilambone	63	B4
Girne = Kyrenia	46	D3
Girona = Gerona	19	B7
Gironde →	12	D3
Girvan	10	D4
Gisborne	64	C8
Gisenyi	57	E5
Gitega	57	E5
Giuba →	49	G3
Giurgiu	22	C5
Giza = El Gîza	52	C5
Giżycko	16	A5
Gjirokastra	23	D3
Gjoa Haven	70	B10
Glace Bay	69	D5
Gladstone, Queens., Australia	61	E9
Gladstone, S. Austral., Australia	62	B2
Glåma → Glomma →	9	G10
Glasgow	10	D4
Glazov	28	D6
Gleiwitz = Gliwice	16	C4
Glen Canyon National Recreation Area	83	B7
Glen Innes	63	A5
Glendale	82	C3
Glenelg	62	B2
Glenelg →	62	C3
Glenmorgan	63	A4
Glenorchy	62	D4
Glenreagh	63	B5
Glenrothes	10	C5
Glens Falls	73	C9
Gliwice	16	C4
Głogów	16	C3
Glomma →	9	G10
Glorieuses, Is.	59	G9
Gloucester, Australia	63	B5
Gloucester, U.K.	11	F5
Glusk	17	B9
Gmünd	15	D8
Gmunden	15	E7
Gniezno	16	B3
Go Cong	36	B3
Goa	43	M8
Goa □	43	M8
Goalen Hd.	63	C5
Goalpara	41	D8
Goba, Ethiopia	49	F2
Goba, Mozam.	59	D6
Gobabis	58	C3
Gobi	35	B6
Gochas	58	C3
Godavari →	40	J4
Godavari Point	40	J4
Godhra	43	H8
Gods →	71	C10
Gods L.	71	C10
Godthåb	68	B5
Godwin Austen = K2	42	B10
Goeie Hoop, Kaap die = Good Hope, C. of	58	E3
Gogra = Ghaghara →	40	E5
Goiânia	93	F4
Goiás	93	F3
Goiás □	93	E4
Goio-Ere	94	A6
Gojra	42	D8
Gökçeada	23	D5
Gokteik	41	F11
Gold Coast	63	A5
Golden Gate	80	G2
Goldsboro	75	B9
Goleniów	16	B2
Golfo Aranci	20	D2
Golspie	10	C5
Goma	57	E5
Gomel = Homyel	17	B10
Gomera	54	C1
Gómez Palacio	84	B4
Gomogomo	39	F4
Gomoh	40	F6
Gompa = Ganta	55	G3
Gonābād	44	C4
Gonaïves	87	C5
Gonbab-e Kāvūs	44	B3
Gonda	40	D3
Gonder	53	F6
Gönen	23	D6
Gonghe	34	C5
Gongolgon	63	B4
Goniri	53	F1
Good Hope, C. of	58	E3
Goodooga	63	A4
Goolgowi	63	B4
Goombalie	63	A4
Goondiwindi	63	A5
Gooray	63	A5
Gop	43	H6
Göppingen	14	D5
Gorakhpur	40	D4
Goražde	20	C7
Gordon →	62	D4
Gore, Australia	63	A5
Goré, Chad	53	G2
Gore, Ethiopia	53	G6
Gore, N.Z.	65	G3
Gorgān	44	B3
Gorgona, I.	90	C3
Gorízia	20	B4
Gorki = Nizhniy Novgorod	24	B5
Gorkiy = Nizhniy Novgorod	24	B5
Gorkovskoye Vdkhr.	24	B5
Görlitz	15	C8
Gorlovka = Horlivka	25	D4
Gorna Dzhumayo = Blagoevgrad	22	C4
Gorna Oryakhovitsa	22	C5
Gorno-Altay □	29	D9
Gorno-Altaysk	29	D9
Gorno Slinkino = Gornopravdinsk	29	C8
Gornopravdinsk	29	C8
Gorodenka = Horodenka	17	D7
Gorodok = Horodok	17	D6
Gorokhov = Horokhiv	17	C7
Gorontalo	39	D2
Gorzów Wielkopolski	16	B2
Gosford	63	B5
Goslar	14	C6
Gospič	20	B5
Göta kanal	9	G11
Göteborg	9	G10
Gotha	14	C6
Gothenburg = Göteborg	9	G10
Gotland	9	G11
Göttingen	14	C5
Gottwaldov = Zlín	16	D3
Gouda	14	B3
Gouin, Rés.	69	D3
Goulburn	63	B4
Gounou-Gaya	53	G2
Gouri	53	E2
Gourma Rharous	55	E4
Gourock Ra.	63	C4
Governador Valadares	93	F5
Gowd-e Zirreh	42	E3
Goya	94	B5
Goyder Lagoon	62	A2
Goyllarisquisga	91	F3
Goz Beïda	53	F3
Gozo	21	F5
Graaff-Reinet	58	E4
Gračac	20	B5
Gracias a Dios, C.	86	D3
Grado	18	A2
Gradule	63	A4
Grafton	63	A5
Graham Bell, Os.	28	A7
Graham Land	96	A3
Grahamstown	59	E5
Grajaú	92	D4
Grajaú →	92	C5
Grampian Highlands = Grampian Mts.	10	C4
Grampian Mts.	10	C4
Gran Canaria	54	C1
Gran Chaco	94	B4
Gran Paradiso	20	B1
Gran Sasso d'Italia	20	C4
Granada, Nic.	85	E7
Granada, Spain	18	D4
Granby	69	D3
Grand Bahama	86	A4
Grand Bassam	55	G4
Grand Canyon	83	B6
Grand Cayman	86	C3
Grand Coulee Dam	80	B4
Grand Forks, Canada	71	D8
Grand Forks, U.S.A.	76	B6
Grand Island	76	E5
Grand Lahou	55	G3
Grand Rapids, Canada	71	C10
Grand Rapids, U.S.A.	72	C3
Grand St.-Bernard, Col du	13	D7
Grand Teton	81	D8
Grande →, Argentina	94	A3
Grande →, Bolivia	91	G6
Grande →, Bahia, Brazil	93	E5

Grande →,
Minas Gerais,
Brazil 93 G3
Grande, B. 95 G3
Grande, Rio → 79 F5
Grande Baleine,
R. de la → .. 69 C3
Grande de
Santiago → 84 C3
Granite City ... 77 F9
Granity 65 D4
Granja 92 C5
Granollers 19 B7
Grantham 11 E6
Granville 12 B3
Gras, L. de .. 70 B8
Grasse 13 E7
Grassmere ... 62 B3
Graulhet 12 E4
's-Gravenhage . 14 B3
Gravesend 63 A5
Graz 15 E8
Great Abaco I. . 86 A4
Great Australian
Bight 60 G4
Great Barrier I. 64 B6
Great Barrier
Reef 61 D8
Great Basin ... 80 F5
Great Bear → . 70 B7
Great Bear L. . 70 B7
Great Belt =
Store Bælt .. 9 G10
Great Dividing
Ra. 61 E8
Great Falls 81 B8
Great Inagua I. 86 B5
Great Indian
Desert = Thar
Desert 42 F7
Great Karoo ... 58 E4
Great Lake 62 D4
Great Ouse → 11 E7
Great Ruaha → 57 F7
Great Saint
Bernard P. =
Grand St.-
Bernard, Col
du 13 D7
Great Salt L. .. 81 E7
Great Salt Lake
Desert 81 E7
Great Sandy
Desert 60 E3
Great Sangi =
Sangihe, P. . 39 D3
Great Slave L. . 70 B8
Great Smoky
Mts. Nat. Pk. 75 B6
Great Victoria
Desert 60 F4
Great Wall .. 35 C5
Great Yarmouth 11 E7
Greater Antilles 86 C5
Greater Sunda
Is. 37 F4
Gredos, Sierra
de 18 B3
Greece ■ 23 E4
Greeley 76 E2
Green → 83 A8
Green Bay 72 B2
Green C. 63 C5
Greenland ■ .. 66 C9
Greenock 10 D4
Greensboro ... 75 A8
Greenville,
Liberia 55 G3
Greenville, Ala.,
U.S.A. 74 D4

Greenville,
Miss., U.S.A. 79 C8
Greenville, N.C.,
U.S.A. 75 B9
Greenville, S.C.,
U.S.A. 75 B6
Greenwood,
Miss., U.S.A. 79 C8
Greenwood,
S.C., U.S.A. . 75 B6
Gregory, L. ... 62 A2
Greifswald 15 A7
Greiz 15 C7
Gremikha 28 C4
Grenada ■ ... 87 D7
Grenfell 63 B4
Grenoble 13 D6
Grey → 65 E4
Grey Ra. 63 A3
Greymouth ... 65 E4
Greytown, N.Z. 65 D6
Greytown,
S. Africa ... 59 D6
Griffin 75 C5
Griffith 63 B4
Grimari 56 C4
Grimaylov =
Hrymayliv ... 17 D8
Grimsby 11 E6
Gris-Nez, C. .. 12 A4
Grodno =
Hrodna 17 B6
Grodzyanka =
Hrodzyanka . 17 B9
Grójec 16 C5
Groningen 14 B4
Grootfontein .. 58 B3
Gross Glockner 15 E7
Grosser Arber . 15 D7
Grosseto 20 C3
Groznyy 25 E6
Grudziądz 16 B4
Gryazi 24 C4
Gryazovets ... 28 D5
Gua 40 F5
Guadalajara,
Mexico 84 C4
Guadalajara,
Spain 18 B4
Guadalcanal .. 64 K11
Guadalete → . 18 D2
Guadalquivir → 18 D2
Guadalupe =
Guadalupe ■ 87 C7
Guadalupe,
Sierra de ... 18 C3
Guadarrama,
Sierra de ... 18 B4
Guadeloupe ■ . 87 C7
Guadiana → .. 18 D2
Guadix 18 D4
Guafo, Boca del 95 E2
Guaíra 94 A6
Guaitecas, Is. . 95 E2
Guajará-Mirim . 91 F5
Guajira, Pen. de
la 90 A4
Gualeguay ... 94 C5
Gualeguaychú . 94 C5
Guam ■ 64 H9
Guamúchil ... 84 B3
Guan Xian ... 34 C5
Guanahani =
San Salvador 86 B5
Guanajuato ... 84 C4
Guandacol ... 94 B3
Guane 86 B3
Guangdong □ . 35 D6
Guanghua 35 C6

Guangxi
Zhuangzu
Zizhiqu □ ... 35 D5
Guangzhou ... 35 D6
Guanipa → ... 90 B6
Guantánamo . 86 B4
Guaporé → .. 91 F5
Guaqui 91 G5
Guarapuava .. 94 B6
Guarda 18 B2
Guardafui, C. =
Asir, Ras ... 49 E5
Guasdualito .. 90 B4
Guasipati 90 B6
Guatemala ... 85 E6
Guatemala ■ .. 85 D6
Guatire 90 A5
Guaviare □ .. 90 C4
Guaviare → .. 90 C5
Guaxupé 93 G4
Guayama 87 C6
Guayaquil 90 D3
Guayaquil, G. de 90 D2
Guaymas 84 B2
Guddu Barrage 42 E6
Gudivada 40 J3
Gudur 43 M11
Guecho 19 A4
Guékédou 55 G2
Guelma 54 A6
Guelph 69 D2
Guéréda 53 F3
Guéret 12 C4
Guernica 19 A4
Guernsey 11 G5
Guidónia-
Montecélio .. 20 C4
Guiglo 55 G3
Guildford 11 F6
Guilin 35 D6
Guimarães,
Brazil 92 C5
Guimarães,
Portugal 18 B1
Guimaras 38 B2
Guinea ■ 55 F2
Guinea, Gulf of 51 F3
Guinea-Bissau ■ 55 F1
Guingamp 12 B2
Guiping 35 D6
Güiria 90 A6
Guiuan 38 B3
Guiyang 35 D5
Guizhou □ ... 35 D5
Gujarat □ 43 H7
Gujranwala ... 42 C9
Gujrat 42 C9
Gulbarga 43 L10
Gulf, The 44 E2
Gulgong 63 B4
Güllük 23 F6
Gulshad 29 E8
Gulu 57 D6
Gum Lake ... 62 B3
Gummi 55 F6
Gümüşhane .. 46 B4
Gumzai 39 F4
Guna 43 G10
Gundagai 63 C4
Gungu 56 F3
Gunnedah ... 63 B5
Gunningbar
Cr. 63 B4
Guntakal 43 M10
Guntur 40 J3
Gunungapi ... 39 F3
Gunungsitoli .. 37 D1
Gunza 56 G2
Gupis 42 A8

Gürchän 46 D7
Gurdaspur 42 C9
Gurgaon 42 E10
Gurkha 40 C5
Gurley 63 A4
Gurupá 92 C3
Gurupá, I.
Grande de .. 92 C3
Gurupi → 92 C4
Guryev =
Atyraū 29 E6
Gusau 55 F6
Güstrow 15 B7
Gütersloh 14 C5
Guthrie 78 B5
Guyana ■ 90 B7
Guyane
française ■ =
French
Guiana ■ ... 92 B3
Guyenne 12 D4
Guyra 63 B5
Gwa 41 J10
Gwaai 59 B5
Gwabegar ... 63 B4
Gwādar 42 G3
Gwalior 42 F11
Gwanda 59 C5
Gweru 59 B5
Gwydir → 63 A4
Gyandzha =
Gäncä 25 E6
Gyaring Hu ... 34 C4
Gydanskiy P-ov. 28 B8
Gympie 63 A5
Gyöngyös ... 16 E4
Györ 16 E3
Gypsumville . 71 C10
Gyula 16 E5
Gyumri 25 E5
Gyzylarbat ... 29 F6

H

Ha 'Arava → . 47 E3
Haarlem 14 B3
Haast → 65 E3
Habaswein ... 57 D7
Hachinohe 33 F12
Hadarba, Ras . 52 D6
Hadd, R'as al .. 45 F4
Hadejia 55 F7
Haden 63 A5
Hadera 46 D3
Hadhramaut =
Ḥaḍramawt . 49 D4
Ḥaḍramawt ... 49 D4
Haeju 35 C7
Haerhpin =
Harbin 35 B7
Hafar al Bāṭin . 47 E6
Hafizabad ... 42 C8
Haflong 41 E9
Haft Gel 47 E7
Hafun, Ras ... 49 E5
Hagen 14 C4
Hagerstown .. 73 E7
Hagi 32 B2
Hagondange-
Briey 13 B7
Hague, C. de la 12 B3
Hague, The = 's-
Gravenhage . 14 B3
Haguenau ... 13 B7
Hai'an 35 C7
Haifa = Ḥefa .. 46 D3

Haikou	35	D6
Ḥāʾil	47	F5
Hailar	35	B6
Haileybury	69	D3
Hailun	35	B7
Hainan □	35	E5
Haiphong	35	D5
Haiti ■	87	C5
Haiya Junction	53	E6
Hajar Bangar	53	F3
Hajdúböszörmény	16	E5
Hajnówka	17	B6
Hakken-Zan	32	B4
Hakodate	32	F12
Hala	42	G6
Ḥalab	46	C4
Ḥalabjah	46	D6
Halaib	52	D6
Halberstadt	15	C6
Halcombe	64	D6
Halcon, Mt.	38	B2
Halden	9	G10
Haldia	40	F7
Haldwani	42	E11
Halifax, Canada	69	D4
Halifax, U.K.	11	E6
Ḥalīl →	44	E4
Hall Beach	68	B2
Halle	15	C6
Hallett	62	B2
Halmahera	39	D3
Halmstad	9	G10
Halq el Oued	52	A1
Hälsingborg = Helsingborg	9	G10
Halul	44	E3
Hamada	32	B3
Hamadān	46	D7
Hamadān □	46	D7
Hamāh	46	D4
Hamamatsu	32	B5
Hamar	8	F10
Hamburg	14	B5
Hameln	14	B5
Hamersley Ra.	60	E2
Hamhung	35	C7
Hami	34	B4
Hamilton, Australia	62	C3
Hamilton, Canada	69	D3
Hamilton, N.Z.	64	B6
Hamilton, U.K.	10	D4
Hamley Bridge	62	B2
Hamlin = Hameln	14	B5
Hamm	14	C4
Hammerfest	8	D12
Hammond	72	D2
Hampden	65	F4
Hampton	73	F7
Hamrat esh Sheykh	53	F4
Han Pijesak	20	B7
Hanamaki	33	G12
Hanau	14	C5
Hancock	77	C7
Handa	49	E5
Handan	35	C6
Handeni	57	F7
Hangayn Nuruu	34	B4
Hangchou = Hangzhou	35	C7
Hangu	35	C6
Hangzhou	35	C7
Hangzhou Wan	35	C7
Ḥanīsh	49	E3
Hanle	42	C11
Hanmer Springs	65	E5
Hanna	71	C8
Hannibal	77	F9
Hannover	14	B5
Hanoi	35	D5
Hanover = Hannover	14	B5
Hanover, I.	95	G2
Hansi	42	E9
Hanson, L.	62	B2
Hantsavichy	17	B8
Hanzhong	35	C5
Haora	40	F7
Haparanda	8	E12
Happy Valley-Goose Bay	69	C4
Hapur	42	E10
Ḥaql	47	E3
Har	39	F4
Har Hu	34	C4
Har Us Nuur	34	B4
Ḥaraḍ	47	F7
Haraisan Plateau = Hurayṣān	47	G6
Harardera	49	G4
Harare	59	B6
Harazé	53	F2
Harbin	35	B7
Harbour Grace	69	D5
Hardangerfjorden	8	F9
Hardap Dam	58	C3
Hardwar = Haridwar	42	E11
Hardy, Pen.	95	H3
Harer	49	F3
Hargeisa	49	F3
Hari →	37	E2
Haridwar	42	E11
Haringhata →	41	G7
Harīrūd	45	C5
Harīrūd →	42	A2
Harlingen	78	F5
Harlow	11	F7
Harney Basin	80	D4
Härnösand	8	F11
Ḥarrat al Kishb	47	G5
Harrat al ʻUwairidh	47	F4
Harris	10	C3
Harris L.	62	B2
Harrisburg	73	D7
Harrison, C.	68	C5
Harrison Bay	71	A4
Harrogate	11	D6
Hart, L.	62	B2
Hartford	73	D9
Hartlepool	11	D6
Harvey	72	D2
Harwich	11	F7
Haryana □	42	E10
Haryn →	17	B8
Harz	14	C6
Hasa	47	F7
Hasan Kuli	29	F6
Hassi Inifel	54	C5
Hassi Messaoud	54	B6
Hastings, N.Z.	64	C7
Hastings, U.K.	11	F7
Hastings, U.S.A.	76	E5
Hastings Ra.	63	B5
Hatay = Antalya	46	C2
Hatfield P.O.	62	B3
Hatgal	34	A5
Hathras	42	F11
Hatia	41	F8
Hattah	62	B3
Hatteras, C.	75	B10
Hattiesburg	79	D9
Hatvan	16	E4
Hau Bon = Cheo Reo	36	B3
Haugesund	9	G9
Hauraki G.	64	B6
Haut Atlas	54	B3
Hautah, Wahāt al	47	G6
Hauts Plateaux	54	B4
Havana = La Habana	86	B3
Havant	11	F6
Havel →	15	B7
Havelock	65	D5
Haverfordwest	11	F4
Havlíčkův Brod	16	D2
Havre	81	A9
Havza	46	B3
Hawaii □	82	J12
Hawaiian Is.	82	H12
Hawea, L.	65	F3
Hawera	64	C6
Hawick	10	D5
Hawke B.	64	C7
Hawker	62	B2
Hawkwood	63	A5
Hay	63	B3
Hay River	70	B8
Haya	39	E3
Hayes →	71	C10
Hayrabolu	22	D6
Haysyn	17	D9
Hayvoron	17	D9
Hazaribag	40	F5
Healesville	63	C4
Hearst	69	D2
Hebei □	35	C6
Hebel	63	A4
Hebron = Al Khalīl	47	E3
Hebron	68	C4
Hechi	35	D5
Hechuan	35	C5
Heerlen	13	A6
Ḥefa	46	D3
Hefei	35	C6
Hegang	35	B8
Heidelberg	14	D5
Heilbron	59	D5
Heilbronn	14	D5
Heilongjiang □	35	B7
Heilunkiang = Heilongjiang □	35	B7
Heinze Is.	41	K11
Hejaz = Al Ḥijāz	47	F4
Hekımhan	46	C4
Hekou	34	D5
Helena, Ark., U.S.A.	79	B8
Helena, Mont., U.S.A.	81	B7
Helensville	64	B6
Helgoland	14	A4
Heligoland = Helgoland	14	A4
Heligoland B. = Deutsche Bucht	14	A5
Hellín	19	C5
Helmand □	42	D3
Helmand →	42	D2
Helmand, Hamun	45	D5
Helmsdale	10	B5
Helsingborg	9	G10
Helsingfors = Helsinki	9	F13
Helsinki	9	F13
Helwân	52	C5
Hemel Hempstead	11	F6
Henan □	35	C6
Henares →	18	B4
Henderson	75	A8
Hendon	63	A5
Hengyang	35	D6
Henrietta Maria, C.	69	C2
Hentiyn Nuruu	35	B5
Henty	63	C4
Henzada	41	J10
Herāt	42	B3
Herāt □	42	B3
Hercegnovi	22	C2
Hereford	11	E5
Herford	14	B5
Hermidale	63	B4
Hermitage	65	E4
Hermite, I.	95	H3
Hermon, Mt. = Ash Shaykh, J.	46	D3
Hermosillo	84	B2
Hernád →	16	E5
Hernandarias	94	B6
Heroica Nogales = Nogales	84	A2
Heron Bay	69	D2
Herrick	62	D4
's-Hertogenbosch	14	C3
Hesse = Hessen □	14	C5
Hessen □	14	C5
Hewett, C.	68	A4
Hexham	11	D5
Heyfield	63	C4
Heywood	62	C3
Hibbing	77	B8
Hibbs B.	62	D4
Hickory	75	B7
Hicks, Pt.	63	C4
Hida-Sammyaku	32	A5
Hidalgo del Parral	84	B3
Higashiōsaka	32	B4
High Atlas = Haut Atlas	54	B3
High Point	75	B8
High River	71	C8
High Tatra = Tatry	16	D4
High Wycombe	11	F6
Hiiumaa	24	B1
Ḥijārah, Ṣaḥrāʼ al	47	E6
Ḥijāz □	48	C2
Hijo = Tagum	38	C3
Hikone	32	B5
Hikurangi	64	A6
Hikurangi, Mt.	64	C7
Hildesheim	14	B5
Hillston	63	B4
Hilo	82	J13
Hilversum	14	B3
Himachal Pradesh □	42	D10
Himalaya	40	C5
Himatnagar	43	H8
Himeji	32	B4
Himi	32	A5
Ḥimṣ	46	D4
Hindmarsh, L.	62	C3
Hindu Kush	42	B7
Hindubagh	42	D5
Hindupur	43	N10

Hinganghat ... 43 J11
Hingoli 43 K10
Hinna = Imi ... 49 F3
Hinojosa del
  Duque 18 C3
Hirakud Dam .. 40 G4
Hiratsuka 32 B6
Hirosaki 33 F12
Hiroshima 32 B3
Hisar 42 E9
Hispaniola 87 C5
Hita 32 C2
Hitachi 32 A7
Hjälmaren 9 G11
Hlyboka 17 D7
Ho 55 G5
Ho Chi Minh
  City = Phanh
  Bho Ho Chi
  Minh ... 36 B3
Hoai Nhon 36 B3
Hoare B. 68 B4
Hobart 62 D4
Hobbs Coast .. 96 B18
Hodgson 71 C10
Hódmezővásárhely
  .... 16 E5
Hodna, Chott el 54 A5
Hodonín 16 D3
Hoek van
  Holland 14 B3
Hof 15 C6
Höfu 32 B2
Hogan Group . 62 C4
Hoggar =
  Ahaggar 54 D6
Hoh Xil Shan . 34 C3
Hohe Rhön .. 14 C5
Hohhot 35 B6
Hoi An 36 A3
Hokianga
  Harbour .. 64 A5
Hokitika 65 E4
Hokkaidō □ .. 32 F12
Holbrook 63 C4
Holguín 86 B4
Hollandia =
  Jayapura .. 39 E6
Holman Island . 70 A8
Holy Cross 71 B4
Holyhead 11 E4
Homalin 41 E10
Hombori 55 E4
Home B. 68 B4
Homer 71 C4
Homs = Ḥimṣ . 46 D4
Homyel 17 B10
Hon Chong ... 36 B2
Honan =
  Henan □ 35 C6
Honda 90 B4
Hondeklipbaai . 58 E3
Honduras ■ ... 85 E7
Honduras, G. de 85 D7
Honfleur 12 B4
Hong Kong ■ . 35 D6
Hongjiang 35 D5
Hongshui He ➙ 35 D5
Hongze Hu 35 C6
Honiara 64 K10
Honolulu 82 H12
Honshū 32 G11
Hood, Mt. 80 C3
Hooghly ➙ =
  Hughli ➙ . 40 G7
Hook of Holland
  = Hoek van
  Holland ... 14 B3
Hoorn 14 B3

Hoover Dam .. 82 B5
Hope, L. 62 A2
Hope, Pt. 71 B3
Hopedale 68 C4
Hopei =
  Hebei □ 35 C6
Hopetoun 62 C3
Hopetown 58 D4
Hoquiam 80 B2
Horlick Mts. ... 96 C1
Horlivka 25 D4
Hormoz 44 E3
Hormoz, Jaz. ye 44 E4
Hormozgān □ . 44 E3
Hormuz, Str. of 44 E4
Horn 15 D8
Horn, Cape =
  Hornos, C. de 95 H3
Hornavan 8 E11
Hornos, C. de . 95 H3
Hornsby 63 B5
Horodenka 17 D7
Horodok,
  Khmelnytskyy,
  Ukraine ... 17 D8
Horodok, Lviv,
  Ukraine ... 17 D6
Horokhiv 17 C7
Horqin Youyi
  Qianqi 35 B7
Horqueta 94 A5
Horsham 62 C3
Horton ➙ 70 B7
Hose, Gunung-
  Gunung ... 37 D4
Hoshangabad . 43 H10
Hoshiarpur ... 42 D9
Hospet 43 M10
Hospitalet de
  Llobregat ... 19 B7
Hoste, I. 95 H3
Hot Springs .. 79 B7
Hotan 34 C2
Houhora Heads 64 A5
Houlton 73 A12
Houma 79 E8
Houston 79 E6
Hovd 34 B4
Hövsgöl Nuur . 34 A5
Howard 63 A5
Howe, C. 63 C5
Howitt, L. 62 A2
Howrah = Haora 40 F7
Höxter 14 C5
Hoy 10 B5
Høyanger 8 F9
Hoyerswerda .. 15 C8
Hpungan Pass . 41 D11
Hradec Králové 16 C2
Hrodna 17 B6
Hrodzyanka ... 17 B9
Hron ➙ 16 E4
Hrvatska =
  Croatia ■ ... 20 B6
Hrymayliv 17 D8
Hsenwi 41 F11
Hsiamen =
  Xiamen ... 35 D6
Hsian = Xi'an . 35 C5
Hsinhailien =
  Lianyungang 35 C6
Hsisha Chuntao 36 A4
Hsüchou =
  Xuzhou ... 35 C6
Hua Hin 36 B1
Huacho 91 F3
Huachón 91 F3
Huai He ➙ ... 35 C6
Huainan 35 C6

Huallaga ➙ .. 90 E3
Huambo 58 A3
Huancabamba . 91 E3
Huancane 91 G5
Huancapi 91 F4
Huancavelica .. 91 F3
Huancayo 91 F3
Huang Hai =
  Yellow Sea .. 35 C7
Huang He ➙ . 35 C6
Huangliu 35 E5
Huangshi 35 C6
Huánuco 91 E3
Huaraz 91 E3
Huarmey 91 F3
Huascarán 91 E3
Huasco 94 B2
Huatabampo .. 84 B3
Huayllay 91 F3
Hubei □ 35 C6
Hubli-Dharwad
  = Dharwad . 43 M9
Huddersfield .. 11 E6
Hudiksvall 8 F11
Hudson ➙ .. 73 D8
Hudson Bay .. 68 B2
Hudson Mts. .. 96 B2
Hudson Str. ... 68 B4
Hue 36 A3
Huelva 18 D2
Huesca 19 A5
Hughenden ... 61 E7
Hughli ➙ 40 G7
Huila, Nevado
  del 90 C3
Huinca Renancó 94 C4
Huize 34 D5
Hukawng Valley 41 D11
Hulan 35 B7
Ḥulayfā' 47 F5
Huld 35 B5
Hull = Kingston
  upon Hull ... 11 E6
Hull 69 D3
Hulun Nur 35 B6
Humahuaca ... 94 A3
Humaitá, Brazil 91 E6
Humaitá,
  Paraguay ... 94 B5
Humber ➙ ... 11 E6
Humboldt 71 C9
Humboldt ➙ . 80 F4
Hume, L. 63 C4
Humenné 16 D5
Hūn 52 C2
Hunan □ 35 D6
Hunedoara ... 17 F6
Hungary ■ 16 E4
Hungerford ... 63 A3
Hŭngnam 35 C7
Hunsrück 14 D4
Hunter I. 62 D3
Hunter Ra. ... 63 B5
Hunterville 64 C6
Huntington ... 72 E4
Huntington
  Beach 82 D3
Huntly, N.Z. ... 64 B6
Huntly, U.K. ... 10 C5
Huntsville 74 B4
Huonville 62 D4
Hupeh =
  Hubei □ 35 C6
Huraysān 47 G6
Huron, L. 72 B4
Hurunui ➙ ... 65 E5
Huși 17 E9
Hutchinson .. 78 A5
Hutton, Mt. .. 63 A4
Hvar 20 C6

Hwang Ho =
  Huang He ➙ 35 C6
Hwange 59 B5
Hyargas Nuur . 34 A4
Hyderabad,
  India 43 L11
Hyderabad,
  Pakistan 42 G6
Hyères 13 E7
Hyères, Is. d' .. 13 E7

I

I-n-Gall 55 E6
Iaco ➙ 91 E5
Ialomiţa ➙ ... 22 B6
Iaşi 17 E8
Iba 38 A2
Ibadan 55 G5
Ibagué 90 C3
Ibar ➙ 22 C3
Ibarra 90 C3
Ibi 55 G6
Ibiá 93 F4
Ibicuy 94 C5
Ibioapaba, Sa.
  da 92 C5
Ibiza 19 C6
Ibonma 39 E4
Ibotirama 93 E5
Ibu 39 D3
Icá 91 F3
Içá ➙ 90 D5
Içana 90 C5
İçel = Mersin .. 46 C3
Iceland ■ 8 B4
Ich'ang =
  Yichang ... 35 C6
Ichchapuram .. 40 H5
Ichihara 32 B7
Ichikawa 32 B6
Ichilo ➙ 91 G6
Ichinomiya ... 32 B5
Idaho □ 81 C6
Idaho Falls 81 D7
Idar-Oberstein . 14 D4
Idd el Ghanam . 53 F3
Iddan 49 F4
Idehan 52 C1
Idehan Marzūq 52 D1
Idelès 54 D6
Idfû 52 D5
Ídhi Óros 23 G5
Ídhra 23 F4
Idi 36 C1
Idiofa 56 E3
Idlip 46 D4
Ierápetra 23 G5
Iesi 20 C4
Ifanadiana 59 J9
Ife 55 G5
Ifni 54 C2
Iforas, Adrar des 54 D5
Igarapava 93 G4
Igarapé Açu ... 92 C4
Igarka 28 C9
Igbetti 55 G5
Iglésias 21 E2
Igli 54 B4
Igloolik 68 B2
Iğneada Burnu 22 D7
Igoumenítsa .. 23 E3
Iguaçu 94 B6
Iguaçu, Cat. del 94 B6
Iguaçu Falls =
  Iguaçu, Cat.
  del 94 B6

Iguala	84	D5
Igualada	19	B6
Iguassu =		
Iguaçu →	94	B6
Iguatu	92	D6
Iguéla	56	E1
Ihosy	59	J9
Iida	32	B5
Iisalmi	8	F13
Iizuka	32	C2
Ijebu-Ode	55	G5
IJsselmeer	14	B3
Ikaría	23	F6
Ikela	56	E4
Iki	32	C1
Ilagan	38	A2
Ilām	46	D6
Iława	16	B4
Île-de-France	12	B5
Ilebo	56	E4
Ilhéus	93	E6
Ili →	29	E8
Ilich	29	E7
Iligan	38	C2
Illampu =		
Ancohuma,		
Nevada	91	G5
Illana B.	38	C2
Illapel	94	C2
Iller →	14	D5
Illimani	91	G5
Illium = Troy	23	E6
Ilmen, Ozero	24	B3
Ilo	91	G4
Iloilo	38	B2
Ilorin	55	G5
Ilwaki	39	F3
Imabari	32	B3
Imandra, Ozero	28	C4
Imari	32	C1
imeni 26		
Bakinskikh		
Komissarov =		
Neftçala	46	C7
Imeri, Serra	90	C5
Imi	49	F3
Imola	20	B3
Imperatriz	92	D4
Impéria	20	C2
Imperial Dam	82	D5
Impfondo	56	D3
Imphal	41	E9
İmroz =		
Gökçeada	23	D5
Imuruan B.	38	B1
In Belbel	54	C5
In Salah	54	C5
Ina	32	B5
Inangahua		
Junction	65	D4
Inanwatan	39	E4
Iñapari	91	F5
Inarijärvi	8	E13
Inca	19	C7
İnce-Burnu	46	B3
Inchon	35	C7
İncirliova	23	F6
Incomáti →	59	D6
Indalsälven →	8	F11
Indaw	41	E11
Independence,		
Kans., U.S.A.	79	A6
Independence,		
Mo., U.S.A.	77	F7
India ■	43	J10
Indian →	75	F7
Indian Harbour	68	C5
Indian Head	71	C9
Indiana □	72	D3
Indianapolis	72	E2
Indiga	28	C5
Indigirka →	31	B12
Indonesia ■	37	E4
Indore	43	H9
Indravati →	40	H3
Indre →	12	C4
Indus →	43	G5
İnebolu	46	B3
Ingende	56	E3
Inglewood,		
Queens.,		
Australia	63	A5
Inglewood, Vic.,		
Australia	62	C3
Inglewood, N.Z.	64	C6
Ingolstadt	15	D6
Ingraj Bazar	40	E7
Ingrid		
Christensen		
Coast	96	A10
Ingulec =		
Inhulec	25	D3
Inhambane	59	C7
Inhaminga	59	B7
Inharrime	59	C7
Inhulec	25	D3
Ining = Yining	34	B3
Inírida →	90	C5
Injune	63	A4
Inle L.	41	G11
Inn →	15	D7
Innamincka	62	A3
Inner Hebrides	10	C3
Inner Mongolia		
= Nei		
Monggol		
Zizhiqu □	35	B6
Innisfail	71	C8
Innsbruck	15	E6
Inongo	56	E3
Inoucdjouac =		
Inukjuak	68	C3
Inowrocław	16	B4
Inquisivi	91	G5
Insein	41	J11
Interlaken	13	C7
International		
Falls	77	A8
Intiyaco	94	B4
Inukjuak	68	C3
Inútil, B.	95	G2
Inuvik	70	B6
Invercargill	65	G3
Inverell	63	A5
Invergordon	10	C4
Inverness	10	C4
Inverurie	10	C5
Investigator		
Group	62	B1
Investigator Str.	62	C2
Inya	29	D9
Inza	24	C6
Ioánnina	23	E3
Iona	77	D8
Iowa City	77	E9
Ipameri	93	F4
Ipatinga	93	F5
Ipiales	90	C3
Ipin = Yibin	35	D5
Ipixuna	91	E4
Ipoh	36	D2
Ippy	56	C4
İpsala	22	D6
Ipswich,		
Australia	63	A5
Ipswich, U.K.	11	E7
Ipu	92	C5
Iqaluit	68	B4
Iquique	91	H4
Iquitos	90	D4
Iracoubo	92	A3
Iráklion	23	G5
Iran ■	44	C3
Iran, Gunung-		
Gunung	37	D4
Iran Ra. = Iran,		
Gunung-		
Gunung	37	D4
Īrānshahr	45	E5
Irapuato	84	C4
Iraq ■	46	D5
Irbid	46	D3
Irebu	56	E3
Ireland ■	11	E3
Irian Jaya □	39	E5
Iringa	57	F7
Iriri →	92	C3
Irish Sea	11	E4
Irkineyeva	30	D7
Irkutsk	30	D8
Iron Baron	62	B2
Iron Gate =		
Portile de Fier	22	B4
Iron Knob	62	B2
Ironwood	77	B9
Irpin	17	C10
Irrara Cr. →	63	A4
Irrawaddy □	41	J10
Irrawaddy →	41	K10
Irtysh →	28	C7
Irumu	57	D5
Irún	19	A5
Irunea =		
Pamplona	19	A5
Irvine	10	D4
Irymple	62	B3
Isabella	38	C2
Isangi	56	D4
Isar →	15	D7
Íschia	21	D4
Ise	32	B5
Ise-Wan	32	B5
Isère →	13	D6
Isérnia	20	D5
Ishikari-Wan	32	F12
Ishim →	29	D8
Ishinomaki	33	G12
Ishkuman	42	A8
Isil Kul	29	D8
Isiolo	57	D7
Isiro	57	D5
İskenderun	46	C4
İskŭr →	22	C5
Islamabad	42	C8
Island L.	71	C10
Island Lagoon	62	B2
Islay	10	D3
Isle →	12	D3
Isle of Wight □	11	F6
Ismail = İzmayil	17	F9
Ismâ'ilîya	47	E3
Isna	52	C5
İsparta	46	C2
İspica	21	F5
Israel ■	46	E3
Issoire	13	D5
Issyk-Kul, Ozero		
= Ysyk-Köl,		
Ozero	29	E8
İstanbul	22	D7
Istiaia	23	E4
Istra	20	B4
İstranca Dağları	22	D6
Istres	13	E6
Istria = Istra	20	B4
Itabaiana	92	D6
Itaberaba	93	E5
Itabira	93	F5
Itabuna	93	E6
Itaipú, Reprêsa		
de	94	B6
Itaituba	92	C2
Itajaí	94	B7
Italy ■	20	C4
Itapecuru-Mirim	92	C5
Itaperuna	93	G5
Itapicuru →,		
Bahia, Brazil	93	E6
Itapicuru →,		
Maranhão,		
Brazil	92	C5
Itapipoca	92	C6
Itaquatiara	90	D7
Itaquí	94	B5
Itatuba	91	E6
Ithaca = Itháki	23	E3
Itháki	23	E3
Ito	32	B6
Itonamas →	91	F6
Ituaçu	93	E5
Ituiutaba	93	F4
Itumbiara	93	F4
Iturbe	94	A3
Itzehoe	14	B5
Ivanava	17	B7
Ivanhoe	63	B3
Ivano-Frankivsk	17	D7
Ivano-Frankovsk		
= Ivano-		
Frankivsk	17	D7
Ivanovo =		
Ivanava	17	B7
Ivanovo	24	B5
Ivatsevichy	17	B7
Ivory Coast ■	55	G3
Ivrea	20	B1
Ivujivik	68	B3
Iwahig	36	C5
Iwaki	33	G12
Iwakuni	32	B3
Iwata	33	H11
Iwate-San	33	G12
Iwo	55	G5
Ixiamas	91	F5
Izhevsk	28	D6
İzmayil	17	F9
İzmir	23	E6
İzmit	25	E2
İznik Gölü	23	D7
Izumi-sano	32	B4
Izumo	32	B3
Izyaslav	17	C8

J

Jabalpur	43	H11
Jablah	46	D4
Jablanica	22	D3
Jablonec	16	C2
Jaboatão	92	D6
Jaca	19	A5
Jacareí	94	A7
Jacarèzinho	94	A7
Jackson,		
Australia	63	A4
Jackson, Mich.,		
U.S.A.	72	C3
Jackson, Miss.,		
U.S.A.	79	C8

Jackson, *Tenn.*,
U.S.A. 74 B3
Jackson B. 65 E3
Jacksons 65 E4
Jacksonville,
Fla., U.S.A. .. 75 D7
Jacksonville, *Ill.*,
U.S.A. 77 F9
Jacksonville,
N.C., U.S.A. . 75 B9
Jacmel 87 C5
Jacobabad ... 42 E6
Jacobina 93 E5
Jacundá -→ ... 92 C3
Jadotville =
Likasi 57 G5
Jādū 52 B1
Jaén, *Peru* ... 91 E3
Jaén, *Spain* .. 18 D4
Jaffa = Tel Aviv-
Yafo 46 D3
Jaffa, C. 62 C2
Jaffna 43 Q12
Jagadhri 42 D10
Jagdalpur 40 H3
Jagraon 42 D9
Jagtial 43 K11
Jaguariaíva .. 94 A7
Jaguaribe -→ . 92 C6
Jahrom 44 D3
Jailolo 39 D3
Jailolo, Selat .. 39 D3
Jaipur 42 F9
Jakarta 37 F3
Jalalabad 42 B7
Jalapa Enríquez 84 D5
Jalas, Jabal al . 47 F4
Jalgaon,
Maharashtra,
India 43 J10
Jalgaon,
Maharashtra,
India 43 J9
Jalna 43 K9
Jalón -→ 19 B5
Jalpaiguri 41 D7
Jalq 45 E5
Jaluit I. 64 J11
Jamaica ■ 86 C4
Jamalpur,
Bangla. 41 E7
Jamalpur, *India* 40 E6
Jamanxim -→ . 92 C2
Jambe 39 E4
Jambi 37 E2
Jambi □ 37 E2
James -→ 76 D6
James B. 69 C2
James Ross I. . 96 A4
Jamestown,
Australia 62 B2
Jamestown,
N. Dak., U.S.A. 76 B5
Jamestown,
N.Y., U.S.A. . 73 C6
Jamkhandi 43 L9
Jammu 42 C9
Jammu &
Kashmir □ .. 42 B10
Jamnagar 43 H7
Jamrud 42 C7
Jamshedpur .. 40 F6
Jand 42 C8
Jandaq 44 C3
Jandowae 63 A5
Janesville 77 D10
Januária 93 F5
Jaora 43 H9
Japan ■ 33 G11

Japan, Sea of . 33 G11
Japen = Yapen 39 E5
Japurá -→ 90 D5
Jaque 90 B3
Jarama -→ ... 18 B4
Jargalant =
Hovd 34 B4
Jarosław 17 C6
Jarso 53 G6
Jarvis I. 65 K15
Jarwa 40 D4
Jäsk 44 E4
Jasło 16 D5
Jasper 74 C4
Jászberény ... 16 E4
Jataí 93 F3
Játiva 19 C5
Jaú 93 G4
Jauja 91 F3
Jaunpur 40 E4
Java = Jawa .. 37 F3
Java Sea 37 E3
Javhlant =
Ulyasutay ... 34 B4
Jawa 37 F3
Jaya, Puncak . 39 E5
Jayanti 41 D7
Jayapura 39 E6
Jayawijaya,
Pegunungan 39 E5
Jaynagar 40 D6
Jean Marie
River 70 B7
Jebba 55 G5
Jebel, Bahr
el -→ 53 G5
Jedburgh 10 D5
Jedda = Jiddah 47 G4
Jędrzejów 16 C5
Jefferson City . 77 F8
Jega 55 F5
Jelenia Góra .. 16 C2
Jelgava 24 B1
Jemaja 37 D3
Jembongan .. 36 C5
Jena 15 C6
Jeparit 62 C3
Jequié 93 E5
Jequitinhonha . 93 F5
Jequitinhonha -→
.... 93 F6
Jerada 54 B4
Jerantut 37 D2
Jérémie 86 C5
Jerez de la
Frontera 18 D2
Jerez de los
Caballeros .. 18 C2
Jerilderie 63 C4
Jersey 11 G5
Jersey City ... 73 D8
Jerusalem ... 47 E3
Jervis B. 63 C5
Jesselton =
Kota Kinabalu 36 C5
Jessore 41 F7
Jeypore 40 H4
Jhal Jhao 42 F4
Jhalawar 43 G10
Jhang Maghiana 42 D8
Jhansi 43 G11
Jharsaguda ... 40 G5
Jhelum 42 C8
Jhelum -→ 42 D8
Jhunjhunu ... 42 E9
Jiamusi 35 B8
Ji'an 35 D6
Jianchuan 34 D4
Jiangling 35 C6

Jiangmen 35 D6
Jiangsu □ 35 C6
Jiangxi □ 35 D6
Jian'ou 35 D6
Jianshui 34 D5
Jiao Xian 35 C7
Jiaozuo 35 C6
Jiaxing 35 C7
Jiayi 35 D7
Jibuti =
Djibouti ■ .. 49 E3
Jiddah 47 G4
Jido 41 C10
Jihlava 16 D2
Jihlava -→ ... 16 D3
Jijel 54 A6
Jijiga 49 F3
Jilin 35 B7
Jilong 35 D7
Jima 53 G6
Jiménez 84 B4
Jinan 35 C6
Jindabyne 63 C4
Jindřichův
Hradeç 16 D2
Jingdezhen ... 35 D6
Jinggu 34 D5
Jingxi 35 D5
Jinhua 35 D6
Jining,
Nei Mongol Zizhiqu,
China 35 B6
Jining,
Shandong,
China 35 C6
Jinja 57 D6
Jinnah Barrage 42 C7
Jinotega 85 E7
Jinshi 35 D6
Jinzhou 35 B7
Jiparaná -→ .. 91 E6
Jipijapa 90 D2
Jisr ash
Shughūr 46 D4
Jiu -→ 22 C4
Jiujiang 35 D6
Jiuquan 34 C4
Jixi 35 B8
Jizzakh 29 E7
Joaçaba 94 B6
João Pessoa .. 92 D7
Joaquín V.
González 94 B4
Jodhpur 42 F8
Jofane 59 C6
Jogjakarta =
Yogyakarta .. 37 F4
Johannesburg . 59 D5
Johnson City .. 75 A6
Johnston Falls
= Mambilima
Falls 57 G5
Johnstown ... 73 D6
Johor Baharu . 37 D2
Joinvile 94 B7
Joinville I. 96 A4
Joliet 72 D1
Joliette 69 D3
Jolo 38 C2
Jome 39 E3
Jonesboro ... 79 B8
Jönköping 9 G10
Jonquière 69 D3
Joplin 79 A6
Jordan ■ 47 E4
Jorhat 41 D10
Jorong 37 E4
Jos 55 G6
Jowzjān □ 45 B6

Juan de Fuca
Str. 80 A1
Juan de Nova . 59 H8
Juan Fernández,
Arch. de 89 G3
Juárez 94 D5
Juàzeiro 93 D5
Juàzeiro do
Norte 92 D6
Jubbulpore =
Jabalpur 43 H11
Juby, C. 54 C2
Júcar -→ 19 C5
Juchitán 85 D5
Jugoslavia =
Yugoslavia ■ 22 B3
Juiz de Fora .. 93 G5
Juli 91 G5
Juliaca 91 G4
Jullundur 42 D9
Jumilla 19 C5
Jumla 40 C4
Jumna =
Yamuna -→ . 40 E3
Junagadh 43 J7
Jundiaí 94 A7
Juneau 71 C6
Junee 63 B4
Jungfrau 13 C7
Junggar Pendi . 34 B3
Junín 94 C4
Junín de los
Andes 95 D2
Jūniyah 46 D3
Jur, Nahr el -→ 53 G4
Jura = Jura,
Mts. du 13 C7
Jura =
Schwäbische
Alb 14 D5
Jura -→ 10 D4
Jura, Mts. du .. 13 C7
Jurado 90 B3
Juruá -→ 90 D5
Juruena -→ ... 91 E7
Juruti 92 C2
Justo Daract .. 94 C3
Jutland =
Jylland 9 G9
Juventud, I. de
la 86 B3
Juwain 42 D2
Jylland 9 G9
Jyväskylä 8 F13

K

K2 .... 42 B10
Kaapkruis 58 C2
Kaapstad =
Cape Town . 58 E3
Kabaena 39 F2
Kabala 55 G2
Kabale 57 E6
Kabalo 57 F5
Kabambare ... 57 E5
Kabanjahe ... 37 D1
Kabara 55 E4
Kabardino-
Balkar
Republic =
Kabardino
Balkaria □ . 25 E5
Kabardino
Balkaria □ . 25 E5
Kabare 39 E4

Kabarega Falls	57	D6
Kabasalan	38	C2
Kabba	55	G6
Kabīr, Zab al →	46	C5
Kabīr Kūh	47	D6
Kabkabiyah	53	F3
Kabompo →	58	A4
Kabongo	57	F5
Kabūd Gonbad	44	B4
Kābul	42	B6
Kābul □	42	B6
Kabul →	42	C8
Kaburuang	39	D3
Kabwe	59	A5
Kachchh, Gulf of	43	H6
Kachchh, Rann of	43	G6
Kachin □	41	D11
Kachiry	29	D8
Kaçkar	46	B5
Kadan Kyun	36	B1
Kadina	62	B2
Kadirli	46	C4
Kadiyevka = Stakhanov	25	D4
Kadoma	59	B5
Kâdugli	53	F4
Kaduna	55	F6
Kaédi	55	E2
Kaelé	53	F1
Kaesŏng	35	C7
Kāf	47	E4
Kafakumba	56	F4
Kafan = Kapan	25	F6
Kafanchan	55	G6
Kaffrine	55	F1
Kafia Kingi	53	G3
Kafirévs, Ákra	23	E5
Kafue →	59	B5
Kafulwe	57	F5
Kaga Bandoro	56	C3
Kagan	29	F7
Kağizman	46	B5
Kagoshima	33	D2
Kagoshima-Wan	33	D2
Kagul = Cahul	17	F9
Kahama	57	E6
Kahayan →	37	E4
Kahemba	56	F3
Kahnūj	44	E4
Kahoolawe	82	H12
Kahramanmaraş	46	C4
Kai, Kepulauan	39	F4
Kai Besar	39	F4
Kai Is. = Kai, Kepulauan	39	F4
Kai Kecil	39	F4
Kaiama	55	G5
Kaiapoi	65	E5
Kaieteur Falls	90	B7
Kaifeng	35	C6
Kaikohe	64	A5
Kaikoura	65	E5
Kaikoura Ra.	65	D5
Kaimana	39	E4
Kaimanawa Mts.	64	C6
Kaingaroa Forest	64	C7
Kainji Res.	55	F5
Kaipara Harbour	64	B6
Kaironi	39	E4
Kairouan	54	A7
Kaiserslautern	14	D4
Kaitaia	64	A5
Kaitangata	65	G3
Kajaani	8	F13
Kajabbi	61	D6
Kajana = Kajaani	8	F13
Kajo Kaji	53	H5
Kaka	53	F5
Kakamas	58	D4
Kakamega	57	D6
Kakanui Mts.	65	F4
Kakhovka	25	D3
Kakhovske Vdskh.	25	D3
Kakinada	40	J4
Kalabagh	42	C7
Kalabahi	39	F2
Kalabo	58	A4
Kalach	24	C5
Kaladan →	41	G9
Kalahari	58	C4
Kalakan	31	D9
Kalámai	23	F4
Kalamata = Kalámai	23	F4
Kalamazoo	72	C3
Kalan	46	C4
Kalao	39	F2
Kalaotoa	39	F2
Kalat	42	E5
Kalce	20	B5
Kale	23	F7
Kalegauk Kyun	41	K11
Kalemie	57	F5
Kalewa	41	F10
Kalgan = Zhangjiakou	35	B6
Kalgoorlie-Boulder	60	G3
Kaliakra, Nos	22	C7
Kalianda	37	F3
Kalibo	38	B2
Kalima	57	E5
Kalimantan	37	E4
Kalimantan Barat □	37	D4
Kalimantan Selatan □	37	E5
Kalimantan Tengah □	37	E4
Kalimantan Timur □	37	D5
Kálimnos	23	F6
Kalinin = Tver	24	B4
Kaliningrad, Kaliningd., Russia	24	C1
Kaliningrad, Moskva, Russia	24	B4
Kalinkavichy	17	B9
Kalinkovichi = Kalinkavichy	17	B9
Kalispell	81	A6
Kalisz	16	C4
Kaliua	57	F6
Kalkrand	58	C3
Kalmar	9	G11
Kalmyk Republic = Kalmykia □	25	D6
Kalmykia □	25	D6
Kalocsa	16	E4
Kalomo	59	B5
Kaluga	24	C4
Kalush	17	D7
Kalutara	43	R11
Kama →	29	D6
Kamaishi	33	G12
Kamaran	49	D3
Kamchatka, P-ov.	31	D13
Kamchatka Pen. = Kamchatka, P-ov.	31	D13
Kamchiya →	22	C6
Kamen	29	D9
Kamenjak, Rt.	20	B4
Kamenka Bugskaya = Kamyanka-Buzka	17	C7
Kamensk Uralskiy	29	D7
Kamin-Kashyrskyy	17	C7
Kamina	57	F5
Kamloops	71	C7
Kampala	57	D6
Kampar →	37	D2
Kampen	14	B3
Kampot	36	B2
Kampuchea = Cambodia ■	36	B3
Kampung →	39	F5
Kampungbaru = Tolitoli	39	D2
Kamrau, Teluk	39	E4
Kamsack	71	C9
Kamyanets-Podilskyy	17	D8
Kamyanka-Buzka	17	C7
Kamyshin	24	C6
Kanaaupscow	69	C3
Kananga	56	F4
Kanash	24	B6
Kanastraíon, Ákra = Palioúrion, Ákra	23	E4
Kanazawa	33	A5
Kanchanaburi	36	B1
Kanchenjunga	40	D7
Kanchipuram	43	N11
Kanda Kanda	56	F4
Kandahar = Qandahār	42	D4
Kandalaksha	28	C4
Kandalu	42	E3
Kandangan	37	E5
Kandi	55	F5
Kandla	43	H7
Kandos	63	B4
Kandy	43	R12
Kangaroo I.	62	C2
Kangāvar	46	D7
Kangean, Kepulauan	37	F5
Kangean Is. = Kangean, Kepulauan	37	F5
Kangiqsualujjuaq		
Kangiqsujuaq	68	C4
Kangirsuk	68	C4
Kangnŭng	35	C7
Kango	56	D2
Kangto	41	D9
Kaniapiskau →	68	C4
Kaniapiskau L.	69	C4
Kanin, Poluostrov	28	C5
Kanin Nos, Mys	28	C5
Kanin Pen. = Kanin, Poluostrov	28	C5
Kaniva	62	C3
Kankakee	72	D2
Kankan	55	F3
Kankendy = Xankändi	25	F6
Kanker	40	G3
Kannapolis	75	B7
Kannauj	42	F11
Kannod	43	H10
Kano	55	F6
Kanowit	37	D4
Kanoya	33	D2
Kanpetlet	41	G9
Kanpur	40	D3
Kansas □	76	F5
Kansas City, Kans., U.S.A.	77	F7
Kansas City, Mo., U.S.A.	77	F7
Kansk	30	D7
Kansu = Gansu □	34	C5
Kantang	36	C1
Kanuma	32	A6
Kanye	59	C5
Kaohsiung = Gaoxiong	35	D7
Kaokoveld	58	B2
Kaolack	55	F1
Kapan	25	F6
Kapanga	56	F4
Kapchagai = Qapshaghay	29	E8
Kapfenberg	15	E8
Kapiri Mposhi	59	A5
Kāpīsā □	45	C7
Kapit	37	D4
Kapoeta	53	H5
Kaposvár	16	E3
Kapuas →	37	E3
Kapuas Hulu, Pegunungan	37	D4
Kapuas Hulu Ra. = Kapuas Hulu, Pegunungan	37	D4
Kapunda	62	B2
Kapuni	64	C6
Kaputar	63	B5
Kara	28	C7
Kara Bogaz Gol, Zaliv = Garabogazköl Aylagy	29	E6
Kara Kalpak Republic □ = Karakalpakstan □	29	E6
Kara Kum	29	F6
Kara Sea	28	B8
Karabiğa	23	D6
Karabük	46	B3
Karaburun	23	E6
Karabutak = Qarabutaq	29	E7
Karacabey	23	D7
Karacasu	23	F7
Karachi	43	G5
Karad	43	L9
Karadeniz Boğazı	22	D7
Karaganda = Qaraghandy	29	E8
Karagayly	29	E8
Karaikal	43	P11
Karaikkudi	43	P11
Karaj	44	C2
Karakalpakstan □	29	E6
Karakas	29	E9
Karakelong	38	D3
Karakitang	39	D3
Karaklis = Vanadzor	25	E5
Karakoram Pass	42	B10
Karakoram Ra.	42	B10
Karalon	30	D9
Karaman	46	C3

Karamay

Karamay 34 B3
Karambu 37 E5
Karamea Bight 65 D4
Karasburg 58 D3
Karasino 28 C9
Karasuk 29 D8
Karatau =
  Qarataū 29 E8
Karatau, Khrebet 29 E7
Karawanken ... 20 A5
Karazhal 29 E8
Karbalā 47 D6
Karcag 16 E5
Kardhítsa 23 E3
Karelia □ 28 C4
Karelian
  Republic □ =
  Karelia □ ... 28 C4
Kargänrüd 46 C7
Kargasok 29 D9
Kargat 29 D9
Kargil 42 B10
Kariba, L. 59 B5
Kariba Dam .. 59 B5
Kariba Gorge . 59 B5
Karibib 58 C3
Karimata,
  Kepulauan .. 37 E3
Karimata, Selat 37 E3
Karimata Is. =
  Karimata,
  Kepulauan .. 37 E3
Karimnagar ... 43 K11
Karimunjawa,
  Kepulauan .. 37 F4
Karin 49 E4
Karkaralinsk =
  Qarqaraly ... 29 E8
Karkinitska
  Zatoka 25 D3
Karkinitskiy Zaliv
  = Karkinitska
  Zatoka 25 D3
Karl-Marx-Stadt
  = Chemnitz .. 15 C7
Karlovac 20 B5
Karlovo 22 C5
Karlovy Vary .. 16 C1
Karlsbad =
  Karlovy Vary 16 C1
Karlskrona 9 G11
Karlsruhe 14 D5
Karlstad 9 G10
Karnal 42 E10
Karnali ─► 40 C3
Karnaphuli Res. 41 F9
Karnataka □ .. 43 N10
Karnische Alpen 20 A4
Kärnten □ 15 E7
Karonga 57 F6
Karoonda 62 C2
Karora 53 E6
Kárpathos 23 G6
Kars 46 B5
Karsakpay 29 E7
Karshi = Qarshi 29 F7
Karsun 24 C6
Karufa 39 E4
Karungu 57 E6
Karviná 16 D4
Karwar 43 M9
Kasai ─► 56 E3
Kasama 57 G6
Kasanga 57 F6
Kasangulu ... 56 E3
Kasaragod 43 N9
Kasba L. 70 B9
Kasempa 59 A5
Kasenga 57 G5

Kāshān 44 C2
Kashi 34 C2
Kashk-e Kohneh 42 B3
Kāshmar 44 C4
Kashun Noerh =
  Gaxun Nur .. 34 B5
Kasimov 24 C5
Kasiruta 39 E3
Kasongo 57 E5
Kasongo Lunda 56 F3
Kásos 23 G6
Kassalâ 53 E6
Kassel 14 C5
Kassue 39 F5
Kastamonu ... 46 B3
Kasulu 57 E6
Kasur 42 D9
Katako Kombe . 56 E4
Katamatite 63 C4
Katanga =
  Shaba □ 57 F4
Katangi 43 J11
Kateríni 23 D4
Katha 41 E11
Kathiawar 43 H7
Katihar 40 E6
Katima Mulilo . 58 B4
Katingan =
  Mendawai ─► 37 E4
Katiola 55 G3
Katmandu 40 D5
Katoomba 63 B5
Katowice 16 C4
Katsina 55 F6
Kattegat 9 G10
Kauai 82 G11
Kaunas 24 C1
Kaura Namoda 55 F6
Kavála 22 D5
Kaw 92 B3
Kawagoe 32 B6
Kawaguchi ... 32 B6
Kawambwa ... 57 F5
Kawardha 40 G3
Kawasaki 32 B6
Kawerau 64 C7
Kawhia Harbour 64 C6
Kawio,
  Kepulauan .. 38 D3
Kawnro 41 F12
Kawthoolei =
  Kawthule □ . 41 H11
Kawthule □ ... 41 H11
Kaya 55 F4
Kayah □ 41 H11
Kayan ─► 37 D5
Kayeli 39 E3
Kayes 55 F2
Kayoa 39 D3
Kayrunnera ... 62 B3
Kayseri 46 C3
Kayuagung ... 37 E2
Kazachye 31 B11
Kazakstan ■ .. 29 E7
Kazan 24 B6
Kazanlŭk 22 C5
Kazatin =
  Kozyatyn ... 17 D9
Kāzerün 44 D2
Kazumba 56 F4
Kazym ─► 28 C7
Ké-Macina 55 F3
Kéa 23 F5
Kebnekaise .. 8 E11
Kebri Dehar ... 49 F3
Kecskemét 16 E4
Kediri 37 F4
Kédougou 55 F2
Keetmanshoop 58 D3

Keewatin □ ... 70 B10
Kefallinía 23 E3
Kefamenanu .. 39 F2
Keffi 55 G6
Keighley 11 E6
Keith 62 C3
Keith Arm 70 B7
Kekri 42 G9
Kël 30 C10
Kelang 37 D2
Kelibia 52 A1
Kellé 56 E2
Kells =
  Ceanannus
  Mor 11 E3
Kélo 53 G2
Kelowna 71 D8
Kelso 65 F3
Keluang 37 D2
Kem 28 C4
Kema 39 D3
Kemah 46 C4
Kemerovo 29 D9
Kemi 8 E12
Kemi älv ─► .. 8 E12
Kemijoki ─► . 8 E12
Kemijoki ─► .. 8 E12
Kemp Land ... 96 A9
Kempsey 63 B5
Kempten 14 E6
Kendal 37 F4
Kendall 63 B5
Kendari 39 E2
Kendawangan . 37 E4
Kende 55 F5
Kendrapara ... 40 G6
Kenema 55 G2
Keng Tawng .. 41 G12
Keng Tung ... 41 G12
Kenge 56 E3
Kenhardt 58 D4
Kenitra 54 B3
Kennedy
  Taungdeik .. 41 F9
Kennewick .. 80 B4
Kenogami ─► . 69 C2
Kenosha 72 C2
Kent Group ... 62 C4
Kent Pen. 70 B9
Kentau 29 E7
Kentucky □ ... 72 F3
Kentville 69 D4
Kenya ■ 57 D7
Kenya, Mt. ... 57 E7
Kepi 39 F5
Kerala □ 43 P10
Kerang 62 C3
Kerch 25 D4
Kerchoual 55 E5
Keren 53 E6
Kericho 57 E7
Kerinci 37 E2
Kerki 29 F7
Kérkira 23 E2
Kermadec Is. .. 64 M13
Kermadec
  Trench 65 N13
Kermān 44 D4
Kermān □ 44 D4
Kermānshāh =
  Bākhtarān .. 46 D6
Kerme Körfezi . 23 F6
Kerrobert 71 C9
Kerulen ─► ... 35 B6
Kerzaz 54 C4
Keşan 22 D6
Kestenga 28 C4
Ket ─► 29 D9
Keta 55 G5

Ketapang 37 E4
Ketchikan 71 C6
Kętrzyn 16 A5
Keweenaw B. . 69 D2
Key West 86 B3
Khabarovo ... 28 C7
Khabarovsk .. 31 E11
Khābūr ─► 46 D5
Khairpur 42 F6
Khakassia □ .. 30 D6
Khakhea 58 C4
Khalkhāl 46 C7
Khalkís 23 E4
Khalmer-Sede =
  Tazovskiy ... 28 C8
Khalmer Yu ... 28 C7
Khalturin 24 B6
Khalūf 48 C6
Khambat, G. of 43 J8
Khambhat 43 H8
Khamir 49 D3
Khānābād 45 B7
Khānaqīn 46 D6
Khandwa 43 J10
Khanewal 42 D7
Khaniá 23 G5
Khaníon, Kólpos 23 G4
Khankendy =
  Xankändi ... 25 F6
Khanty-
  Mansiysk ... 29 C7
Khapcheranga . 30 E9
Kharagpur ... 40 F6
Kharan Kalat . 42 E4
Khārānaq 44 C3
Kharda 43 K9
Khârga, El
  Wâhât el ... 52 C5
Khargon 43 J9
Khārk, Jazireh . 47 E7
Kharkiv 24 D4
Kharkov =
  Kharkiv ... 24 D4
Kharovsk 24 B5
Kharsānīya ... 47 F7
Kharta 22 D7
Khartoum = El
  Khartûm ... 53 E5
Khāsh 42 E2
Khashm el Girba 53 F6
Khaskovo 22 D5
Khatanga 30 B8
Khatanga ─► . 30 B8
Khaybar, Harrat 47 F5
Khed Brahma . 43 G8
Khemmarat ... 36 A3
Khenchela ... 54 A6
Khenifra 54 B3
Kherson 25 D3
Kheta ─► 30 B8
Khilok 30 D9
Khíos 23 E6
Khiuma =
  Hiiumaa ... 24 B1
Khiva 29 E7
Khīyāv 46 C6
Khmelnik 17 D8
Khmelnitskiy =
  Khmelnytskyy 17 D8
Khmelnytskyy . 17 D8
Khmer Rep. =
  Cambodia ■ . 36 B3
Khodoriv 17 D7
Khodzent =
  Khudzhand .. 29 E7
Khojak P. 42 D5
Kholm, Afghan. 45 B7
Kholm, Russia . 24 B3
Khomeyn 44 C2

Khon Kaen 36 A2
Khoper → 29 E5
Khóra Sfakíon . 23 G5
Khorat =
  Nakhon
  Ratchasima . 36 B2
Khorramābād . 46 D7
Khorrāmshahr . 47 E7
Khotyn 17 D8
Khouribga 54 B3
Khowai 41 E8
Khoyniki 17 C9
Khu Khan 36 B2
Khudzhand ... 29 E7
Khūgiānī 42 D4
Khulna 41 F7
Khulna □ 41 F7
Khūr 44 C4
Khurays 47 F7
Khūriyā Mūriyā,
  Jazā 'ir 48 D6
Khush 42 C3
Khushab 42 C8
Khust 17 D6
Khuzdar 42 F5
Khvor 44 C3
Khvormūj 44 D2
Khvoy 46 C6
Khyber Pass .. 42 B7
Kiama 63 B5
Kiamba 38 C2
Kiangsi =
  Jiangxi □ ... 35 D6
Kiangsu =
  Jiangsu □ ... 35 C6
Kibangou 56 E2
Kibombo 57 E5
Kibondo 57 E6
Kibwesa 57 F5
Kibwezi 57 E7
Kicking Horse
  Pass 71 C8
Kidal 55 E5
Kidnappers, C. . 64 C7
Kiel 14 A6
Kiel Canal =
  Nord-Ostsee-
  Kanal → ... 14 A5
Kielce 16 C5
Kieler Bucht ... 14 A6
Kiev = Kyyiv .. 17 C10
Kiffa 55 E2
Kifri 46 D6
Kigali 57 E6
Kigoma-Ujiji .. 57 E5
Kihee 62 A3
Kii-Suidō 32 C4
Kikinda 22 B3
Kikládhes ... 23 F5
Kikwit 56 E3
Kilcoy 63 A5
Kilimanjaro .. 57 E7
Kilindini 57 E7
Kilis 46 C4
Kiliya 17 F9
Kilkenny 11 E3
Kilkís 23 D4
Killarney,
  Australia 63 A5
Killarney, Ireland 11 E2
Killeen 78 D5
Killíni 23 F4
Kilmarnock .. 10 D4
Kilmore 63 C3
Kilosa 57 F7
Kilrush 11 E2
Kilwa Kivinje . 57 F7
Kimaam 39 F5
Kimba 62 B2

Kimberley 58 D4
Kimberley
  Plateau 60 D4
Kimry 24 B4
Kinabalu,
  Gunong 36 C5
Kindia 55 G2
Kindu 57 E5
Kineshma 24 B5
King George B. 95 G4
King George I. . 96 A4
King George Is. 69 C2
King I. = Kadan
  Kyun 36 B1
King I. 62 C3
King William I. . 70 B10
King William's
  Town 59 E5
Kingaroy 63 A5
Kingoonya .. 62 B2
Kings Canyon
  National Park 82 B3
King's Lynn .. 11 E7
Kingscote ... 62 C2
Kingsport 75 A6
Kingston,
  Canada 69 D3
Kingston,
  Jamaica 86 C4
Kingston, N.Z. . 65 F3
Kingston South
  East 62 C2
Kingston upon
  Hull 11 E6
Kingstown ... 87 D7
Kingsville ... 78 F5
Kınık 23 E6
Kinkala 56 E2
Kinleith 64 C6
Kinsale 11 F2
Kinshasa ... 56 E3
Kinston 75 B9
Kintampo ... 55 G4
Kintap 37 E5
Kiparissía .. 23 F3
Kiparissiakós
  Kólpos 23 F3
Kipembawe .. 57 F6
Kipili 57 F6
Kipushi 59 A5
Kirghizia ■ =
  Kyrgyzstan ■ 29 E8
Kirghizstan =
  Kyrgyzstan ■ 29 E8
Kiri 56 E3
Kiribati ■ 64 K12
Kırıkkale 46 C3
Kirillov 24 B4
Kirin = Jilin .. 35 B7
Kiritimati 65 J15
Kirkcaldy ... 10 C5
Kirkcudbright .. 11 D4
Kirkee 43 K8
Kirkenes 8 E14
Kirkland Lake . 69 D2
Kırklareli ... 22 D6
Kirkūk 46 D6
Kirkwall 10 B5
Kirov 28 D5
Kirovabad =
  Gäncä 25 E6
Kirovakan =
  Vanadzor .. 25 E5
Kirovograd =
  Kirovohrad .. 25 D3
Kirovohrad ... 25 D3
Kirovsk =
  Babadayhan . 29 F7
Kirovsk 28 C4

Kirsanov 24 C5
Kırşehir 46 C3
Kirteh 45 C5
Kirthar Range . 42 F5
Kiruna 8 F12
Kirundu 57 E5
Kiryū 32 A6
Kisangani 57 D5
Kisar 39 F3
Kiselevsk ... 29 D9
Kishanganj ... 40 D7
Kishangarh ... 42 F7
Kishinev =
  Chişinău 17 E9
Kishiwada ... 32 B4
Kishtwar 42 C9
Kisii 57 E6
Kisiju 57 F7
Kiska I. 71 C1
Kiskörös ... 16 E4
Kiskunfélegyháza
  .... 16 E4
Kiskunhalas ... 16 E4
Kislovodsk 25 E5
Kiso-Sammyaku 32 B5
Kissidougou .. 55 G2
Kisumu 57 E6
Kita 55 F3
Kitab 29 F7
Kitakami-
  Gawa → ... 33 G12
Kitakyūshū ... 32 C2
Kitale 57 D7
Kitchener 69 D2
Kitega = Gitega 57 E5
Kitgum 57 D6
Kíthira 23 F4
Kíthnos 23 F5
Kitikmeot □ ... 70 B9
Kittakittaooloo,
  L. 62 A2
Kitui 57 E7
Kitwe 59 A5
Kivertsi ... 17 C7
Kivu, L. ... 57 E5
Kiyev = Kyyiv . 17 C10
Kiyevskoye
  Vdkhr. =
  Kyyivske
  Vdskh. 17 C10
Kızıl Irmak → . 25 E4
Kizlyar 25 E6
Kizyl-Arvat =
  Gyzylarbat .. 29 F6
Kladno 16 C2
Klagenfurt ... 15 E8
Klaipeda ... 24 B1
Klamath Falls . 80 D3
Klamath Mts. .. 80 E2
Klarälven → .. 9 G10
Klatovy 16 D1
Klawer 58 E3
Klerksdorp .. 59 D5
Kletsk = Klyetsk 17 B8
Klin 28 D4
Kłodzko ... 16 C3
Klondike 70 B6
Klouto 55 G5
Kluane L. ... 70 B6
Kluczbork .. 16 C4
Klyetsk ... 17 B8
Knossós ... 23 G5
Knox Coast .. 96 A12
Knoxville 75 B6
Koartac =
  Quaqtaq ... 68 B4
Koba, Aru,
  Indonesia .. 39 F4
Koba, Bangka,
  Indonesia ... 37 E3

Kobarid 20 A4
Kobdo = Hovd 34 B4
Kōbe 32 B4
København .. 9 G10
Koblenz 14 C4
Kobroor,
  Kepulauan .. 39 F4
Kobryn 17 B7
Kocaeli = İzmit 25 E2
Kočani 22 D4
Koch Bihar ... 41 D7
Kōchi 32 C3
Kochiu = Gejiu 34 D5
Kodiak 71 C4
Kodiak I. 71 C4
Koes 58 D3
Kofiau 39 E3
Koforidua 55 G4
Kōfu 32 B6
Kogan 63 A5
Koh-i-Bābā .. 42 B5
Kohat 42 C7
Kohima 41 E10
Kohkīlūyeh va
  Būyer
  Aḥmadi □ .. 44 D2
Kohler Ra. ... 96 B1
Kokand =
  Qŭqon 29 E8
Kokas 39 E4
Kokchetav =
  Kökshetaū .. 29 D7
Koko Kyunzu . 41 K9
Kokomo 72 D2
Kokonau 39 E5
Kökshetaū ... 29 D7
Koksoak → .. 68 C4
Kokstad ... 59 E5
Kola, Indonesia 39 F4
Kola, Russia .. 28 C4
Kola Pen. =
  Kolskiy
  Poluostrov .. 28 C4
Kolaka 39 E2
Kolar 43 N11
Kolar Gold
  Fields 43 N11
Kolayat 42 F8
Kolchugino =
  Leninsk-
  Kuznetskiy .. 29 D9
Kolda 55 F2
Kole 56 E4
Kolepom = Yos
  Sudarso,
  Pulau 39 F5
Kolguyev,
  Ostrov 28 C5
Kolhapur 43 L9
Kolín 16 C2
Köln 14 C4
Koło 16 B4
Kołobrzeg ... 16 A2
Kolokani ... 55 F3
Kolomna ... 24 B4
Kolomyya ... 17 D7
Kolonodale .. 39 E2
Kolosib 41 E9
Kolpashevo .. 29 D9
Kolpino ... 24 B3
Kolskiy
  Poluostrov .. 28 C4
Kolwezi 57 G5
Kolyma → ... 31 C13
Komárno 16 E4
Komatsu 32 A5
Komi □ 28 C6
Kommunarsk =
  Alchevsk 25 D4

Kommunizma, Pik

Kommunizma,
Pik 29 F8
Komodo 39 F1
Komono 56 E2
Komoran, Pulau 39 F5
Komotini 22 D5
Kompong Cham 36 B3
Kompong
Chhnang 36 B2
Kompong Som 36 B2
Komrat =
Comrat 17 E9
Komsomolets,
Ostrov 30 A7
Konarhá □ 42 B7
Konch 42 G11
Kondoa 57 E7
Konduga 53 F1
Köneürgench .. 29 E6
Kong 55 G4
Kong, Koh 36 B2
Konglu 41 D11
Kongolo 57 F5
Kongor 53 G5
Königsberg =
Kaliningrad . 24 C1
Konin 16 B4
Konjic 20 C6
Konosha 28 C5
Konotop 24 C3
Konqi He ➤ .. 34 B4
Końskie 16 C5
Konstanz 14 E5
Kontagora 55 F6
Kontum 36 B3
Konya 46 C3
Konya Ovası .. 46 C3
Konza 57 E7
Kooloonong ... 62 B3
Koondrook 63 C3
Koonibba 62 B1
Koorawatha ... 63 B4
Kootenay L. .. 71 D8
Kopaonik 22 C3
Koper 20 B4
Kopeysk 29 D7
Kopi 62 B2
Koprivnica ... 20 A6
Kopychyntsi .. 17 D7
Korab 22 D3
Korça 23 D3
Korce = Korça . 23 D3
Korčula 20 C6
Kordestan 46 D5
Kordestān □ .. 46 D6
Kordofân 53 F4
Korea, North ■ 35 C7
Korea, South ■ 35 C7
Korea Bay ... 35 C7
Korea Strait ... 35 C7
Korets 17 C8
Korhogo 55 G3
Korim 39 E5
Korinthiakós
Kólpos 23 E4
Kórinthos 23 F4
Kōriyama 33 G12
Korneshty =
Corneşti 17 E9
Koro, Ivory C. . 55 G3
Koro, Mali ... 55 F4
Korogwe 57 F7
Koroit 62 C3
Koror 38 C4
Körös ➤ 16 E5
Korosten 17 C9
Korostyshev .. 17 C9
Korti 53 E5
Kortrijk 14 C2

Kos 23 F6
Kościan 16 B3
Kosciusko, Mt. . 63 C4
Kosha 52 D5
K'oshih = Kashi 34 C2
Košice 16 D5
Kosovo □ 22 C3
Kosovska-
Mitrovica =
Titova-
Mitrovica ... 22 C3
Kôstî 53 F5
Kostopil 17 C8
Kostroma 24 B5
Kostrzyn 16 B2
Koszalin 16 A3
Kota 43 G9
Kota Baharu .. 36 C2
Kota Belud ... 36 C5
Kota Kinabalu . 36 C5
Kota Tinggi ... 37 D2
Kotaagung ... 37 F2
Kotabaru 37 E5
Kotabumi 37 E2
Kotamobagu .. 39 D2
Kotawaringin .. 37 E4
Kotelnich 24 B6
Kotelnyy, Ostrov 31 B11
Kotka 9 F13
Kotlas 28 C5
Kotli 42 C8
Kotor 22 C2
Kotovsk 17 E9
Kotri 42 G6
Kottayam 43 Q10
Kotturu 43 M10
Kotuy ➤ 30 B8
Kotzebue 71 B3
Kouango 56 D4
Koudougou ... 55 F4
Kouilou ➤ ... 56 E2
Kouki 56 C3
Koula Moutou . 56 E2
Koulen 36 B2
Koulikoro 55 F3
Koumra 53 G2
Kounradskiy .. 29 E8
Kouroussa ... 55 F3
Kousseri 53 F1
Koutiala 55 F3
Kovel 17 C7
Kovrov 24 B5
Kowloon 35 D6
Köyceğiz 23 F7
Koyuk 71 B3
Koyukuk ➤ ... 71 B4
Kozan 46 C3
Kozáni 23 D3
Kozhikode =
Calicut 43 P9
Kozyatyn 17 D9
Kpalimé 55 G5
Kra, Isthmus of
= Kra, Kho
Khot 36 B1
Kra, Kho Khot . 36 B1
Kragujevac ... 22 B3
Krajina 20 B6
Krakatau =
Rakata, Pulau 37 F3
Kraków 16 C4
Kraljevo 22 C3
Kramatorsk ... 25 D4
Kranj 20 A5
Kraśnik 16 C6
Krasnoarmeysk 29 D5
Krasnodar ... 25 D4
Krasnoperekopsk
.... 25 D3

Krasnoselkupsk 28 C9
Krasnoturinsk . 28 D7
Krasnoufimsk . 29 D6
Krasnouralsk .. 29 D7
Krasnovodsk =
Krasnowodsk 29 E6
Krasnowodsk . 29 E6
Krasnoyarsk ... 30 D7
Krasnyy Luch . 25 D4
Krasnyy Yar ... 25 D6
Kratie 36 B3
Krau 39 E6
Krefeld 14 C4
Kremen 20 B5
Kremenchug =
Kremenchuk . 24 D3
Kremenchuk . 24 D3
Kremenchuksk
Vdskh. 24 D3
Kremenets 17 C7
Krems 15 D8
Kribi 56 D1
Krichev =
Krychaw 17 B10
Krishna ➤ .. 40 K3
Krishnanagar .. 41 F7
Kristiansand .. 9 G9
Kristiansund .. 8 F9
Kriti 23 G5
Krivoy Rog =
Kryvyy Rih .. 25 D3
Krk 20 B5
Kronprins Olav
Kyst 96 A9
Kronshtadt ... 24 B2
Kroonstad ... 59 D5
Kropotkin,
Irkutsk, Russia 30 D9
Kropotkin,
Krasnodar,
Russia 25 D5
Krosno 16 D5
Krotoszyn 16 C3
Krugersdorp .. 59 D5
Krung Thep =
Bangkok ... 36 B2
Krupki 17 A9
Kruševac 22 C3
Krychaw 17 B10
Krymskiy
Poluostrov =
Krymskyy
Pivostriv 25 D3
Krymskyy
Pivostriv 25 D3
Kryvyy Rih 25 D3
Ksar el Boukhari 54 A5
Ksar el Kebir .. 54 B3
Ksar es Souk =
Ar Rachidiya 54 B4
Kuala 37 D3
Kuala Kubu
Baharu 37 D2
Kuala Lipis ... 37 D2
Kuala Lumpur . 37 D2
Kuala
Terengganu . 36 C2
Kualajelai 37 E4
Kualakapuas .. 37 E4
Kualakurun ... 37 E4
Kualapembuang 37 E4
Kualasimpang . 37 D1
Kuandang 39 D2
Kuangchou =
Guangzhou . 35 D6
Kuantan 37 D2
Kuba = Quba . 25 E6
Kuban ➤ 25 D4
Kucing 37 D4

Kuda 43 H7
Kudat 36 C5
Kudymkar ... 28 D6
Kueiyang =
Guiyang 35 D5
Kufra Oasis = Al
Kufrah 52 D3
Kufstein 15 E7
Küh-e 'Alījūq .. 44 D2
Küh-e-Hazārān 44 D4
Küh-e Sorkh .. 44 C4
Küh-e Taftān . 45 D5
Kūhak 42 F3
Kūhhā-ye-
Bashäkerd .. 45 E4
Kūhpāyeh ... 44 C3
Kuito 58 A3
Kujū-San 32 C2
Kukawa 53 F1
Kukësi 22 C3
Kula 23 E7
Kulasekarappattinam
.... 43 Q11
Kuldja = Yining 34 B3
Kŭlob 29 F7
Kulsary 29 E6
Kulunda 29 D8
Kulwin 62 C3
Kulyab = Kŭlob 29 F7
Kum Tekei 29 E8
Kuma ➤ 25 E6
Kumagaya ... 32 A6
Kumai 37 E4
Kumamba,
Kepulauan .. 39 E5
Kumamoto ... 32 C2
Kumanovo ... 22 C3
Kumara 65 E4
Kumasi 55 G4
Kumayri =
Gyumri 25 E5
Kumba 56 D1
Kumbakonam . 43 P11
Kumbarilla ... 63 A5
Kumo 55 F7
Kumon Bum .. 41 D11
Kunama 63 C4
Kungala 63 A5
Kungrad =
Qŭnghirot ... 29 E6
Kungur 29 D6
Kunlong 41 F12
Kunlun Shan .. 34 C3
Kunming 34 D5
Kunsan 35 C7
Kunya-Urgench
Köneürgench 29 E6
Kuopio 8 F13
Kupa ➤ 20 B7
Kupyansk 28 E4
Kuqa 34 B3
Kür ➤ 25 E6
Kura = Kür ➤ . 25 E6
Kurashiki 32 B3
Kurayoshi 32 B3
Kŭrdzhali 22 D5
Kure 32 B3
Kurgaldzhinskiy 29 D8
Kurgan 29 D7
Kuria Maria Is.
= Khūrīyā
Mūrīyā, Jazā
'ir 48 D6
Kurigram 41 E7
Kuril Is. =
Kurilskiye
Ostrova 31 E12
Kurilskiye
Ostrova 31 E12

Kurmuk 53 F5
Kurnool 43 L11
Kurow 65 F4
Kurrajong 63 B5
Kurri Kurri ... 63 B5
Kursk 24 C4
Kuruktag 34 B3
Kuruman 58 D4
Kurume 32 C2
Kus Gölü 23 D6
Kuşadası 23 F6
Kushiro 32 F12
Kushtia 41 F7
Kuskokwim → 71 B3
Kuskokwim B. . 71 C3
Kustanay =
  Qostanay ... 29 D7
Kütahya 46 C2
Kutaisi 25 E5
Kutaraja =
  Banda Aceh . 36 C1
Kutch, Gulf of =
  Kachchh, Gulf
  of 43 H6
Kutch, Rann of
  = Kachchh,
  Rann of 43 G6
Kutno 16 B4
Kutu 56 E3
Kutum 53 F3
Kuujjuaq 68 C4
Kuwait = Al
  Kuwayt 47 E7
Kuwait ■ 47 E6
Kuwana 32 B5
Kuybyshev =
  Samara 24 C7
Kuybyshev ... 29 D8
Kuybyshevskoye
  Vdkhr. 24 B6
Küysanjaq ... 46 C6
Kuyumba 30 C7
Kuzey Anadolu
  Dağları 46 B3
Kuznetsk 24 C6
Kvarner 20 B5
Kvarnerič 20 B5
Kwakoegron .. 92 A2
Kwamouth 56 E3
Kwando → 58 B4
Kwangju 35 C7
Kwangsi-Chuang
  = Guangxi
  Zhuangzu
  Zizhiqu □ ... 35 D5
Kwangtung =
  Guangdong □ 35 D6
Kwatisore ... 39 E4
Kweichow =
  Guizhou □ .. 35 D5
Kwekwe 59 B5
Kwidzyn 16 B4
Kwoka 39 E4
Kyabé 53 G2
Kyabra Cr. → . 63 A3
Kyabram 63 C4
Kyancutta ... 62 B2
Kyangin 41 H10
Kyaukpadaung 41 G10
Kyaukpyu 41 H9
Kyaukse 41 G11
Kyle Dam 59 C6
Kyneton 63 C3
Kyō-ga-Saki .. 32 B4
Kyoga, L. 57 D6
Kyogle 63 A5
Kyongpyaw .. 41 J10
Kyōto 32 B4
Kyrenia 46 D3

Kyrgyzstan ■ . . 29 E8
Kystatyam 30 C10
Kytal Ktakh ... 31 C10
Kyunhla 41 F10
Kyūshū 32 C2
Kyūshū-Sanchi 32 C2
Kyustendil ... 22 C4
Kyusyur 30 B10
Kywong 63 B4
Kyyiv 17 C10
Kyyivske Vdskh. 17 C10
Kyzyl Kum 29 E7
Kyzyl-Kyya ... 29 E8
Kzyl-Orda =
  Qyzylorda ... 29 E7

L

La Albufera ... 19 C5
La Alcarria ... 19 B4
La Asunción .. 90 A6
La Banda 94 B4
La Calera 94 C2
La Carlota 94 C4
La Ceiba 85 D7
La Chaux de
  Fonds 13 C7
La Cocha 94 B3
La Coruña ... 18 A1
La Crosse 77 D9
La Dorada ... 90 B4
La Estrada ... 18 A1
La Grange ... 74 C5
La Guaira 90 A5
La Güera 54 D1
La Habana ... 86 B3
La Línea de la
  Concepción . 18 D3
La Mancha ... 19 C4
La Palma,
  Canary Is. ... 54 C1
La Palma,
  Panama 86 E4
La Palma del
  Condado 18 D2
La Paragua ... 90 B6
La Paz,
  Entre Ríos,
  Argentina ... 94 C5
La Paz, San Luis,
  Argentina ... 94 C3
La Paz, Bolivia . 91 G5
La Paz, Mexico . 84 C2
La Pedrera ... 90 D5
La Plata 94 D5
La Quiaca 94 A3
La Rioja 94 B3
La Rioja □ ... 19 A4
La Robla 18 A3
La Roche-sur-
  Yon 12 C3
La Rochelle ... 12 C3
La Roda 19 C4
La Romana ... 87 C6
La Serena 94 B2
La Seyne-sur-
  Mer 13 E6
La Spézia 20 B2
La Tortuga 87 D6
La Tuque 69 D3
La Unión 95 E2
La Urbana ... 90 B5
La Vega 87 C5
Labe = Elbe → 14 B5
Labé 55 F2
Labis 37 D2

Laboulaye 94 C4
Labrador, Coast
  of □ 69 C4
Lábrea 91 E6
Labuan, Pulau . 36 C5
Labuha 39 E3
Labuhanbajo . 39 F2
Labuk, Telok .. 36 C5
Labyrinth, L. .. 62 B2
Labytnangi ... 28 C7
Lac La Biche .. 71 C8
Lac la Martre .. 70 B8
Lacanau 12 D3
Laccadive Is. =
  Lakshadweep
  Is. 26 H11
Lacepede B. .. 62 C2
Lachine 69 D3
Lachlan → 63 B3
Lackawanna .. 73 C6
Lacombe 71 C8
Ladakh Ra. ... 42 B10
Lādīz 45 D5
Ladoga, L. =
  Ladozhskoye
  Ozero 28 C4
Ladozhskoye
  Ozero 28 C4
Ladysmith,
  Canada 71 D7
Ladysmith,
  S. Africa 59 D5
Lae 61 B8
Lafayette 79 D7
Lafia 55 G6
Laghmān □ .. 45 C7
Laghouat 54 B5
Lagonoy Gulf . 38 B2
Lagos, Nigeria . 55 G5
Lagos, Portugal 18 D1
Laguna 94 B7
Lagunas, Chile 91 H5
Lagunas, Peru . 90 E3
Lahad Datu .. 38 C1
Lahat 37 E2
Lahewa 37 D1
Lāhījān 46 C7
Lahn → 14 C4
Lahore 42 D9
Lahti 8 F13
Lahtis = Lahti . 8 F13
Laï 53 G2
Laibin 35 D5
Laidley 63 A5
Laila = Laylá . 47 G6
Lairg 10 B4
Lajere 55 F7
Lajes 94 B6
Lake Cargelligo 63 B4
Lake Charles . 79 D7
Lake City 75 D6
Lake Harbour . 68 B4
Lake Worth ... 75 F7
Lakeland 75 E7
Lakes Entrance 63 C4
Lakewood 72 D5
Lakonikós
  Kólpos 23 F4
Lakor 39 F3
Lakota 55 G3
Lakshadweep Is. 26 H11
Lala Ghat 41 E9
Lalibela 53 F6
Lalín 18 A1
Lalitapur =
  Patan 40 D5
Lamaing 41 K11
Lamas 91 E3
Lambaréné ... 56 E2

Lambert Glacier 96 B10
Lame 55 F6
Lamego 18 B2
Lameroo 62 C3
Lamía 23 E4
Lamon Bay ... 38 B2
Lampa 91 G4
Lampedusa ... 21 G4
Lampione ... 21 G4
Lampung □ .. 37 F2
Lamu 57 E8
Lanai I. 82 H12
Lanak La 42 B11
Lanak'o
  Shank'ou =
  Lanak La ... 42 B11
Lanao, L. 38 C2
Lancaster, U.K. 11 D5
Lancaster,
  U.S.A. 73 D7
Lancaster Sd. . 68 A2
Lanchow =
  Lanzhou ... 35 C5
Lanciano ... 20 C5
Landeck 14 E6
Landes 12 D3
Landi Kotal .. 42 B7
Land's End ... 11 F4
Landshut ... 15 D7
La'nga Co ... 40 B3
Langkawi, P. . 36 C1
Langkon 36 C5
Langres 13 C6
Langres, Plateau
  de 13 C6
Langsa 36 D1
Languedoc ... 13 E5
Langzhong ... 35 C5
Lannion 12 B2
Lansdowne ... 63 B5
Lansing 72 C3
Lanusei 21 E2
Lanzarote ... 54 C2
Lanzhou 35 C5
Laoag 38 A2
Laoang 38 B3
Laon 13 B5
Laos ■ 36 A3
Lapa 94 B7
Laparan 38 C1
Lapland =
  Lappland ... 8 E12
Lappland ... 8 E12
Lapseki 23 D6
Laptev Sea .. 30 B10
L'Aquila 20 C4
Lär 44 E3
Larache 54 A3
Laramie 76 E2
Larantuka ... 39 F2
Larap 38 B2
Larat 39 F4
Laredo 78 F4
Lariang 39 E1
Lárisa 23 E4
Larnaca 46 D3
Larne 11 D4
Larrimah 60 D5
Larsen Ice Shelf 96 A3
Larvik 9 G9
Laryak 29 C8
Las Anod 49 F4
Las Cruces ... 83 D9
Las Flores ... 94 D5
Las Khoreh .. 49 E4
Las Lajas 94 D2
Las Lomitas .. 94 A4
Las Palmas ... 54 C1
Las Pipinas ... 94 D5

Las Plumas

Las Plumas ... 95 E3	Leh 42 B10	Levskigrad =	Limoeiro do
Las Rosas 94 C4	Leicester 11 E6	Karlovo 22 C5	Norte 92 D6
Las Varillas .. 94 C4	Leiden 14 B3	Lewis 10 B3	Limoges 12 D4
Las Vegas 82 B5	Leie → 14 C2	Lewisporte 69 D5	Limón 86 E3
Lashio 41 F11	Leine → 14 B5	Lewiston, Idaho,	Limousin 12 D4
Lassen Pk. 80 E3	Leinster □ 11 E3	U.S.A. 80 B5	Limoux 12 E5
Lastoursville .. 56 E2	Leipzig 15 C7	Lewiston,	Limpopo → .. 59 D6
Lastovo 20 C6	Leiria 18 C1	Maine, U.S.A. 73 B10	Limuru 57 E7
Latacunga 90 D3	Leitrim 11 D2	Lexington, Ky.,	Linares, Chile . 94 D2
Latakia = Al	Leizhou Bandao 35 D6	U.S.A. 72 E3	Linares, Mexico 84 C5
Lādhiqīyah .. 46 D3	Leksula 39 E3	Lexington, N.C.,	Linares, Spain . 18 C4
Latina 20 D4	Lékva Ori 23 G5	U.S.A. 75 B7	Linchuan 35 D6
Latium =	Leleque 95 E2	Leyte 38 B2	Lincoln,
Lazio □ 20 C4	Léman, L. 13 C7	Lezha 22 D2	Argentina ... 94 C4
Latrobe 62 D4	Lemhi Ra. 81 C7	Lhasa 34 D4	Lincoln, N.Z. .. 65 E5
Latvia ■ 24 B1	Lena → 30 B10	Lhazê 34 D3	Lincoln, U.K. .. 11 E6
Lauchhammer . 15 C7	Leninabad =	Lhokkruet 36 D1	Lincoln, U.S.A. 77 E6
Launceston ... 62 D4	Khudzhand .. 29 E7	Lhokseumawe . 36 C1	Linden 90 B7
Laurel 79 D9	Leninakan =	Lhuntsi Dzong . 41 D8	Líndhos 23 F7
Lauria 21 D5	Gyumri 25 E5	Lianga 38 C3	Lindi 57 F7
Laurium 69 D2	Leningrad =	Lianyungang .. 35 C6	Lingayen 38 A2
Lausanne 13 C7	Sankt-	Liaoning □ ... 35 B7	Lingayen G. ... 38 A2
Laut 36 D3	Peterburg ... 24 B3	Liaoyuan 35 B7	Lingen 14 B4
Laut Kecil,	Leninogorsk =	Liard → 70 B7	Lingga 37 E2
Kepulauan .. 37 E5	Ridder 29 D9	Libau = Liepāja 24 B1	Lingga,
Laval 12 B3	Leninsk 25 D6	Libenge 56 D3	Kepulauan .. 37 E2
Lávrion 23 F5	Leninsk-	Liberal 78 A3	Lingga Arch. =
Lawas 36 D5	Kuznetskiy .. 29 D9	Liberec 16 C2	Lingga,
Lawele 39 F2	Lenmalu 39 E4	Liberia ■ 55 G3	Kepulauan .. 37 E2
Lawng Pit 41 E11	Lens 13 A5	Lîbîya, Sahrâ' . 52 C4	Linguéré 55 E1
Lawrence, N.Z. 65 F3	Lentini 21 F5	Libobo, Tanjung 39 E3	Lingyun 35 D5
Lawrence,	Leoben 15 E8	Libonda 58 A4	Linhai 35 D7
U.S.A. 77 F7	Leodhas =	Libourne 12 D3	Linhares 93 F5
Lawton 78 B4	Lewis 10 B3	Libreville 56 D1	Linköping 9 G11
Laylá 47 G6	León, Mexico .. 84 C4	Libya ■ 52 C2	Linosa, I. 21 G4
Lazio □ 20 C4	León, Nic. 85 E7	Libyan Desert =	Lins 94 A7
Le Creusot 13 C6	León, Spain ... 18 A3	Lîbîya, Sahrâ' 52 C4	Linville 63 A5
Le Havre 12 B4	León, Montañas	Licantén 94 D2	Linxia 34 C5
Le Mans 12 C4	de 18 A2	Licata 21 F4	Linz 15 D8
Le Mont-St.-	Leongatha 63 C4	Lichinga 59 A7	Lion, G. du .. 13 E6
Michel 12 B3	Léopold II, Lac	Lichtenburg .. 59 D5	Lions, G. of =
Le Moule 87 C7	= Mai-	Lida 17 B7	Lion, G. du = 13 E6
Le Puy-en-Velay 13 D5	Ndombe, L. . 56 E3	Liechtenstein ■ 14 E5	Lipa 38 B2
Le Touquet-	Léopoldville =	Liège 14 C3	Lípari 21 E5
Paris-Plage .. 12 A4	Kinshasa 56 E3	Liegnitz =	Lípari, Is. =
Le Tréport 12 A4	Leova 17 E9	Legnica 16 C3	Eólie, Ís. 21 E5
Le Verdon-sur-	Lepel = Lyepyel 24 C2	Lienyünchiangshih	Lipcani 17 D8
Mer 12 D3	Lepikha 31 C10	=	Lipetsk 24 C4
Leamington Spa	Léré 53 G1	Lianyungang 35 C6	Lipkany =
= Royal	Lérida 19 B6	Lienz 15 E7	Lipcani 17 D8
Leamington	Lerwick 10 A6	Liepāja 24 B1	Lipovets 17 D9
Spa 11 E6	Les Cayes 86 C5	Liffey → 11 E3	Lippe → 14 C4
Leandro Norte	Les Sables-	Lifford 11 D3	Liptrap C. 63 C4
Alem 94 B5	d'Olonne ... 12 C3	Lightning Ridge 63 A4	Lira 57 D6
Leavenworth .. 77 F7	Lesbos =	Liguria □ 20 B2	Liria 19 C5
Lebak 38 C2	Lésvos 23 E6	Ligurian Sea .. 20 C2	Lisala 56 D4
Lebanon ■ 46 D3	Leskovac 22 C3	Lijiang 34 D5	Lisboa 18 C1
Łębork 16 A3	Lesotho ■ 59 D5	Likasi 57 G5	Lisbon = Lisboa 18 C1
Lebrija 18 D2	Lesser Antilles . 87 D7	Lille 13 A5	Lisburn 11 D3
Lebu 94 D2	Lesser Sunda Is. 39 F2	Lillehammer .. 8 F10	Lisburne, C. ... 71 B3
Lecce 23 D2	Lésvos 23 E6	Lilongwe 59 A6	Lishui 35 D6
Lecco 20 B2	Leszno 16 C3	Liloy 38 C2	Lisichansk =
Lech → 14 D6	Lethbridge 71 D8	Lim → 20 C7	Lysychansk . 25 D4
Łęczyca 16 B4	Leti, Kepulauan 39 F3	Lima, Indonesia 39 E3	Lisieux 12 B4
Leduc 71 C8	Leti Is. = Leti,	Lima, Peru 91 F3	Liski 24 C4
Leeds 11 E6	Kepulauan .. 39 F3	Lima, U.S.A. .. 72 D3	Lismore 63 A5
Leer 14 B4	Leticia 90 D4	Lima → 18 B1	Lister, Mt. 96 B15
Leeton 63 B4	Letpadan 41 J10	Limassol 46 D3	Liston 63 A5
Leeuwarden ... 14 B3	Letpan 41 H10	Limay Mahuida 94 D3	Listowel 11 E2
Leeuwin, C. ... 60 G2	Letterkenny ... 11 D3	Limbang 36 D5	Litang 38 C1
Leeward Is. ... 87 C7	Leuser, G. 37 D1	Limbe 56 D1	Litani → 46 D3
Leganés 18 B4	Leuven 14 C3	Limbri 63 B5	Lithgow 63 B5
Legazpi 38 B2	Levádhia 23 E4	Limburg 14 C5	Líthinon, Ákra . 23 G5
Leghorn =	Levice 16 D4	Limeira 94 A7	Lithuania ■ ... 24 B1
Livorno 20 C3	Levin → 64 D6	Limerick 11 E2	Litoměřice 16 C2
Legionowo 16 B5	Lévis 69 D3	Limfjorden 9 G9	Little Barrier I. . 64 B6
Legnago 20 B3	Levkás 23 E3	Limia =	Little Laut Is. =
Legnica 16 C3	Levkôsia =	Lima → 18 B1	Laut Kecil,
Legume 63 A5	Nicosia 46 D3	Límnos 23 E5	Kepulauan .. 37 E5

Little
  Missouri → . 76 B3
Little River .. 65 E5
Little Rock 79 B7
Liukang
  Tenggaja ... 39 F1
Liuwa Plain ... 58 A4
Liuzhou 35 D5
Liverpool,
  Australia 63 B5
Liverpool,
  Canada 69 D4
Liverpool, U.K. . 11 E5
Liverpool Plains 63 B5
Liverpool Ra. .. 63 B5
Livingstone ... 59 B5
Livingstonia ... 57 G6
Livny 24 C4
Livorno 20 C3
Livramento ... 94 C5
Liwale 57 F7
Ljubljana 20 A5
Ljungan → ... 8 F11
Ljusnan → ... 8 F11
Llancanelo,
  Salina 94 D3
Llanelli 11 F4
Llanes 18 A3
Llano Estacado 78 C2
Llanos 90 C4
Lleida = Lérida 19 B6
Llobregat → . 19 B7
Lloret de Mar . 19 B7
Lluchmayor ... 19 C7
Llullaillaco,
  Volcán 94 A3
Loa → 94 A2
Lobatse 59 D5
Lobería 94 D5
Lobito 58 A2
Locarno 13 C8
Loch Garman =
  Wexford 11 E3
Loches 12 C4
Lock 62 B2
Lodhran 42 E7
Lodi, Italy 20 B2
Lodi, U.S.A. ... 80 F3
Lodja 56 E4
Lodwar 57 D7
Łódź 16 C4
Lofoten 8 E10
Logan 81 E8
Logan, Mt. 70 B5
Logone → 53 F2
Logroño 19 A4
Lohardaga 40 F5
Loi-kaw 41 H11
Loir → 12 C3
Loire → 12 C2
Loja, Ecuador . 90 D3
Loja, Spain ... 18 D3
Loji 39 E3
Lokandu 57 E5
Lokitaung 57 D7
Lokoja 55 G6
Lokolama 56 E3
Loliondo 57 E7
Lom 22 C4
Lomami → 56 D4
Lomblen 39 F2
Lombok 37 F5
Lomé 55 G5
Lomela 56 E4
Lomela → 56 E4
Lomié 56 D2

Lomond, L. ... 10 C4
Lompobatang . 39 F1
Łomza 16 B6
Loncoche 95 D2
Londa 43 M9
London 11 F6
Londonderry .. 11 D3
Londonderry, C. 60 C4
Londonderry, I. 95 H2
Londrina 94 A6
Long Beach ... 82 D3
Long I.,
  Bahamas ... 86 B4
Long I., U.S.A. . 73 D9
Long Xuyen ... 36 B3
Longford,
  Australia 62 D4
Longford,
  Ireland 11 E3
Longiram 37 E5
Longlac 69 D2
Longnawan ... 37 D4
Longreach 61 E7
Longview, Tex.,
  U.S.A. 79 C6
Longview,
  Wash., U.S.A. 80 B2
Lons-le-Saunier 13 C6
Lop Nor = Lop
  Nur 34 B4
Lop Nur 34 B4
Lopez, C. 56 E1
Lora → 42 D4
Lora, Hamun-i- 42 E4
Lora Cr. → ... 62 A2
Lora del Río .. 18 D3
Lorain 72 D4
Loralai 42 D6
Lorca 19 D5
Lorestān □ 46 D6
Loreto 92 D4
Lorient 12 C2
Lorne 62 C3
Lorraine 13 B7
Los Alamos ... 83 C5
Los Andes 94 C2
Los Angeles,
  Chile 94 D2
Los Angeles,
  U.S.A. 82 C3
Los Blancos ... 94 A4
Los Hermanos . 90 A6
Los Mochis ... 84 B3
Los Roques ... 90 A5
Los Testigos .. 90 A6
Los Vilos 94 C2
Lošinj 20 B5
Lot → 12 D4
Lota 94 D2
Loubomo 56 E2
Louga 55 E1
Louis Trichardt 59 C5
Louis XIV, Pte. . 69 C3
Louisiade Arch. 61 C9
Louisiana □ ... 79 D8
Louisville 72 E3
Loulé 18 D1
Lourdes 12 E3
Louth, Australia 63 B4
Louth, U.K. ... 11 E6
Louvain =
  Leuven 14 C3
Lovech 22 C5
Low Tatra =
  Nízké Tatry .. 16 D4
Lowell 73 C10
Lower California
  = Baja
  California ... 84 A1

Lower Hutt 65 D6
Lower Saxony =
  Niedersachsen □
  .... 14 B5
Lower Tunguska
  = Tunguska,
  Nizhnyaya → 30 C6
Lowestoft 11 E7
Łowicz 16 B4
Lowyar □ 45 C7
Loxton 62 B3
Loyalty Is. =
  Loyauté, Is. . 64 M11
Loyang =
  Luoyang ... 35 C6
Loyauté, Is. ... 64 M11
Loyev = Loyew 17 C10
Loyew 17 C10
Luachimo 56 F4
Luacono 56 G4
Lualaba → ... 57 D5
Luanda 56 F2
Luangwa 59 B6
Luangwa → .. 59 A6
Luanshya 59 A5
Luapula → ... 57 F5
Luarca 18 A2
Luashi 56 G4
Luau 56 G4
Lubalo 56 F3
Lubang Is. ... 38 B2
Lubbock 78 C3
Lübeck 15 B6
Lubefu 56 E4
Lubero = Luofu 57 E5
Lubin 16 C3
Lublin 17 C6
Lubuagan 38 A2
Lubuk Antu ... 37 D4
Lubuklinggau . 37 E2
Lubuksikaping . 37 D2
Lubumbashi .. 59 A5
Lubutu 57 E5
Lucca 20 C3
Lucena, Phil. . 38 B2
Lucena, Spain . 18 D3
Lučenec 16 D4
Lucerne =
  Luzern 13 C8
Lucira 58 A2
Luckenwalde .. 15 B7
Lucknow 40 D3
Lüda = Dalian . 35 C7
Lüderitz 58 D3
Ludhiana 42 D9
Ludwigsburg .. 14 D5
Ludwigshafen . 14 D5
Luebo 56 F4
Lufira → 57 F5
Lufkin 79 D6
Luga 24 B2
Lugano 13 C8
Lugansk =
  Luhansk 25 D4
Lugh Ganana .. 49 G3
Lugo, Italy 20 B3
Lugo, Spain .. 18 A2
Lugoj 17 F5
Lugovoy 29 E8
Luhansk 25 D4
Luiana 58 B4
Luimneach =
  Limerick 11 E2
Luís Correia .. 92 C5
Luitpold Coast . 96 B5
Luiza 56 F4
Luján 94 C5
Lukanga Swamp 59 A5
Lukenie → ... 56 E3

Lukolela 56 E3
Łuków 17 C6
Lule älv → ... 8 E12
Luleå 8 E12
Lüleburgaz 22 D6
Lulonga → ... 56 D3
Lulua → 56 E4
Luluabourg =
  Kananga 56 F4
Lumai 58 A4
Lumbala
  N'guimbo ... 58 A4
Lumsden 65 F3
Lumut, Tg. 37 E3
Lundazi 59 A6
Lundu 37 D3
Lüneburg 14 B6
Lüneburg Heath
  = Lüneburger
  Heide 14 B6
Lüneburger
  Heide 14 B6
Lunéville 13 B7
Lunglei 41 F9
Luni → 42 G8
Luni → 43 G7
Luninets =
  Luninyets ... 17 B8
Luninyets 17 B8
Luofu 57 E5
Luoyang 35 C6
Luozi 56 E2
Luremo 56 F3
Lurgan 11 D3
Lusaka 59 B5
Lusambo 56 E4
Lushnja 23 D2
Lushoto 57 E7
Luta = Dalian . 35 C7
Luton 11 F6
Lutong 36 D4
Lutsk 17 C7
Lützow
  Holmbukta .. 96 A8
Luwuk 39 E2
Luxembourg . 13 B7
Luxembourg ■ 14 D4
Luxor = El
  Uqsur 52 C5
Luza 24 A6
Luzern 13 C8
Luzhou 35 D5
Luziânia 93 F4
Luzon 38 A2
Lviv 17 D7
Lvov = Lviv ... 17 D7
Lyakhavichy ... 17 B8
Lyakhovskiye,
  Ostrova 31 B12
Lyallpur =
  Faisalabad .. 42 D8
Lydenburg 59 D6
Lydia 23 E7
Lyell 65 D5
Lyepyel 24 C2
Łyna → 16 A5
Lynchburg ... 73 F6
Lynd Ra. 63 A4
Lyndhurst 62 B2
Lynn Lake 71 C9
Lyon 13 D6
Lyonnais 13 D6
Lyons = Lyon . 13 D6
Lys = Leie → . 14 C2
Lysychansk ... 25 D4
Lyttelton 65 E5
Lyubertsy 24 B4
Lyuboml 17 C7

Ma'alah

M

Ma'alah 47 F6
Ma'ān 47 E3
Ma'anshan 35 C6
Ma'arrat an
  Nu'mān 46 D4
Maas → 14 C3
Maastricht 13 A6
Mabrouk 55 E4
Macaé 93 G5
McAllen 78 F4
Macao =
  Macau ■ .. 35 D6
Macapá 92 B3
Macau 92 D6
Macau ■ 35 D6
M'Clintock Chan. 70 A9
McComb 79 D8
Macdonnell Ras. 60 E5
McDouall Peak 62 A1
Macdougall L. . 70 B10
Macedonia =
  Makedhonía □ 23 D4
Macedonia ■ .. 22 D3
Maceió 93 D6
Macenta 55 G3
Macerata 20 C4
Macfarlane, L. . 62 B2
Macgillycuddy's
  Reeks 11 F2
McGregor Ra. . 62 A3
Mach 42 E5
Machado =
  Jiparaná → . 91 E6
Machakos 57 E7
Machala 90 D3
Machilipatnam 40 J3
Machiques 90 A4
Machupicchu .. 91 F4
Macintyre → . 63 A5
Mackay 61 E8
Mackay, L. 60 E4
Mackenzie → . 70 B6
Mackenzie City
  = Linden .. 90 B7
Mackenzie Mts. 70 B6
McKinley, Mt. . 71 B4
McKinney 79 C5
Macksville ... 63 B5
Maclean 63 A5
Maclear 59 E5
Macleay → ... 63 B5
McMurdo Sd. . 96 B15
Mâcon, France 13 C6
Macon, U.S.A. . 75 C6
Macondo 58 A4
McPherson Ra. 63 A5
Macquarie
  Harbour 62 D4
MacRobertson
  Land 96 B10
Madagali 53 F1
Madagascar ■ . 59 J9
Madā'in Sālih . 47 F4
Madama 52 D1
Madang 61 B8
Madaoua 55 F6
Madaripur ... 41 F8
Madauk 41 J11
Madaya 41 F11
Maddalena ... 20 D2
Madeira 54 B1
Madeira → ... 90 D7
Madha 43 L9
Madhya
  Pradesh □ .. 43 H10
Madikeri 43 N9

Madimba 56 E3
Madīnat ash
  Sha'b 49 E3
Madingou 56 E2
Madison 77 D10
Madiun 37 F4
Madras = Tamil
  Nadu □ 43 P10
Madras 43 N12
Madre, Laguna 78 F5
Madre, Sierra . 38 A2
Madre de
  Dios → 91 F5
Madre de Dios,
  I. 95 G1
Madre
  Occidental,
  Sierra 84 B3
Madrid 18 B4
Madurai 43 Q11
Madurantakam 43 N11
Mae Sot 36 A1
Maebashi 32 A6
Maestrazgo,
  Mts. del 19 B5
Maevatanana .. 59 H9
Mafeking =
  Mafikeng .. 59 D5
Maffra 63 C4
Mafia I. 57 F7
Mafikeng 59 D5
Mafra, Brazil .. 94 B7
Mafra, Portugal 18 C1
Magadan 31 D13
Magadi 57 E7
Magallanes,
  Estrecho de . 95 G2
Magangué 90 B4
Magburaka ... 55 G2
Magdalena,
  Argentina ... 94 D5
Magdalena,
  Bolivia 91 F6
Magdalena,
  Malaysia 36 D5
Magdalena → 90 A4
Magdeburg ... 15 B6
Magelang 37 F4
Magellan's Str.
  = Magallanes,
  Estrecho de . 95 G2
Maggiore, L. .. 20 B2
Magnetic Pole
  (South) =
  South
  Magnetic Pole 96 A13
Magnitogorsk . 29 D6
Magosa =
  Famagusta .. 46 D3
Maguarinho, C. 92 C4
Mağusa =
  Famagusta .. 46 D3
Magwe 41 G10
Mahābād 46 C6
Mahabo 59 J8
Mahagi 57 D6
Mahajanga ... 59 H9
Mahakam → . 37 E5
Mahalapye ... 59 C5
Maḥallāt 44 C2
Mahanadi → . 40 G6
Mahanoro ... 59 H9
Maharashtra □ 43 J9
Mahbubnagar . 43 L10
Mahdia 52 A1
Mahenge 57 F7
Maheno 65 F4
Mahia Pen. ... 64 C7
Mahesana ... 43 H8
Mahia Pen. ... 64 C7

Mahilyow 17 B10
Mahón 19 C8
Mai-Ndombe, L. 56 E3
Maicurú → ... 92 C3
Maidstone ... 11 F7
Maiduguri 53 F1
Maijdi 41 F8
Maikala Ra. .. 40 G3
Main → 14 D5
Maine 12 C3
Maingkwan ... 41 D11
Mainit, L. 38 C3
Mainland,
  Orkney, U.K. 10 B5
Mainland, Shet.,
  U.K. 10 A6
Maintirano 59 H8
Mainz 14 C5
Maipú 94 D5
Maiquetía 90 A5
Mairabari 41 D9
Maitland,
  N.S.W.,
  Australia 63 B5
Maitland,
  S. Austral.,
  Australia 62 B2
Maizuru 32 B4
Majene 39 E1
Maji 53 G6
Majorca =
  Mallorca ... 19 C7
Maka 55 F2
Makale 39 E1
Makari 56 B2
Makarikari =
  Makgadikgadi
  Salt Pans .. 59 C5
Makasar =
  Ujung
  Pandang 39 F1
Makasar, Selat 39 E1
Makasar, Str. of
  = Makasar,
  Selat 39 E1
Makat 29 E6
Makedhonía □ 23 D4
Makeni 55 G2
Makeyevka =
  Makiyivka ... 25 D4
Makgadikgadi
  Salt Pans .. 59 C5
Makhachkala . 25 E6
Makian 39 D3
Makindu 57 E7
Makinsk 29 D8
Makiyivka 25 D4
Makkah 47 G4
Makó 16 E5
Makokou 56 D2
Makoua 56 E3
Makrai 43 H10
Makurdi 55 G6
Makran 45 E5
Makran Coast
  Range 42 G4
Maksimkin Yar 29 D9
Mākū 46 C6
Makumbi 56 F4
Makurazaki ... 33 D2
Makurdi 55 G6
Malabang ... 38 C2
Malabar Coast 43 P9
Malabo = Rey
  Malabo 56 D1
Malacca, Str. of 36 D1
Maladzyechna . 17 A8
Málaga 18 D3

Malakâl 53 G5
Malakand 42 B7
Malang 37 F4
Malanje 56 F3
Mälaren 9 G11
Malargüe 94 D3
Malaryta 17 C7
Malatya 46 C4
Malawi ■ 59 A6
Malawi, L. 59 A6
Malaybalay ... 38 C3
Malāyer 46 D7
Malaysia ■ ... 36 D4
Malazgirt 46 C5
Malbooma ... 62 B1
Malbork 16 A4
Malden I. 65 K15
Maldives ■ ... 26 J11
Maldonado ... 94 C6
Malé Karpaty .. 16 D3
Maléa, Ákra .. 23 F4
Malegaon 43 J9
Malema 59 A7
Malha 53 E4
Mali ■ 55 E4
Mali → 41 E11
Malik 39 E2
Malili 39 E2
Malin Hd. 10 D3
Malindi 57 E8
Malines =
  Mechelen ... 14 C3
Malino 39 D2
Malita 38 C3
Malkara 22 D6
Mallacoota ... 63 C4
Mallacoota Inlet 63 C4
Mallaig 10 C4
Mallawi 52 C5
Mallorca 19 C7
Mallow 11 E2
Malmö 9 G10
Malolos 38 B2
Malpelo 90 C2
Malta ■ 21 G5
Maltahöhe ... 58 C3
Maluku 39 E3
Maluku □ 39 E3
Maluku Sea =
  Molucca Sea 39 E2
Malvan 43 L8
Malvinas, Is. =
  Falkland Is. □ 95 G5
Malyn 17 C9
Malyy
  Lyakhovskiy,
  Ostrov 31 B12
Mamahatun ... 46 C5
Mamanguape . 92 D6
Mamasa 39 E1
Mamberamo → 39 E5
Mambilima Falls 57 G5
Mamburao ... 38 B2
Mamfe 55 G6
Mamoré → ... 91 F5
Mamou 55 F2
Mamuju 39 E1
Man 55 G3
Man, I. of 11 D4
Man Na 41 F11
Mana 92 A3
Manaar, G. of =
  Mannar, G. of 43 Q11
Manacapuru .. 90 D6
Manacor 19 C7
Manado 39 D2
Managua 85 E7
Manakara 59 J9
Manama = Al
  Manāmah ... 44 E2

Place	Map	Grid
Mananara	59	H9
Mananjary	59	J9
Manaos =		
Manaus	90	D7
Manapouri	65	F2
Manapouri, L.	65	F2
Manas	34	B3
Manas →	41	D8
Manasir	44	E2
Manaung	41	H9
Manaus	90	D7
Manay	38	C3
Manchester, U.K.	11	E5
Manchester, U.S.A.	73	C10
Mand →	44	D3
Manda	57	G6
Mandal	9	G9
Mandalay	41	F11
Mandale =		
Mandalay	41	F11
Mandalī	46	D6
Mandar, Teluk	39	E1
Mandaue	38	B2
Mandi	42	D10
Mandimba	59	A7
Mandioli	39	E3
Mandla	40	F3
Mandritsara	59	H9
Mandsaur	43	G9
Mandvi	43	H6
Mandya	43	N10
Manfalût	52	C5
Manfred	63	B3
Manfredónia	20	D5
Mangalia	22	C7
Mangalore	43	N9
Mangaweka	64	C6
Manggar	37	E3
Manggawitu	39	E4
Mangkalihat, Tanjung	39	D1
Mangla Dam	42	C8
Mangnai	34	C4
Mango	55	F5
Mangoche	59	A7
Mangole	39	E3
Mangonui	64	A5
Mangueigne	53	F3
Mangueira, L. da	94	C6
Mangyshlak Poluostrov	29	E6
Manhuaçu	93	G5
Manica	59	B6
Manicoré	91	E6
Manicouagan →	69	D4
Manīfah	47	F7
Manihiki	65	L14
Manila	38	B2
Manila B.	38	B2
Manilla	63	B5
Manipur □	41	E9
Manipur →	41	F10
Manisa	23	E6
Manitoba □	71	C10
Manitoba, L.	71	C10
Manitowoc	72	B2
Manizales	90	B3
Manja	59	J8
Manjacaze	59	C6
Manjhand	42	G6
Manjil	46	C7
Manjra →	43	K10
Mankono	55	G3
Manly	63	B5
Manmad	43	J9
Manna	37	E2
Mannahill	62	B3
Mannar	43	Q11
Mannar, G. of	43	Q11
Mannar I.	43	Q11
Mannheim	14	D5
Mannum	62	B2
Mano	55	G2
Manokwari	39	E4
Manombo	59	J8
Manono	57	F5
Manosque	13	E6
Manresa	19	B6
Mansa	57	G5
Mansel I.	68	B2
Mansfield, Australia	63	C4
Mansfield, U.K.	11	E6
Mansfield, U.S.A.	72	D4
Manta	90	D2
Mantalingajan, Mt.	36	C5
Mantes-la-Jolie	12	B4
Manthani	43	K11
Mantiqueira, Serra da	93	G5
Mántova	20	B3
Mantua = Mántova	20	B3
Manu	91	F4
Manuae	65	L15
Manuel Alves →	93	E4
Manui	39	E2
Manyara, L.	57	E7
Manych-Gudilo, Ozero	25	D5
Manyoni	57	F6
Manzai	42	C7
Manzanares	18	C4
Manzanillo, Cuba	86	B4
Manzanillo, Mexico	84	D4
Manzhouli	35	B6
Mao	53	F2
Maoke, Pegunungan	39	E5
Maoming	35	D6
Mapam Yumco	40	B3
Mapia, Kepulauan	39	D4
Mapuera →	90	D7
Maputo	59	D6
Maqnā	47	E3
Maquela do Zombo	56	F3
Maquinchao	95	E3
Mar, Serra do	94	B7
Mar Chiquita, L.	94	C4
Mar del Plata	94	D5
Mar Menor	19	D5
Maraã	90	D5
Marabá	92	D4
Maracá, I. de	92	B3
Maracaibo	90	A4
Maracaibo, L. de	90	B4
Maracay	90	A5
Marādah	52	C2
Maradi	55	F6
Marāgheh	46	C6
Marāh	47	F6
Marajó, I. de	92	C4
Maralal	57	D7
Marama	62	C3
Marampa	55	G2
Marand	46	C6
Maranguape	92	C6
Maranhão = São Luís	92	C5
Maranhão □	92	D4
Maranoa →	63	A4
Marañón →	90	D4
Marão	59	C6
Maraş = Kahramanmaraş	46	C4
Maratua	39	D1
Marbella	18	D3
Marburg	14	C5
Marche	12	C4
Marchena	18	D3
Mardan	42	B8
Mardin	46	C5
Marek = Stanke Dimitrov	22	C4
Marek	39	E2
Margarita, I. de	90	A6
Margate	11	F7
Margelan = Marghilon	29	E8
Marghilon	29	E8
Mari El □	24	B6
Mari Republic □ = Mari El □	24	B6
Maria I.	62	D4
Maria van Diemen, C.	64	A5
Mariana Trench	64	H9
Marianao	86	B3
Ma'rib	49	D4
Maribor	20	A5
Marîdî	53	H4
Marie Byrd Land	96	B18
Marie-Galante	87	C7
Mariecourt = Kangiqsujuaq	68	B3
Mariental	58	C3
Mariinsk	29	D9
Marília	93	G3
Marín	18	A1
Marinduque	38	B2
Maringá	94	A6
Marion	79	A9
Mariscal Estigarribia	91	H6
Maritime Alps = Maritimes, Alpes	13	D7
Maritimes, Alpes	13	D7
Maritsa = Évros →	22	D6
Mariupol	25	D4
Marīvān	46	D6
Markazī □	44	C2
Markham, Mt.	96	C15
Marks	24	C6
Marla	62	A1
Marmagao	43	M8
Marmara	23	D6
Marmara, Sea of = Marmara Denizi	22	D7
Marmara Denizi	22	D7
Marmaris	23	F7
Marmolada, Mte.	20	A3
Marne →	12	B5
Maroantsetra	59	H9
Marondera	59	B6
Maroni →	92	A3
Maroochydore	63	A5
Maroona	62	C3
Maroua	53	F1
Marovoay	59	H9
Marquesas Is. = Marquises, Is.	65	K17
Marquette	69	D2
Marquises, Is.	65	K17
Marrakech	54	B3
Marrawah	62	D3
Marree	62	A2
Marromeu	59	B7
Marrowie Cr. →	63	B4
Marrupa	59	A7
Marsá Matrûh	52	B4
Marsá Susah	52	B3
Marsabit	57	D7
Marsala	21	F4
Marsden	63	B4
Marseille	13	E6
Marseilles = Marseille	13	E6
Marshall, Liberia	55	G2
Marshall, U.S.A.	79	C6
Marshall Is. ■	64	J12
Marshalltown	77	D8
Martaban	41	J11
Martaban, G. of	41	J11
Martapura, Kalimantan, Indonesia	37	E4
Martapura, Sumatera, Indonesia	37	E2
Marte	53	F1
Martigny	13	C7
Martigues	13	E6
Martin	16	D4
Martina Franca	21	D6
Martinborough	65	D6
Martinique ■	87	D7
Martinique Passage	87	C7
Martinsville	75	A8
Marton	64	D6
Martos	18	D4
Marudi	37	D4
Marugame	32	B3
Marulan	63	B5
Marwar	42	G8
Mary	29	F7
Maryborough = Port Laoise	11	E3
Maryborough, Queens., Australia	63	A5
Maryborough, Vic., Australia	62	C3
Maryland □	73	E7
Maryvale	63	A5
Marzūq	52	C1
Masaka	57	E6
Masalembo, Kepulauan	37	F4
Masalima, Kepulauan	37	F5
Masamba	39	E2
Masan	35	C7
Masandam, Ras	44	E4
Masasi	57	G7
Masaya	85	E7
Masbate	38	B2
Mascara	54	A5
Masela	39	F3
Maseru	59	D5
Mashābih	47	F4
Mashhad	44	B4
Mashike	32	F12
Mashkel, Hamun-i-	42	E3
Mashki Chāh	42	E3
Masi Manimba	56	E3
Masindi	57	D6
Masisea	91	E4
Masjed Soleyman	47	E7

Mask, L.

Mask, L. 11 E2
Masohi 39 E3
Mason City ... 77 D8
Masqat 45 F4
Massa 20 B3
Massachusetts □ 73 C9
Massaguet 53 F2
Massakory 53 F2
Massangena . 59 C6
Massawa = Mitsiwa 53 E6
Massena 73 B8
Massénya 53 F2
Massif Central . 13 D5
Massinga 59 C7
Masson I. 96 A11
Mastanli = Momchilgrad 22 D5
Masterton 65 D6
Mastuj 42 A8
Mastung 42 E5
Masty 17 B7
Masuda 32 B2
Masvingo 59 C6
Mataboor 39 E5
Matad 35 B6
Matadi 56 F2
Matagalpa 85 E7
Matagami, L. .. 69 D3
Matak, P. 37 D3
Matakana 63 B4
Matam 55 E2
Matamoros, Coahuila, Mexico 84 B4
Matamoros, Tamaulipas, Mexico 84 B5
Ma'ţan as Sarra 52 D3
Matane 69 D4
Matanzas 86 B3
Matapan, C. = Taínaron, Ákra 23 F4
Mataram 37 F5
Matarani 91 G4
Mataró 19 B7
Mataura 65 G3
Matehuala 84 C4
Matera 21 D6
Mathura 42 F10
Mati 38 C3
Matiri Ra. 65 D5
Matmata 54 B6
Mato Grosso □ 93 E3
Mato Grosso, Planalto do .. 93 F3
Mato Grosso do Sul □ 93 F3
Matochkin Shar 28 B6
Matosinhos .. 18 B1
Matrah 45 F4
Matsue 32 B3
Matsumoto ... 32 A6
Matsusaka ... 32 B5
Matsuyama ... 32 C3
Mattagami → . 69 C2
Mattancheri .. 43 Q10
Matterhorn ... 13 D7
Matthew's Ridge 90 B6
Matucana 91 F3
Matun 42 C6
Maturín 90 B6
Mau Ripur ... 43 G11
Maubeuge ... 13 A5
Maude 63 B3
Maudin Sun .. 41 K10
Maués 90 D7
Mauganj 40 E3

Maui 82 H12
Maulamyaing = Moulmein ... 41 J11
Maumere 39 F2
Maun 58 B4
Maungmagan Kyunzu ... 41 K11
Mauritania ■ .. 55 D2
Mauritius ■ ... 51 H9
Mavinga 58 B4
Mawk Mai ... 41 G11
Mawlaik 41 F10
Mawson Coast 96 A10
Maxixe 59 C7
Mayaguana ... 87 B5
Mayagüez 87 C6
Maydena 62 D4
Mayenne → .. 12 C3
Maykop 25 E5
Mayo 70 B6
Mayo L. 70 B6
Mayon Volcano 38 B2
Mayor I. 64 B7
Mayu 39 D3
Mazabuka 59 B5
Mazagán = El Jadida 54 B3
Mazagão 92 C3
Mazán 90 D4
Māzandarān □ 44 B3
Mazar-e Sharif . 45 B6
Mazara del Vallo 21 F4
Mazarredo ... 95 F3
Mazarrón 19 D5
Mazaruni → . 90 B7
Mazatlán 84 C3
Māzhān 44 C4
Mazīnān 44 B4
Mazoe → 59 B6
Mazurian Lakes = Mazurski, Pojezierze ... 16 B5
Mazurski, Pojezierze ... 16 B5
Mazyr 17 C9
Mbabane 59 D6
Mbaïki 56 D3
Mbala 57 F6
Mbale 57 D6
Mbalmayo 56 D2
Mbamba Bay .. 57 G6
Mbandaka 56 D3
Mbanza Congo 56 F2
Mbanza Ngungu 56 F2
Mbarara 57 E6
Mbeya 57 F6
Mbini □ 56 D2
Mbour 55 F1
Mbout 55 E2
Mbuji-Mayi ... 56 F4
Mbulu 57 E7
Mchinji 59 A6
Mead, L. 82 B5
Mearim → ... 92 C5
Meaux 13 B5
Mecca = Makkah 47 G4
Mechelen 14 C3
Mecheria 54 B4
Mecklenburg .. 15 B6
Mecklenburger Bucht 15 A6
Meconta 59 A7
Medan 37 D1
Medanosa, Pta. 95 F3
Medéa 54 A5
Medellín 90 B3
Mederdra 55 E1
Medford 80 D2

Medgidia 22 B7
Mediaş 17 E7
Medicine Bow Ra. 81 E10
Medicine Hat .. 71 C8
Medina = Al Madīnah ... 47 F4
Medina del Campo 18 B3
Medina-Sidonia 18 D3
Medinipur 40 F6
Mediterranean Sea 50 C5
Médoc 12 D3
Medveditsa → 24 D5
Medvezhyegorsk 28 C4
Meekatharra .. 60 F2
Meerut 42 E10
Mega 53 H6
Mégara 23 F4
Meghalaya □ .. 41 E8
Mei Xian 35 D6
Meiganga 56 C2
Meiktila 41 G10
Meissen 15 C7
Mejillones 94 A2
Mékambo 56 D2
Mekdela 53 F6
Mekhtar 42 D6
Meknès 54 B3
Mekong → ... 36 C3
Mekongga ... 39 E2
Mekvari = Kür → 25 E6
Melagiri Hills .. 43 N10
Melaka 37 D2
Melalap 36 C5
Melanesia ... 64 K10
Melbourne, Australia 63 C3
Melbourne, U.S.A. 75 E7
Mélèzes → ... 68 C3
Melfi 53 F2
Melfort 71 C9
Melilla 19 E4
Melitopol 25 D4
Melk 15 D8
Melo 94 C6
Melolo 39 F2
Melrose 63 B4
Melun 13 B5
Melut 53 F5
Melville 71 C9
Melville I. 60 C5
Melville Pen. .. 68 B2
Memba 59 G8
Memboro 39 F1
Memel = Klaipėda 24 B1
Memmingen .. 14 E6
Mempawah ... 37 D3
Memphis, Tenn., U.S.A. 74 B3
Memphis, Tex., U.S.A. 78 B3
Ménaka 55 E5
Menan = Chao Phraya → .. 36 B2
Menate 37 E4
Mendawai → . 37 E4
Mende 13 D5
Mendocino, C. . 80 E1
Mendoza 94 C3
Mene Grande . 90 B4
Menemen ... 23 E6
Menggala 37 E3
Mengzi 34 D5
Menindee 62 B3

Menindee L. .. 62 B3
Meningie 62 C2
Menominee ... 72 B2
Menongue ... 58 A3
Menorca 19 C7
Mentawai, Kepulauan .. 37 E1
Menton 13 E7
Meppel 14 B4
Merabéllou, Kólpos 23 G5
Meran = Merano 20 A3
Merano 20 A3
Merauke 39 F6
Merbein 62 B3
Merca 49 G3
Merced 82 B2
Mercedes, Buenos Aires, Argentina ... 94 C5
Mercedes, Corrientes, Argentina ... 94 B5
Mercedes, San Luis, Argentina ... 94 C3
Mercedes, Uruguay ... 94 C5
Merceditas ... 94 B2
Mercer 64 B6
Mercy C. 68 B4
Meredith, C. ... 95 G4
Merga = Nukheila 53 E4
Mergui Arch. = Myeik Kyunzu 36 B1
Mérida, Mexico 85 C7
Mérida, Spain . 18 C2
Mérida, Venezuela .. 90 B4
Mérida, Cord. de 87 E5
Meridian 74 C3
Meriruma 92 B3
Merowe 53 E5
Merriwa 63 B5
Merriwagga .. 63 B4
Merrygoen ... 63 B4
Mersa Fatma .. 49 E3
Merseburg 15 C6
Mersin 46 C3
Mersing 37 D2
Merthyr Tydfil . 11 F5
Mértola 18 D2
Meru 57 D7
Mesa 83 D7
Meshed = Mashhad ... 44 B4
Meshra er Req . 53 G4
Mesolóngion .. 23 E3
Mesopotamia = Al Jazirah ... 46 D5
Messina, Italy . 21 E5
Messina, S. Africa 59 C6
Messina, Str. di 21 E5
Messíni 23 F4
Messiniakós Kólpos 23 F4
Mesta → 22 D5
Meta → 90 B5
Metairie 79 E8
Metán 94 B4
Metangula ... 59 A6
Metema 53 F6
Methven 65 E4
Metil 59 B7
Metlakatla ... 71 C6
Mettur Dam ... 43 P10

Metz	13	B7
Meulaboh	37	D1
Meureudu	36	C1
Meuse →	13	A6
Mexiana, I.	92	C4
Mexicali	84	A1
México	84	D5
México □	84	D5
Mexico ■	84	C4
Mexico, G. of	85	B7
Meymaneh	42	B4
Mezen	28	C5
Mezen →	28	C5
Mézene	13	D6
Mezökövesd	16	E5
Mezötúr	16	E5
Mhow	43	H9
Miami	75	G7
Miami Beach	75	G7
Mīāndowāb	46	C6
Miandrivazo	59	H9
Mīāneh	46	C6
Mianwali	42	C7
Miaoli	35	D7
Miarinarivo	59	H9
Miass	29	D7
Michalovce	17	D5
Michigan □	72	C3
Michigan, L.	72	C2
Michipicoten	69	D2
Michurin	22	C6
Michurinsk	24	C5
Micronesia,		
Federated		
States of ■	64	J10
Middelburg	58	E5
Middlesbrough	11	D6
Midi, Canal		
du →	12	E4
Midland	78	D3
Midyat	46	C5
Midžŏr	22	C4
Międzychód	16	B2
Międzyrzec		
Podlaski	17	C6
Mielec	16	C5
Miercurea Ciuc	17	E7
Mieres	18	A3
Miguel Alves	92	C5
Mihara	32	B3
Mikhaylovgrad	22	C4
Míkonos	23	F5
Mikura-Jima	32	C6
Milagro	90	D3
Milan = Milano	20	B2
Milang	62	B2
Milano	20	B2
Milâs	23	F6
Milazzo	21	E5
Mildura	62	B3
Miles	63	A5
Miletus	23	F6
Milford Haven	11	F4
Milford Sd.	65	F2
Milḥ, Baḥr al	47	D5
Miliana	54	C5
Mill I.	96	A12
Millau	13	D5
Millicent	62	C3
Millinocket	73	B11
Millmerran	63	A5
Milne Inlet	68	A2
Mílos	23	F5
Milparinka P.O.	62	A3
Milton	65	G3
Milton Keynes	11	E6
Miltou	53	F2
Milwaukee	72	C2
Min Chiang →	35	D6
Min Jiang →	35	D5
Mina Pirquitas	94	A3
Mīnā Su'ud	47	E7
Mīnā'al Aḥmadī	47	E7
Mīnāb	44	E4
Minamata	32	C2
Minas	94	C5
Minas Gerais □	93	F4
Minbu	41	G10
Mindanao	38	C2
Mindanao Sea =		
Bohol Sea	38	C2
Mindanao		
Trench	38	B3
Minden	14	B5
Mindiptana	39	F6
Mindoro	38	B2
Mindoro Str.	38	B2
Mindouli	56	E2
Mingäçevir Su		
Anbarı	25	E6
Mingan	69	C4
Mingechaurskoye		
Vdkhr. =		
Mingäçevir Su		
Anbarı	25	E6
Mingin	41	F10
Mingt'iehkaitafan		
= Mintaka		
Pass	42	A9
Minho =		
Miño →	18	B1
Minho →	18	B1
Minna	55	G6
Minneapolis	77	C8
Minnedosa	71	C10
Minnesota □	77	B7
Minnipa	62	B2
Miño →	18	B1
Minorca =		
Menorca	19	C7
Minore	63	B4
Minot	76	A4
Minsk	24	C2
Mińsk		
Mazowiecki	16	B5
Mintaka Pass	42	A9
Minto	71	B5
Minutang	41	C11
Minvoul	56	D2
Mir	53	F1
Mira	20	B4
Miraj	43	L9
Miram Shah	42	C7
Miranda	93	G2
Miranda de Ebro	19	A4
Miranda do		
Douro	18	B2
Miri	36	D4
Mirim, L.	94	C6
Mirpur Khas	42	G6
Mirzapur	40	E4
Mirzapur-cum-		
Vindhyachal =		
Mirzapur	40	E4
Mish'āb, Ra's		
al	47	E7
Mishan	35	B8
Miskīn	44	F4
Miskolc	16	D5
Misool	39	E4
Misrātah	52	B2
Misriç	46	C5
Missinaibi →	69	C2
Mississippi □	79	C9
Mississippi →	79	E9
Mississippi River		
Delta	79	E9
Missoula	81	B6
Missouri □	77	F8
Missouri →	77	F9
Mistassini L.	69	C3
Misurata =		
Misrātah	52	B2
Mitchell,		
Australia	63	A4
Mitchell, U.S.A.	76	D5
Mitchell →	61	D7
Mitilíni	23	E6
Mito	32	A7
Mitrovica =		
Titova-		
Mitrovica	22	C3
Mitsinjo	59	H9
Mitsiwa	53	E6
Mittagong	63	B5
Mitú	90	C4
Mitumba,		
Chaîne des	57	F5
Mitumba Mts. =		
Mitumba,		
Chaîne des	57	F5
Mitwaba	57	F5
Mitzic	56	D2
Miyake-Jima	32	B6
Miyako	33	G12
Miyakonojō	32	D2
Miyazaki	32	D2
Miyet, Bahr el =		
Dead Sea	47	E3
Mizal	47	G6
Mizdah	52	B1
Mizoram □	41	F9
Mjøsa	8	F10
Mladá Boleslav	16	C2
Mława	16	B5
Mljet	20	C6
Mmabatho	59	D5
Moa	39	F3
Moabi	56	E2
Moalie Park	62	A3
Moba	57	F5
Mobaye	56	D4
Mobayi	56	D4
Mobile	74	D3
Mobutu Sese		
Seko, L. =		
Albert L.	57	D6
Moçambique	59	H8
Moçâmedes =		
Namibe	58	B2
Mochudi	59	C5
Mocimboa da		
Praia	57	G8
Mocoa	90	C3
Mocuba	59	B7
Modane	13	D7
Módena	20	B3
Modesto	82	B2
Módica	21	F5
Moe	63	C4
Moengo	92	A3
Mogadishu =		
Muqdisho	49	G4
Mogador =		
Essaouira	54	B3
Mogami →	33	G12
Mogaung	41	E11
Mogi das Cruzes	94	A7
Mogi-Mirim	94	A7
Mogilev =		
Mahilyow	17	B10
Mogilev-		
Podolskiy =		
Mohyliv-		
Podilskyy	17	D8
Mogoi	39	E4
Mogok	41	F11
Mohács	16	F4
Moḥammadābād		
	44	B4
Mohoro	57	F7
Mohyliv-		
Podilskyy	17	D8
Mointy	29	E8
Moisie	69	C4
Moisie →	69	C4
Moïssala	53	G2
Mojave Desert	82	C4
Mokai	64	C6
Mokokchung	41	D10
Mokra Gora	22	C3
Molchanovo	29	D9
Moldavia ■ =		
Moldova ■	17	E9
Molde	8	F9
Moldova ■	17	E9
Moldoveana	17	F7
Molepolole	59	C5
Molfetta	20	D6
Moline	77	E9
Moliro	57	F6
Mollendo	91	G4
Molodechno =		
Maladzyechna	17	A8
Molokai	82	H12
Molong	63	B4
Molopo →	58	D4
Molotov = Perm	29	D6
Moloundou	56	D3
Molu	39	F4
Molucca Sea =	39	E2
Moluccas =		
Maluku	39	E3
Moma	59	B7
Mombasa	57	E7
Momchilgrad	22	D5
Mompós	90	B4
Mon →	41	G10
Mona, Canal de		
la	87	C6
Monaco ■	13	E7
Monahans	78	D2
Monastir =		
Bitola	22	D3
Monastir	52	A1
Moncayo, Sierra		
del	19	B5
Mönchengladbach		
	14	C4
Monchique	18	D1
Monclova	84	B4
Mondego →	18	B1
Mondeodo	39	E2
Mondovì	20	B1
Monforte de		
Lemos	18	A2
Mong Hsu	41	G12
Mong Kung	41	G11
Mong Nai	41	G11
Mong Pawk	41	F12
Mong Ton	41	G12
Mong Wa	41	G13
Mong Yai	41	F12
Mongalla	53	G5
Monghyr =		
Munger	40	E6
Mongibello =		
Etna	21	F5
Mongo	53	F2
Mongolia ■	35	B5
Mongororo	53	F3
Mongu	58	B4
Monkoto	56	E4
Monópoli	21	D6
Monqoumba	56	D3

Monroe

Monroe, La.,
 U.S.A. 79 C7
Monroe, N.C.,
 U.S.A. 75 B7
Monrovia 55 G2
Mons 14 C2
Monse 39 E2
Mont-de-Marsan 12 E3
Mont-St.-Michel,
 Le = Le Mont-
 St.-Michel ... 12 B3
Montalbán 19 B5
Montana □ ... 81 B9
Montargis 13 C5
Montauban ... 12 D4
Montbéliard .. 13 C7
Montceau-les-
 Mines 13 C6
Monte Alegre . 92 C3
Monte Azul .. 93 F5
Monte-Carlo .. 20 C1
Monte Caseros 94 C5
Monte Comán . 94 C3
Monte Santu, C.
 di 21 D2
Montecristi ... 90 D2
Montecristo .. 20 C3
Montego Bay .. 86 C4
Montélimar ... 13 D6
Montemorelos . 84 B5
Montenegro □ 22 C2
Montepuez 59 A7
Monterey 82 B2
Montería 90 B3
Monterrey 84 B4
Montes Claros . 93 F5
Montesilvano
 Marina 20 C5
Montevideo ... 94 C5
Montgomery =
 Sahiwal 42 D8
Montgomery .. 74 C4
Montijo 18 C2
Montilla 18 D3
Montluçon ... 13 C5
Montmorillon . 12 C4
Montoro 18 C3
Montpellier ... 13 E5
Montréal 69 D3
Montreux 13 C7
Montrose, U.K. 10 C5
Montrose,
 U.S.A. 83 A9
Montserrat ■ . 87 C7
Monveda 56 D4
Monywa 41 F10
Monza 20 B2
Monze 59 B5
Monze, C. 43 G5
Monzón 19 B6
Moolawatana . 62 A2
Moomin Cr. ⟶ 63 A4
Moonie 63 A5
Moonie ⟶ ... 63 A4
Moonta 62 B2
Mooroopna ... 63 C4
Moose Jaw ... 71 C9
Moosomin 71 C9
Moosonee 69 C2
Mopeia Velha . 59 B7
Mopti 55 F4
Moquegua 91 G4
Mora 8 F10
Moradabad ... 42 E11
Morafenobe .. 59 H8
Moramanga ... 59 H9
Moratuwa 43 R11
Morava ⟶,
 Serbia, Yug. . 22 B3

Morava ⟶,
 Slovak Rep. . 16 D3
Moravian Hts. =
 Ceskomoravská
 Vrchovina ... 16 D2
Morawhanna .. 90 B7
Moray Firth .. 10 C5
Morden 71 D10
Mordovian
 Republic □ =
 Mordvinia □ 24 C5
Mordvinia □ .. 24 C5
Morea 62 C3
Moree 63 A4
Morelia 84 D4
Morella 19 B5
Morena, Sierra 18 C3
Moreton I. 63 A5
Morgan 62 B2
Morgan City .. 79 E8
Morioka 33 G12
Morlaix 12 B2
Mornington .. 63 C4
Mornington, I. . 95 F1
Moro G. 38 C2
Morocco ■ ... 54 B3
Morococha ... 91 F3
Morogoro 57 F7
Morombe 59 J8
Morón 86 B4
Mörön ⟶ 35 B6
Morón de la
 Frontera ... 18 D3
Morondava ... 59 J8
Morotai 39 D3
Moroto 57 D6
Morphou 46 D3
Morrinhos 93 F4
Morrinsville .. 64 B6
Morristown ... 75 A6
Morrumbene .. 59 C7
Morshansk ... 24 C5
Morteros 94 C4
Mortes, R.
 das ⟶ 93 E3
Mortlake 62 C3
Morundah 63 B4
Moruya 63 C5
Morvan 13 C6
Morven 63 A4
Morwell 63 C4
Moscos Is. ... 41 L11
Moscow =
 Moskva 24 B4
Moscow 80 B5
Mosel ⟶ 13 A7
Moselle =
 Mosel ⟶ ... 13 A7
Mosgiel 65 F4
Moshi 57 E7
Moskva 24 B4
Moskva ⟶ ... 24 B4
Moson-
 magyaróvár . 16 E3
Mosquera 90 C3
Moss Vale 63 B5
Mossaka 56 E3
Mossburn 65 F3
Mosselbaai ... 58 E4
Mossendjo ... 56 E2
Mossgiel 63 B3
Mossoró 92 D6
Mossuril 59 G8
Most 16 C1
Mostaganem .. 54 A5
Mostar 20 C6
Mostardas ... 94 C6
Mostiska =
 Mostyska ... 17 D6

Mosty = Masty 17 B7
Mostyska 17 D6
Mosul = Al
 Mawşil 46 C5
Motihari 40 D5
Motril 18 D4
Motueka 65 D5
Motueka ⟶ .. 65 D5
Mouanda 56 E2
Moúdhros ... 23 E5
Moudjeria ... 55 E2
Mouila 56 E2
Moulamein ... 63 C3
Moulins 13 C5
Moulmein ... 41 J11
Moundou 53 G2
Mount Barker . 62 C2
Mount Darwin . 59 B6
Mount Eba ... 62 B2
Mount Gambier 62 C3
Mount Hope,
 N.S.W.,
 Australia 63 B4
Mount Hope,
 S. Austral.,
 Australia 62 B2
Mount Howitt . 62 A3
Mount Isa 61 E6
Mount Lofty Ra. 62 B2
Mount McKinley
 National Park 71 B4
Mount Margaret 63 A3
Mount
 Maunganui .. 64 B7
Mount Olympus
 = Uludağ .. 23 D7
Mount Perry .. 63 A5
Mount Vernon . 73 D9
Mountain View 82 B1
Moura, Brazil .. 90 D6
Moura, Portugal 18 C2
Mourdi,
 Dépression du 53 E3
Mourdiah 55 F3
Moussoro 53 F2
Moutohara ... 64 C7
Moutong 39 D2
Moyale 57 D7
Moyamba 55 G2
Moyen Atlas .. 54 B3
Moyo 57 F5
Moyyero ⟶ .. 30 C8
Mozambique =
 Moçambique 59 H8
Moçambique ■ 59 B7
Mozambique
 Chan. 59 H8
Mozdok 25 E5
Mozyr = Mazyr 17 C9
Mpanda 57 F6
Mpika 59 A6
Mpwapwa ... 57 F7
Msaken 52 A1
Msoro 59 A6
Mu Us Shamo . 35 C5
Muaná 92 C4
Muar 37 D2
Muarabungo .. 37 E2
Muaraenim ... 37 E2
Muarajuloi ... 37 E4
Muarakaman .. 37 E5
Muaratebo ... 37 E2
Muaratembesi . 37 E2
Muaratewe ... 37 E4
Mubarraz = Al
 Mubarraz ... 47 F7
Mubende 57 D6
Mubi 53 F1
Muckadilla ... 63 A4

Muconda 56 G4
Mucuri 93 F6
Mudanjiang ... 35 B7
Mudanya 23 D7
Mudgee 63 B4
Mufulira 59 A5
Muğla 23 F7
Mugu 40 C4
Muhammad Qol 52 D6
Mühlhausen .. 14 C6
Mühlig Hofmann
 fjell 96 B7
Mukachere .. 17 D6
Mukachevo =
 Mukacheve .. 17 D6
Mukah 37 D4
Mukden =
 Shenyang ... 35 B7
Mukomuko ... 37 E2
Muktsar 42 D9
Mukur 42 C5
Mula 19 C5
Mulchén 94 D2
Mulde ⟶ 15 C7
Mulgathing ... 62 B1
Mulgrave 69 D4
Mulhacén 18 D4
Mulhouse 13 C7
Mull 10 C4
Mullengudgery 63 B4
Muller,
 Pegunungan 37 D4
Mullingar 11 E3
Mullumbimby . 63 A5
Multan 42 D7
Mulwala 63 C4
Mumbai =
 Bombay 43 K8
Mumbwa 59 A5
Muna 39 F2
München 15 D6
Munchen-
 Gladbach =
 Mönchengladbach
 .... 14 C4
Muncie 72 D3
Mundala 39 E6
Münden 14 C5
Mundo Novo .. 93 E5
Mungallala ... 63 A4
Mungallala
 Cr. ⟶ 63 A4
Mungbere 57 D5
Munger 40 E6
Mungindi 63 A4
Munhango ... 58 A3
Munich =
 München ... 15 D6
Muñoz Gamero,
 Pen. 95 G2
Münster 14 C4
Munster □ ... 11 E2
Muntok 37 E3
Mupa 58 B3
Muqdisho 49 G4
Murallón,
 Cuerro 95 F2
Murang'a 57 E7
Murashi 24 B6
Muratlı 22 D6
Murchison ⟶ . 60 F1
Murchison, Mt. 96 B15
Murchison Falls
 = Kabarega
 Falls 57 D6
Murcia 19 C5
Murcia □ 19 D5
Mureş ⟶ 16 E5

134

Mureşul =
Mureş → ... 16 E5
Murfreesboro . 74 B4
Murgab =
Murghob .. 29 F8
Murghob 29 F8
Murgon 63 A5
Müritz-see ... 15 B7
Murmansk 28 C4
Murom 24 B5
Muroran 32 F12
Muroto-Misaki . 32 C4
Murray → 62 C2
Murray Bridge . 62 C2
Murree 42 C8
Murrumbidgee →
.... 62 B3
Murrumburrah 63 B4
Murrurundi ... 63 B5
Murtoa 62 C3
Murwara 40 F3
Murwillumbah . 63 A5
Mürzzuschlag . 15 E8
Muş 46 C5
Mûsa, G. 47 E3
Musa Khel 42 D6
Mûsá Qal'eh .. 42 C4
Musaffargarh .. 42 D7
Musala, Bulgaria 22 C4
Musala,
Indonesia ... 37 D1
Musay'id 44 E2
Muscat =
Masqat 45 F4
Muscat & Oman
= Oman ■ .. 48 C6
Muscatine 77 E9
Musgrave Ras. . 60 F5
Mushie 56 E3
Musi → 37 E2
Muskogee 79 B6
Musmar 53 E6
Musoma 57 E6
Mussoorie 42 D11
Mustafakemalpaşa
.... 23 D7
Mustang 40 C4
Musters, L. ... 95 F3
Muswellbrook . 63 B5
Mût, Egypt 52 C4
Mut, Turkey ... 46 C3
Mutare 59 B6
Muting 39 F6
Mutoray 30 C8
Muxima 66 F2
Muynak 29 E6
Muzaffarabad . 42 B8
Muzaffarnagar . 42 E10
Muzaffarpur .. 40 D5
Muzhi 28 C7
Muztag 34 C3
Mvuma 59 B6
Mwanza,
Tanzania .. 57 E6
Mwanza, Zaïre . 57 F5
Mweka 56 E4
Mwenezi 59 C6
Mwenga 57 E5
Mweru, L. ... 57 F5
Mwinilunga .. 58 A4
My Tho 36 B3
Myanaung .. 41 H10
Myanmar ■ =
Burma ■ .. 41 G11
Myaungmya .. 41 J10
Mycenæ 23 F4
Myeik Kyunzu . 36 B1
Myingyan 41 G10
Myitkyina 41 E11

Mykolayiv 25 D3
Mymensingh .. 41 E8
Mysia 23 E6
Mysore =
Karnataka □ . 43 N10
Mysore 43 N10
Myszków 16 C4
Mytishchi 24 B4

N

Na Hearadh =
Harris 10 C3
Naab → 15 D7
Naberezhnyye
Chelny 29 D6
Nabeul 52 A1
Nabire 39 E5
Nablus =
Nābulus 46 D3
Nābulus 46 D3
Nachingwea . 57 G7
Nackara 62 B2
Nacogdoches . 79 D6
Nacozari 84 A3
Nadiad 43 H8
Nadūshan 44 C3
Nadvirna 17 D7
Nadvornaya =
Nadvirna ... 17 D7
Nadym 28 C8
Nadym → 28 C8
Nafada 55 F7
Nafūd ad Dahy 47 G6
Nafud Desert =
An Nafūd ... 47 E5
Naga 38 B2
Nagaland □ .. 41 E10
Nagano 32 A6
Nagaoka 33 G11
Nagappattinam 43 P11
Nagar Parkar . 43 G7
Nagasaki 32 C1
Nagaur 42 F8
Nagercoil 43 Q10
Nagīneh 44 C4
Nagoya 32 B5
Nagpur 43 J11
Nagykanizsa . 16 E3
Nagykörös 16 E4
Nahanni Butte . 70 B7
Nahāvand 46 D7
Nā'ifah 48 D5
Nain, Canada .. 68 C4
Nā'īn, Iran 44 C3
Nairn 10 C5
Nairobi 57 E7
Naivasha 57 E7
Najafābād 44 C2
Najd 47 F5
Najibabad 42 E11
Nakamura 32 C3
Nakfa 53 E6
Nakhichevan =
Naxçıvan ... 25 F6
Nakhichevan
Republic □ =
Naxçıvan □ . 25 F6
Nakhodka 31 E11
Nakhon Phanom 36 A2
Nakhon
Ratchasima . 36 B2
Nakhon Sawan 36 A2
Nakhon Si
Thammarat . 36 C2
Nakina 69 C2

Nakuru 57 E7
Nal → 42 G4
Nalchik 25 E5
Nalgonda 43 L11
Nallamalai Hills 43 M11
Nālūt 52 B1
Nam Co 34 C4
Nam-Phan 36 B3
Namacurra ... 59 B7
Namak,
Daryācheh-ye 44 C3
Namak, Kavir-e 44 C4
Namaland 58 C3
Namangan ... 29 E8
Namapa 59 A7
Namber 39 E4
Nambour 63 A5
Nambucca
Heads 63 B5
Nameh 37 D5
Namib Desert =
Namibwoestyn
.... 58 C2
Namibe 58 B2
Namibia ■ ... 58 C3
Namibwoestyn 58 C2
Namlea 39 E3
Namoi → 63 B4
Nampa 80 D5
Nampula 59 B7
Namrole 39 E3
Namse Shankou 40 C4
Namtu 41 F11
Namur 14 C3
Namutoni 58 B3
Namwala 59 B5
Nanaimo 71 D7
Nanango 63 A5
Nanao 32 A5
Nanchang ... 35 D6
Nanching =
Nanjing ... 35 C6
Nanchong ... 35 C5
Nancy 13 B7
Nanda Devi .. 42 D11
Nanded 43 K10
Nandewar Ra. . 63 B5
Nandurbar ... 43 J9
Nandyal 43 M11
Nanga-Eboko .. 56 D2
Nanga Parbat . 42 B9
Nangapinoh .. 37 E4
Nangarhár □ .. 42 B7
Nangatayap .. 37 E4
Nanjing 35 C6
Nanking =
Nanjing ... 35 C6
Nanning 35 D5
Nanping 35 D6
Nansei-Shotō =
Ryūkyū-rettō . 35 D7
Nantes 12 C3
Nanuque 93 F5
Nanusa,
Kepulauan . 38 D3
Nanyang 35 C6
Nanyuki 57 D7
Nao, C. de la .. 19 C6
Naoetsu 32 A6
Napa 80 F2
Napier 64 C7
Naples = Nápoli 21 D5
Naples 75 F7
Napo → 90 D4
Nápoli 21 D5
Nappa Merrie . 62 A3
Nara, Japan .. 32 B4
Nara, Mali ... 55 E3
Naracoorte 62 C3

Naradhan 63 B4
Narasapur 40 J3
Narathiwat ... 36 C2
Narayanganj . 41 F8
Narayanpet ... 43 L10
Narbonne 13 E5
Nardò 23 D2
Narew → 16 B5
Narin 42 A6
Narmada → .. 43 J8
Narooma 63 C5
Narrabri 63 B4
Narran → 63 A4
Narrandera .. 63 B4
Narromine ... 63 B4
Narsimhapur . 43 H11
Narva 24 B2
Narvik 8 E11
Naryan-Mar .. 28 C6
Naryilco 62 A3
Narym 29 D9
Narymskoye .. 29 E9
Naryn 29 E8
Nasarawa ... 55 G6
Naseby 65 F4
Naser, Buheirat
en 52 D5
Nashville 74 A4
Nasik 43 K8
Nasirabad ... 42 F9
Nassau 86 A4
Nassau, B. .. 95 H3
Nasser, L. =
Naser,
Buheirat en . 52 D5
Nat Kyizin .. 41 K11
Nata 59 C5
Natagaima .. 90 C3
Natal, Brazil .. 92 D6
Natal, Indonesia 37 D1
Naţanz 44 C2
Natashquan .. 69 C4
Natashquan → 69 C4
Natchez 79 D8
Nathalia 63 C4
Nathdwara .. 43 G8
Natitingou ... 55 F5
Natron, L. ... 57 E7
Natuna Besar,
Kepulauan . 37 D3
Natuna Is. =
Natuna Besar,
Kepulauan . 37 D3
Natuna Selatan,
Kepulauan . 37 D3
Naturaliste, C. . 62 D4
Naumburg ... 15 C6
Nauru ■ 64 K11
Naushahra =
Nowshera .. 42 B8
Nauta 90 D4
Nautanwa ... 40 D4
Navahrudak ... 17 B7
Navalmoral de
la Mata 18 C3
Navarino, I. ... 95 H3
Navarra □ ... 19 A5
Năvodari 22 B7
Navoi = Nawoiy 29 E7
Navojoa 84 B3
Návpaktos .. 23 E3
Návplion 23 F4
Navsari 43 J8
Nawabshah .. 42 F6
Nawakot 40 D5
Nawalgarh .. 42 F9
Nawāsif, Harrat 47 G5
Nawoiy 29 E7
Naxçıvan 25 F6

Naxçıvan

Naxçıvan □ ... 25 F6
Náxos 23 F5
Näy Band 44 E3
Nazas → 84 B4
Nazilli 23 F7
Nazir Hat 41 F8
Ncheu 59 A6
Ndalatando ... 56 F2
Ndélé 56 C4
Ndendé 56 E2
Ndjamena 53 F1
Ndjolé 56 E2
Ndola 59 A5
Neagh, Lough . 11 D3
Near Is. 71 C1
Neath 11 F5
Nebine Cr. → . 63 A4
Nebitdag 29 F6
Nebraska □ ... 76 E5
Nébrodi, Monti 21 F5
Neckar → 14 D5
Necochea 94 D5
Neemuch =
  Nimach 43 G9
Neepawa 71 C10
Nefta 54 B6
Neftçala 46 C7
Negapatam =
  Nagappattinam
  .... 43 P11
Negele 49 F2
Negombo 43 R11
Negotin 22 B4
Negra Pt. 38 A2
Negrais, C. =
  Maudin Sun . 41 K10
Negro →,
  Argentina ... 95 E4
Negro →, Brazil 90 D6
Negro →,
  Uruguay 94 C5
Negros 38 C2
Nehbandān ... 45 D5
Nei Monggol
  Zizhiqu □ ... 35 B6
Neijiang 35 D5
Neiva 90 C3
Nejd = Najd .. 47 F5
Nekemte 53 G6
Nellore 43 M11
Nelson, Canada 71 D8
Nelson, N.Z. .. 65 D5
Nelson → ... 71 C10
Nelson, C. 62 C3
Nelson, Estrecho 95 G2
Nelspruit 59 D6
Néma 55 E3
Neman → 24 B1
Nemunas =
  Neman → .. 24 B1
Nemuro 32 F13
Nemuro-Kaikyō 32 F13
Nenagh 11 E2
Nenana 71 B5
Nenjiang 35 B7
Nepal ■ 40 D5
Nepalganj 40 C3
Neretva → ... 20 C6
Ness, L. 10 C4
Nesterov 17 C6
Nesvizh =
  Nyasvizh 17 B8
Netherlands ■ . 14 B3
Netherlands
  Antilles ■ ... 87 D6
Nettilling L. ... 68 B3
Neubrandenburg
  .... 15 B7
Neuchâtel 13 C7

Neuchâtel, Lac
  de 13 C7
Neumünster .. 14 A5
Neunkirchen .. 14 D4
Neuquén 94 D3
Neuruppin ... 15 B7
Neustrelitz ... 15 B7
Neva → 24 B3
Nevada □ 80 F5
Nevada, Sierra,
  Spain 18 D4
Nevada, Sierra,
  U.S.A. 80 F3
Nevers 13 C5
Nevertire 63 B4
Nevinnomyssk 25 E5
Nevis 87 C7
Nevşehir 46 C3
New Amsterdam 90 B7
New Angledool 63 A4
New Bedford . 73 D10
New Bern 75 B9
New Brighton . 65 E5
New Britain,
  Papua N. G. . 64 K10
New Britain,
  U.S.A. 73 D9
New
  Brunswick □ 69 D4
New
  Caledonia ■ . 64 M11
New Delhi 42 E10
New England
  Ra. 63 B5
New Glasgow . 69 D4
New Guinea .. 64 K9
New
  Hampshire □ 73 C10
New Haven ... 73 D9
New Hebrides =
  Vanuatu ■ .. 64 L11
New Ireland .. 64 K10
New Jersey □ . 73 E9
New London .. 73 D9
New Mexico □ 83 C9
New Norfolk .. 62 D4
New Orleans .. 79 E8
New Plymouth 64 C6
New Providence 86 A4
New Siberian Is.
  =
  Novosibirskiye
  Ostrova 31 B12
New South
  Wales □ 63 B4
New
  Westminster . 71 D7
New York □ ... 73 C8
New York City . 73 D9
New Zealand ■ 64 D7
Newala 57 G7
Newark 73 D8
Newburgh ... 73 D8
Newbury 11 F6
Newcastle,
  Australia 63 B5
Newcastle,
  S. Africa 59 D5
Newcastle-upon-
  Tyne 11 D6
Newfoundland □
  .... 69 C5
Newman 60 E2
Newport,
  I. of W., U.K. 11 F6
Newport, Newp.,
  U.K. 11 F5
Newport, U.S.A. 73 D10
Newport News . 73 F7

Newquay 11 F4
Newry 11 D3
Newton Boyd . 63 A5
Neya 24 B5
Neyrīz 44 D3
Neyshābūr ... 44 B4
Nezhin = Nizhyn 24 C3
Ngabang 37 D3
Ngabordamlu,
  Tanjung 39 F4
Ngami
  Depression .. 58 C4
Nganglong
  Kangri 40 A3
Ngaoundéré .. 56 C2
Ngapara 65 F4
Ngoring Hu ... 34 C4
Ngudu 57 E6
Nguigmi 53 F1
Nguru 55 F7
Nha Trang ... 36 B3
Nhill 62 C3
Niafounké ... 55 E4
Niagara Falls,
  Canada 69 D3
Niagara Falls,
  U.S.A. 73 C6
Niah 37 D4
Niamey 55 F5
Niangara 57 D5
Nias 37 D1
Nicaragua ■ . 85 E7
Nicaragua, L. de 85 E7
Nicastro 21 E6
Nice 13 E7
Nichinan 33 D2
Nicobar Is. ... 27 J13
Nicosia 46 D3
Nicoya, Pen. de 85 F7
Niedersachsen □
  .... 14 B5
Niemen =
  Neman → .. 24 B1
Nienburg 14 B5
Nieuw
  Amsterdam . 92 A2
Nieuw Nickerie 92 A2
Niğde 46 C3
Niger ■ 55 E6
Niger → 55 G6
Nigeria ■ 55 G6
Nightcaps ... 65 F3
Nii-Jima 32 B6
Niigata 33 G11
Niihama 32 C3
Niihau 82 H11
Nijmegen 14 C3
Nikiniki 39 F2
Nikki 55 G5
Nikolayev =
  Mykolayiv ... 25 D3
Nikolayevsk .. 24 C6
Nikopol 25 D3
Nikshahr 45 E5
Nikšić 22 C2
Nîl, Nahr en → 52 B5
Nîl el Abyad → 53 E5
Nîl el Azraq → 53 E5
Nile = Nîl, Nahr
  en → 52 B5
Nimach 43 G9
Nîmes 13 E6
Nimfaíon, Ákra
  = Pínnes,
  Ákra 23 D5
Nimmitabel ... 63 C4
Nīmrūz □ 45 D5
Nimule 57 D6
Nīnawá 46 C5

Nindigully 63 A4
Nineveh =
  Nīnawá 46 C5
Ningbo 35 D7
Ningde 35 D6
Ningpo =
  Ningbo 35 D7
Ningsia Hui A.R.
  = Ningxia
  Huizu
  Zizhiqu □ ... 35 C5
Ningxia Huizu
  Zizhiqu □ ... 35 C5
Niobrara → .. 76 D5
Nioro du Sahel 55 E3
Niort 12 C3
Nipawin 71 C9
Nipigon, L. ... 69 D2
Niquelândia .. 93 E4
Nirmal 43 K11
Nirmali 40 D6
Niš 22 C3
Nişāb, Si. Arabia 47 E6
Nişāb, Yemen . 49 E4
Nishinomiya .. 32 B4
Nistru =
  Dnister → .. 17 E10
Niţā' 47 F7
Niterói 93 G5
Nitra 16 D4
Nitra → 16 E4
Niue 65 L14
Niut 37 D4
Nivernais 13 C5
Nizamabad ... 43 K11
Nizamghat ... 41 C10
Nizhne-Vartovsk 29 C8
Nizhneyansk .. 31 B11
Nizhniy
  Novgorod ... 24 B5
Nizhniy Tagil .. 29 D6
Nizhyn 24 C3
Nizip 46 C4
Nízké Tatry ... 16 D4
Njombe 57 F6
Nkambe 55 G7
Nkawkaw 55 G4
Nkhata Bay ... 57 G6
Nkhota Kota . 59 A6
Nkongsamba . 56 D1
Nmai → 41 E11
Noakhali =
  Maijdi 41 F8
Noatak 71 B3
Nobeoka 32 C2
Nocera Inferiore 21 D5
Nockatunga .. 62 A3
Nogales, Mexico 84 A2
Nogales, U.S.A. 83 E7
Nōgata 32 C2
Noginsk 30 C7
Noire, Mts. ... 12 B2
Noirmoutier, I.
  de 12 C2
Nok Kundi ... 42 E3
Nokhtuysk ... 30 C9
Nokomis 71 C9
Nola 56 D3
Nome 71 B3
Nong Khai ... 36 A2
Noondoo 63 A4
Nóqui 56 F2
Noranda 69 D3
Nord-Ostsee-
  Kanal → ... 14 A5
Nordegg 71 C8
Norderney ... 14 B4
Norderstedt .. 14 B5

Nordfriesische Inseln 14 A5
Nordhausen ... 14 C6
Nordkapp 8 D13
Nordrhein-Westfalen □ . 14 C4
Nordvik 30 B9
Norfolk 73 F7
Norfolk I. 64 M11
Norilsk 30 C6
Norley 63 A3
Norman 78 B5
Norman Wells . 70 B7
Normandie ... 12 B4
Normandy = Normandie ... 12 B4
Normanton ... 61 D7
Norquinco 95 E2
Norrköping ... 9 G11
Norrland 8 F11
Norseman 60 G3
North Battleford 71 C9
North Bay ... 69 D3
North Bend ... 80 D1
North C. 64 A5
North Cape = Nordkapp ... 8 D13
North Carolina □ . . 75 B7
North Channel, Canada 69 D2
North Channel, U.K. 11 D4
North Dakota □ 76 B4
North Frisian Is. = Nordfriesische Inseln 14 A5
North I. 64 B5
North Korea ■ . 35 C7
North Lakhimpur . . 41 D10
North Minch .. 10 B4
North Ossetia □ 25 E5
North Pagai, I. = Pagai Utara . 37 E2
North Platte .. 76 E4
North Platte → 76 E4
North Rhine Westphalia □ = Nordrhein-Westfalen . 14 C4
North Saskatchewan → 71 C8
North Sea 10 D8
North Sporades = Voríai Sporádhes .. 23 E4
North Taranaki Bight 64 C6
North Thompson → 71 C7
North Uist ... 10 C3
North West C. . 60 E1
North West Frontier □ .. 42 B8
North West Highlands ... 10 C4
North West River .. 69 C4
North West Territories □ 70 B9
Northampton .. 11 E6
Northern Circars 40 J4
Northern Ireland □ .. 11 D3
Northern Marianas ■ . 64 H9

Northern Territory □ .. 60 D5
Northland □ ... 64 A5
Northumberland, C. 62 C3
Northumberland Str. 69 D4
Norton Sd. 71 B3
Norway ■ 8 F10
Norway House 71 C10
Norwegian Sea 6 B5
Norwich 11 E7
Nosok 28 B9
Noşratābād .. 45 D4
Nossob → ... 58 D4
Nosy Bé 59 G9
Nosy Boraha .. 59 H9
Nosy Mitsio .. 59 G9
Nosy Varika .. 59 J9
Noteć → 16 B2
Noto-Hanto ... 33 G11
Notre Dame B. 69 D5
Notre Dame de Koartac = Quaqtaq 68 B4
Notre Dame d'Ivugivic = Ivujivik 68 B3
Nottaway → .. 69 C3
Nottingham ... 11 E6
Nouâdhibou .. 54 D1
Nouâdhibou, Ras 54 D1
Nouakchott .. 55 E1
Nouméa 64 M11
Noupoort 58 E4
Nouveau Comptoir = Wemindji ... 69 C3
Nouvelle-Calédonie = New Caledonia ■ . 64 M11
Nova Casa Nova 93 D5
Nova Cruz 92 D6
Nova Friburgo . 93 G5
Nova Gaia = Cambundi-Catembo 56 G3
Nova Iguaçu .. 93 G5
Nova Iorque .. 92 D5
Nova Lima 93 F5
Nova Mambone 59 C7
Nova Scotia □ . 69 D4
Nova Sofala ... 59 C6
Nova Venécia . 93 F5
Nova Zagora .. 22 C5
Novara 20 B2
Novaya Ladoga 24 A3
Novaya Lyalya 29 D7
Novaya Zemlya 28 B6
Nové Zámky .. 16 D4
Novgorod 24 B3
Novgorod-Severskiy = Novhorod-Siverskyy = 24 C3
Novhorod-Siverskyy .. 24 C3
Novi Lígure .. 20 B2
Novi Pazar ... 22 C3
Novi Sad 22 B2
Novo Mesto .. 20 B5
Novo Remanso 93 D5
Novoataysk ... 29 D9
Novocherkassk 25 D5
Novogrudok = Navahrudak . 17 B7

Novohrad-Volynskyy ... 17 C8
Novokazalinsk = Zhangaqazaly 29 E7
Novokuybyshevsk 24 C6
Novokuznetsk . 29 D9
Novomoskovsk 24 C4
Novorossiysk .. 25 E4
Novorybnoye . 30 B8
Novoselytsya .. 17 D8
Novoshakhtinsk 25 D4
Novosibirsk ... 29 D9
Novosibirskiye Ostrova 31 B12
Novotroitsk ... 29 D6
Novouzensk .. 24 C6
Novovolynsk .. 17 C7
Novska 20 B6
Novyy Port ... 28 C8
Now Shahr 44 B2
Nowa Sól 16 C2
Nowgong 41 D9
Nowra 63 B5
Nowshera 42 B8
Nowy Sącz ... 16 D5
Nowy Targ ... 16 D5
Nowy Tomyśl . 16 B3
Noyon 13 B5
Nsanje 59 B7
Nsawam 55 G4
Nsukka 55 G6
Nubian Desert = Nûbîya, Es Sahrâ En .. 52 D5
Nûbîya, Es Sahrâ En .. 52 D5
Nuboai 39 E5
Nueltin L. 70 B10
Nueva Imperial 94 D2
Nueva Rosita . 84 B4
Nuéve de Julio 94 D4
Nuevitas 86 B4
Nuevo, G. 95 E4
Nuevo Laredo . 84 B5
Nugget Pt. 65 G3
Nuhaka 64 C7
Nukey Bluff .. 62 B2
Nukheila 53 E4
Nukus 29 E6
Nulato 71 B4
Nullarbor Plain 60 G4
Numalla, L. .. 63 A3
Numan 53 G1
Numazu 32 B6
Numfoor 39 E4
Numurkah ... 63 C4
Nuneaton 11 E6
Nunivak I. ... 71 B3
Nunkun 42 C10
Núoro 21 D2
Nuremberg = Nürnberg .. 15 D6
Nuriootpa ... 62 B2
Nürnberg ... 15 D6
Nurran, L. = Terewah, L. . 63 A4
Nusa Tenggara Barat □ 37 F5
Nusa Tenggara Timur □ 39 F2
Nushki 42 E5
Nutak 68 C4
Nuuk = Godthåb 68 B5
Nuwakot 40 C4
Nuweveldberge 58 E4
Nuyts Arch. ... 62 B1
Nyah West 62 C3

Nyahanga 57 E6
Nyahururu 57 D7
Nyainqentanglha Shan ... 34 C4
Nyâlâ 53 F3
Nyasa, L. = Malawi, L. .. 59 A6
Nyasvizh 17 B8
Nyda 28 C8
Nyeri 57 E7
Nyíregyháza .. 17 E5
Nylstroom ... 59 C5
Nymagee 63 B4
Nyngan 63 B4
Nyoman = Neman → .. 24 B1
Nysa 16 C3
Nysa → 16 B2
Nyurbe 30 C9
Nzega 57 E6
N'Zérékoré .. 55 G3
Nzeto 56 F2

O

Ō-Shima 32 B6
Oahe, L. 76 C4
Oahu 82 H12
Oak Ridge ... 75 A5
Oakbank 62 B3
Oakey 63 A5
Oakland 82 B1
Oamaru 65 F4
Oates Land .. 96 A15
Oaxaca 84 D5
Ob → 28 C7
Oba 69 D2
Oban 10 C4
Obbia 49 F4
Oberhausen .. 14 C4
Oberon 63 B4
Obi, Kepulauan 39 E3
Obi Is. = Obi, Kepulauan .. 39 E3
Óbidos 92 C2
Obihiro 32 F12
Obilatu 39 E3
Obo 57 C5
Obozerskaya = Obozerskiy = 28 C5
Obozerskiy ... 28 C5
Obskaya Guba . 28 C8
Obuasi 55 G4
Ocala 75 E6
Ocaña 18 C4
Occidental, Cordillera ... 90 C3
Ocean I. = Banaba 64 K11
Oceanside ... 82 D4
Ocniţa 17 D8
Ocumare del Tuy 90 A5
Odate 33 F12
Odawara 32 B6
Oddur 49 G3
Ödemiş 23 E7
Odense 9 G10
Oder → 15 B8
Odesa 25 D3
Odessa = Odesa 25 D3
Odessa 78 D2
Odienné 55 G3
Odintsovo ... 24 B4
Odorheiu Secuiesc 17 E7

Odra

Odra = Oder → 15 B8
Odra → 16 B2
Oeiras 92 D5
Ofanto → 20 D6
Offa 55 G5
Offenbach 14 C5
Offenburg 14 D4
Oga-Hantō 33 G11
Ōgaki 32 B5
Ogbomosho 55 G5
Ogden 81 E8
Ogdensburg 73 B8
Oglio → 20 B3
Ogooué → 56 E1
Ogowe = Ogooué → 56 E1
Ohai 65 F2
Ohakune 64 C6
Ohanet 54 C6
Ohau, L. 65 F3
Ohio □ 72 D4
Ohio → 72 F1
Ohre → 16 C2
Ohrid 22 D3
Ohridsko Jezero 23 D3
Oise → 12 B5
Ōita 32 C2
Oiticica 92 D5
Ojos del Salado, Cerro 94 B3
Oka → 28 D5
Okaba 39 F5
Okahandja 58 C3
Okahukura 64 C6
Okandja 56 E2
Okanogan → 80 A4
Okara 42 D8
Okarito 65 E4
Okaukuejo 58 B3
Okavango Swamps 58 B4
Okaya 32 A6
Okayama 32 B3
Okazaki 32 B5
Okeechobee, L. 75 F7
Okefenokee Swamp 75 D6
Okhotsk, Sea of 31 D12
Oki-Shotō 32 A3
Okiep 58 D3
Okinawa-Jima 35 D7
Oklahoma □ 78 B5
Oklahoma City 78 B5
Okmulgee 79 B6
Oknitsa = Ocniţa 17 D8
Okrika 55 H6
Oktabrsk = Oktyabrsk 29 E6
Oktyabrsk 29 E6
Oktyabrskiy = Aktsyabrski 17 B9
Oktyabrskoy Revolyutsii, Os. 30 B7
Oktyabrskoye 28 C7
Okuru 65 E3
Okushiri-Tō 32 F11
Öland 9 G11
Olary 62 B3
Olavarría 94 D4
Oława 16 C3
Ólbia 20 D2
Old Crow 70 B6
Oldenburg 14 B5
Oldham 11 E5
Olekma → 31 C10
Olekminsk 31 C10
Oleksandriya 17 C8

Olenek 30 C9
Olenek → 30 B10
Oléron, I. d' 12 D3
Oleśnica 16 C3
Olevsk 17 C8
Olhão 18 D2
Olifants → 59 C6
Ólimbos, Óros 23 D4
Olinda 92 D7
Oliveira 93 G5
Olivenza 18 C2
Ollagüe 91 H5
Olomouc 16 D3
Olongapo 38 B2
Olot 19 A7
Olsztyn 16 B5
Olt → 22 C5
Olteniţa 22 B6
Oltu 46 B5
Olympia, Greece 23 F3
Olympia, U.S.A. 80 B2
Olympic Mts. 80 B2
Olympus, Mt. = Ólimbos, Óros 23 D4
Om → 29 D8
Omagh 11 D3
Omaha 77 E7
Oman ■ 48 C6
Oman, G. of 45 E4
Omaruru 58 C3
Omate 91 G4
Ombai, Selat 39 F2
Omboué 56 E1
Ombrone → 20 C3
Omdurmân 53 E5
Ometepec 84 D5
Ōmiya 32 B6
Omo → 53 G6
Omsk 29 D8
Omul, Vf. 17 F7
Ōmura 32 C1
Omuramba Omatako → 58 B4
Ōmuta 32 C2
Onang 39 E1
Oncócua 58 B2
Onda 19 C5
Ondangua 58 B3
Ondo 55 G5
Öndörhaan 35 B6
Onega 28 C4
Onega → 8 F15
Onega, G. of = Onezhskaya Guba 28 C4
Onega, L. = Onezhskoye Ozero 28 C4
Onehunga 64 B6
Oneşti 17 E8
Onezhskaya Guba 28 C4
Onezhskoye Ozero 28 C4
Ongarue 64 C6
Onilahy → 59 J8
Onitsha 55 G6
Onoda 32 B2
Ontario □ 69 D2
Ontario, L. 73 C7
Oodnadatta 62 A2
Oostende 14 C2
Ootacamund 43 P10
Opala 56 E4
Opava 16 D3
Opole 16 C3
Oporto = Porto 18 B1
Opotiki 64 C7
Opua 64 A6

Opunake 64 C5
Oradea 17 E5
Orai 42 G11
Oral = Zhayyq → 29 E6
Oral 24 C7
Oran, Algeria 54 A4
Oran, Argentina 94 A4
Orange = Oranje → 58 D3
Orange, Australia 63 B4
Orange, France 13 D6
Orange, U.S.A. 79 D7
Orange, C. 92 B3
Orangeburg 75 C7
Oranienburg 15 B7
Oranje → 58 D3
Oranjemund 58 D3
Oras 38 B3
Oraşul Stalin = Braşov 17 F7
Orbetello 20 C3
Orbost 63 C4
Orchila, I. 90 A5
Ordos = Mu Us Shamo 35 C5
Ordu 46 B4
Ordzhonikidze = Vladikavkaz 25 E5
Ore Mts. = Erzgebirge 15 C7
Örebro 9 G11
Oregon □ 80 C3
Orekhovo-Zuyevo 24 B4
Orel 24 C4
Ören 23 F6
Orenburg 29 D6
Orense 18 A2
Orepuki 65 G2
Øresund 9 G10
Orgaz 18 C4
Orgeyev = Orhei 17 E9
Orgūn 45 C7
Orhaneli 23 E7
Orhangazi 23 D7
Orhei 17 E9
Orhon Gol → 35 A5
Orient 62 A3
Oriental, Cordillera 90 B4
Orihuela 19 C5
Orinoco → 90 B6
Orissa □ 40 G5
Oristano 21 E2
Oristano, G. di 21 E2
Orizaba 84 D5
Orkney Is. 10 B5
Orlando 75 E7
Orléanais 12 C5
Orléans 12 C4
Ormara 42 G4
Ormoc 38 B2
Ormond 64 C7
Örnsköldsvik 8 F11
Orocué 90 C4
Orol Dengizi = Aral Sea 29 E6
Oroquieta 38 C2
Orós 92 D6
Orosháza 16 E5
Orroroo 62 B2
Orsha 24 C3
Orsk 29 D6
Orşova 22 B4
Ortaca 23 F7
Ortegal, C. 18 A2

Orthez 12 E3
Ortigueira 18 A2
Ortles 20 A3
Ortón → 91 F5
Orūmīyeh 46 C6
Orūmīyeh, Daryācheh-ye 46 C6
Oruro 91 G5
Oruzgān □ 42 C5
Orvieto 20 C4
Oryakhovo 22 C4
Ōsaka 32 B4
Ösel = Saaremaa 24 B1
Osh 29 E8
Oshawa 69 D3
Oshkosh 76 E3
Oshmyany = Ashmyany 17 A7
Oshogbo 55 G5
Oshwe 56 E3
Osijek 20 B7
Osipenko = Berdyansk 25 D4
Osipovichi = Asipovichy 17 B9
Oskarshamn 9 G11
Öskemen 29 E9
Oslo 9 G10
Oslob 38 C2
Oslofjorden 9 G10
Osmanabad 43 K10
Osmaniye 46 C4
Osnabrück 14 B5
Osorio 94 B6
Osorno 95 E2
Ossa, Mt. 62 D4
Óssa, Oros 23 E4
Ostend = Oostende 14 C2
Oster 17 C10
Österdalälven 8 F10
Östersund 8 F10
Ostfriesische Inseln 14 B4
Ostrava 16 D4
Ostróda 16 B4
Ostroh 17 C8
Ostrołęka 16 B5
Ostrów Mazowiecka 16 B5
Ostrów Wielkopolski 16 C3
Ostrowiec-Świętokrzyski 16 C5
Ostuni 21 D6
Ōsumi-Kaikyō 33 D2
Osuna 18 D3
Oswego 73 C7
Oświęcim 16 C4
Otago □ 65 F3
Otago Harbour 65 F4
Ōtake 32 B3
Otaki 64 D6
Otaru 32 F12
Otaru-Wan = Ishikari-Wan 32 F12
Otavalo 90 C3
Otavi 58 B3
Otira Gorge 65 E4
Otjiwarongo 58 C3
Otorohanga 64 C6
Otranto 23 D2
Otranto, C. d' 23 D2
Otranto, Str. of 23 D2
Ōtsu 32 B4
Ottawa 69 D3
Ottawa Is. 68 C2
Ottumwa 77 E8

Oturkpo 55 G6
Otway, B. 95 G2
Otway, C. 62 C3
Otwock 16 B5
Ouachita Mts. . 79 B6
Ouadâne 54 D2
Ouadda 56 C4
Ouagadougou . 55 F4
Ouahran = Oran 54 A4
Ouallene 54 D5
Ouanda Djallé . 56 C4
Ouango 56 D4
Ouargla 54 B6
Ouarzazate 54 B3
Oubangi → 56 E3
Oudtshoorn ... 58 E4
Ouessant, I. d' . 12 B1
Ouesso 56 D3
Ouezzane 54 B3
Ouidah 55 G5
Oujda 54 B4
Oujeft 55 D2
Ouled Djellal .. 54 B6
Oulu 8 E13
Oulujärvi 8 F13
Oulujoki → 8 E13
Oum Chalouba 53 E3
Ounianga-Kébir 53 E3
Ounianga Sérir 53 E3
Ourense =
Orense 18 A2
Ouricuri 92 D5
Ouro Prêto 93 G5
Ouse 62 D4
Outer Hebrides 10 C3
Outjo 58 C3
Ouyen 62 C3
Ovalle 94 C2
Ovamboland .. 58 B3
Oviedo 18 A3
Ovruch 17 C9
Owaka 65 G3
Owambo =
Ovamboland 58 B3
Owase 32 B5
Owbeh 42 B3
Owen Sound .. 69 D2
Owen Stanley
Ra. 61 B8
Owendo 56 D1
Owo 55 G6
Owyhee → 80 D5
Oxford, N.Z. ... 65 E5
Oxford, U.K. ... 11 F6
Oxley 63 B3
Oxnard 82 C3
Oxus =
Amudarya → 29 E6
Oya 37 D4
Oyama 32 A6
Oyem 56 D2
Oyo 55 G5
Ozamiz 38 C2
Ozark Plateau . 79 A8
Ózd 16 D5

P

Pa-an 41 J11
Paarl 58 E3
Pab Hills 42 F5
Pabianice 16 C4
Pabna 41 E7
Pacaja → 92 C3
Pacaraima,
Sierra 90 C6

Pacasmayo ... 91 E3
Pachpadra 42 G8
Pachuca 84 C5
Padaido,
Kepulauan .. 39 E5
Padang 37 E2
Padangpanjang 37 E2
Padangsidempuan
.... 37 D1
Paderborn ... 14 C5
Padloping Island 68 B4
Pádova 20 B3
Padua = Pádova 20 B3
Paducah 72 F1
Paeroa 64 B6
Pafúri 59 C6
Pag 20 B5
Pagadian 38 C2
Pagai Selatan, P. 37 E2
Pagai Utara ... 37 E2
Pagalu =
Annobón ... 51 G4
Pagastikós
Kólpos 23 E4
Pagatan 37 E5
Pahiatua 64 D6
Painan 37 E2
Paint Hills =
Wemindji ... 69 C3
Painted Desert . 83 C7
País Vasco □ .. 19 A4
Paisley 10 D4
Paita 90 E2
Pajares, Puerto
de 18 A3
Pakaraima Mts. 90 B6
Pakistan ■ ... 42 E7
Pakokku 41 G10
Pakse 36 A3
Paktiā □ 42 C6
Pala 53 G2
Palagruža 20 C6
Palam 43 K10
Palampur 42 C10
Palana 62 C4
Palanan 38 A2
Palanan Pt. ... 38 A2
Palangkaraya .. 37 E4
Palani Hills ... 43 P10
Palanpur 43 G8
Palanro 39 E1
Palapye 59 C5
Palau ■ 38 C4
Palawan 36 C5
Palayankottai .. 43 Q10
Paleleh 39 D2
Palembang ... 37 E2
Palencia 18 A3
Palermo 21 E4
Palestine 79 D6
Paletwa 41 G9
Palghat 43 P10
Pali 42 G8
Palioúrion, Ákra 23 E4
Palitana 43 J7
Palk Bay 43 Q11
Palk Strait 43 Q11
Pallanza =
Verbánia ... 20 B2
Palm Springs .. 82 D4
Palma 57 G8
Palma → 93 E4
Palma, B. de .. 19 C7
Palma de
Mallorca 19 C7
Palmares 92 D6
Palmas, C. 55 H3
Pálmas, G. di .. 21 E2

Palmeira dos
Índios 93 D6
Palmeirinhas,
Pta. das 56 F2
Palmer 71 B5
Palmer Arch. .. 96 A3
Palmer Land .. 96 B3
Palmerston ... 65 F4
Palmerston
North 64 D6
Palmi 21 E5
Palmira 90 C3
Palmyra =
Tudmur 46 D4
Palopo 39 E2
Palos, C. de ... 19 D5
Palu, Indonesia 39 E1
Palu, Turkey .. 46 C5
Paluan 38 B2
Pama 55 F5
Pamiers 12 E4
Pamirs 29 F8
Pamlico Sd. ... 75 B10
Pampa 78 B3
Pampa de las
Salinas 94 C3
Pampanua 39 E2
Pampas,
Argentina ... 94 D4
Pampas, Peru . 91 F4
Pamplona,
Colombia ... 90 B4
Pamplona,
Spain 19 A5
Panaji 43 M8
Panamá 85 F9
Panama ■ ... 86 E4
Panamá, G. de 86 E4
Panama City .. 74 D5
Panão 91 E3
Panay 38 B2
Panay, G. 38 B2
Pančevo 22 B3
Pandan 38 B2
Pandharpur ... 43 L9
Pando 94 C5
Pando, L. =
Hope, L. ... 62 A2
Panevežys 24 B1
Panfilov 29 E8
Pang-Long ... 41 F12
Pang-Yang ... 41 F12
Pangani 57 F7
Pangfou =
Bengbu 35 C6
Pangkajene ... 39 E1
Pangkalanbrandan
.... 37 D1
Pangkalanbuun 37 E4
Pangkalansusu 37 D1
Pangkalpinang 37 E3
Pangkoh 37 E4
Pangnirtung ... 68 B4
Pangutaran
Group 38 C2
Panjgur 42 F4
Panjim = Panaji 43 M8
Panjinad
Barrage 42 E7
Panorama 94 A6
Pantar 39 F2
Pante Macassar 39 F2
Pantelleria 21 F3
Pánuco 84 C5
Panyam 55 G6
Paoting =
Baoding 35 C6
Paot'ou =
Baotou 35 B6

Paoua 56 C3
Pápa 16 E3
Papakura 64 B6
Papantla 84 C5
Papar 36 C5
Papua New
Guinea ■ ... 61 B8
Papudo 94 C2
Papun 41 H11
Pará = Belém . 92 C4
Pará □ 92 C3
Paracatu 93 F4
Paracel Is. =
Hsisha
Chuntao ... 36 A4
Parachilna 62 B2
Parachinar ... 42 C7
Paradip 40 G6
Parado 39 F1
Paragua → ... 90 B6
Paraguaçu → . 93 E6
Paraguaná, Pen.
de 90 A4
Paraguarí 94 B5
Paraguay ■ ... 94 A5
Paraguay → .. 94 B5
Paraíba = João
Pessoa 92 D7
Paraíba □ 92 D6
Paraíba do
Sul → 93 G5
Parakou 55 G5
Paramaribo ... 92 A2
Paraná,
Argentina ... 94 C4
Paraná, Brazil . 93 E4
Paraná □ 94 A6
Paraná → 94 C5
Paranaguá 94 B7
Paranaíba → . 93 G3
Paranapanema →
.... 94 A6
Paranapiacaba,
Serra do 94 A7
Parang, Jolo,
Phil. 38 C2
Parang,
Mindanao,
Phil. 38 C2
Paratinga 93 E5
Paratoo 62 B2
Parattah 62 D4
Parbhani 43 K10
Parchim 15 B6
Pardo →,
Bahia, Brazil . 93 F6
Pardo →,
Mato Grosso,
Brazil 93 G3
Pardo →,
São Paulo,
Brazil 93 G4
Pardubice 16 C2
Parecis, Serra
dos 91 F7
Parepare ■ ... 39 E1
Párga 23 E3
Pariaguán 90 B6
Pariaman 37 E2
Parigi 39 E2
Parika 90 B7
Parima, Serra .. 90 C6
Parinari 90 D4
Parîngul Mare . 17 F6
Parintins 92 C2
Paris, France .. 12 B5
Paris, U.S.A. .. 79 C6
Park Range ... 81 F10

Parkersburg ... 72 E5
Parkes 63 B4
Parla 18 B4
Parma 20 B3
Parnaguá 93 E5
Parnaíba, *Piauí, Brazil* 92 C5
Parnaíba, *São Paulo, Brazil* 93 F3
Parnaíba → . 92 C5
Parnassós 23 E4
Parnu 24 B1
Paroo → 63 B3
Páros 23 F5
Parral 94 D2
Parramatta .. 63 B5
Parry Sound .. 69 D3
Partinico 21 E4
Paru → 92 C3
Paruro 91 F4
Parván □ 42 B6
Parvatipuram .. 40 H4
Pasadena, *Calif., U.S.A.* 82 C3
Pasadena, *Tex., U.S.A.* 79 E6
Pasaje 90 D3
Pascagoula ... 79 D9
Paşcani 17 E8
Pasco 80 B4
Pasco, Cerro de 91 F3
Pashmakli = Smolyan 22 D5
Pašman 20 C5
Pasni 42 G3
Paso de Indios 95 E3
Passau 15 D7
Passero, C. 21 F5
Passo Fundo .. 94 B6
Passos 93 G4
Pastaza → ... 90 D3
Pasto 90 C3
Patagonia 95 F3
Patan, *India* .. 43 H8
Patan, *Nepal* .. 40 D5
Patani 39 D3
Patchewollock . 62 C3
Patea 64 C6
Pategi 55 G6
Paternò 21 F5
Paterson 73 D8
Pathankot 42 C9
Patiala 42 D10
Patkai Bum ... 41 D10
Pátmos 23 F6
Patna 40 E5
Patos, L. dos .. 94 C6
Patos de Minas 93 F4
Patquía 94 C3
Pátrai 23 E3
Pátraikós Kólpos 23 E3
Patras = Pátrai 23 E3
Patrocínio ... 93 F4
Pattani 36 C2
Patuakhali 41 F8
Pau 12 E3
Pauini → 90 D6
Pauk 41 G10
Paulis = Isiro . 57 D5
Paulistana 92 D5
Paulo Afonso . 93 D6
Pavia 20 B2
Pavlodar 29 D8
Pavlograd = Pavlohrad ... 25 D4
Pavlohrad ... 25 D4
Pavlovo 24 B5
Pavlovsk 24 C5

Pawtucket 73 D10
Paxoí 23 E3
Payakumbuh .. 37 E2
Payne Bay = Kangirsuk ... 68 C4
Paysandú 94 C5
Paz, B. la 84 C2
Pazar 46 B5
Pazardzhik ... 22 C5
Peace → 70 C8
Peak Hill 63 B4
Peake 62 C2
Peake Cr. → .. 62 A2
Pebane 59 B7
Pebas 90 D4
Peć 22 C3
Pechenga 28 C4
Pechenizhyn .. 17 D7
Pechora → ... 28 C6
Pechorskaya Guba 28 C6
Pecos → 78 E3
Pécs 16 E4
Pedder, L. 62 D4
Pedirka 62 A2
Pedra Azul ... 93 F5
Pedreiras 92 C5
Pedro Afonso . 93 D4
Pedro Juan Caballero ... 94 A5
Peebinga 62 B3
Peel →, *Australia* 63 B5
Peel →, *Canada* 70 B6
Peera Peera Poolanna L. .. 62 A2
Pegasus Bay .. 65 E5
Pegu 41 J11
Pegu Yoma ... 41 H10
Pehuajó 94 D4
Peine 14 B6
Peip'ing = Beijing 35 C6
Peipus, L. = Chudskoye, Oz. 24 B2
Peixe 93 E4
Pekalongan ... 37 F3
Pekanbaru ... 37 D2
Pekin 77 E10
Peking = Beijing 35 C6
Pelagie, Is. 21 G4
Pelaihari 37 E4
Peleaga, Vf. .. 17 F6
Peleng 39 E2
Pelješac 20 C6
Pelly → 70 B6
Pelly Bay 68 B2
Pelly L. 70 B9
Peloponnese = Pelopónnisos □ 23 F4
Pelopónnisos □ 23 F4
Peloro, C. 21 E5
Pelorus Sd. ... 65 D5
Pelotas 94 C6
Pelvoux, Massif du 13 D7
Pematangsiantar 37 D1
Pemba I. 57 F7
Pembroke, *Canada* 69 D3
Pembroke, *U.K.* 11 F4
Penang = Pinang 36 C2
Penápolis 94 A6
Peñarroya-Pueblonuevo . 18 C3
Peñas, C. de .. 18 A3

Penas, G. de .. 95 F2
Pench'i = Benxi 35 B7
Pend Oreille L. 80 A5
Pendembu 55 G2
Pendleton 80 C4
Penedo 93 E6
Penguin 62 D4
Peniche 18 C1
Penida 37 F5
Peninsular Malaysia □ .. 37 D2
Penmarch, Pte. de 12 C1
Pennines 11 D5
Pennsylvania □ 73 D7
Penola 62 C3
Penong 60 G5
Penrith 63 B5
Pensacola 74 D4
Pensacola Mts. 96 C4
Penshurst 62 C3
Penticton 71 D8
Pentland Firth . 10 B5
Penza 24 C6
Penzance 11 F4
Peoria 77 E10
Perabumulih .. 37 E2
Perche, Collines du 12 B4
Perdido, Mte. .. 19 A6
Perdu, Mt. = Perdido, Mte. 19 A6
Pereira 90 C3
Perekerten 62 B3
Pereyaslav-Khmelnytskyy 24 C3
Pergamino 94 C4
Péribonca → . 69 D3
Perico 94 A3
Périgueux 12 D4
Perijá, Sierra de 90 B4
Perlas, Arch. de las 86 E4
Perm 29 D6
Pernambuco = Recife 92 D7
Pernatty Lagoon 62 B2
Pernik 22 C4
Perpendicular Pt. 63 B5
Perpignan 13 E5
Persepolis 44 D3
Pershotravensk 17 C8
Persian Gulf = Gulf, The ... 44 E2
Perth, *Australia* 60 G2
Perth, *U.K.* 10 C5
Perth Amboy .. 73 D8
Peru ■ 90 C3
Perúgia 20 C4
Pervomaysk ... 25 D3
Pervouralsk ... 29 D6
Pésaro 20 C4
Pescara 20 C5
Peshawar 42 B7
Peshkopi 22 D3
Pesqueira 92 D6
Petah Tiqwa .. 46 D3
Petauke 59 A6
Peter I.s Øy ... 96 A2
Peterborough, *Australia* 62 B2
Peterborough, *Canada* 69 D3
Peterborough, *U.K.* 11 E6
Peterhead 10 C6
Petersburg, *Alaska, U.S.A.* 71 C6

Petersburg, *Va., U.S.A.* 73 F7
Petitsikapau, L. 69 C4
Petlad 43 H8
Peto 85 C7
Petone 65 D6
Petrich 22 D4
Petrikov = Pyetrikaw ... 17 B9
Petrograd = Sankt-Peterburg ... 24 B3
Petrolândia .. 93 D6
Petrolina 93 D5
Petropavl 29 D7
Petropavlovsk = Petropavl ... 29 D7
Petropavlovsk-Kamchatskiy 31 D13
Petrópolis ... 93 G5
Petroşani 17 F6
Petrovaradin .. 22 B2
Petrovsk 24 C6
Petrozavodsk .. 28 C4
Peureulak 36 D1
Pforzheim 14 D5
Phagwara 42 D9
Phalodi 42 F8
Phan Rang ... 36 B3
Phangan, Ko .. 36 C2
Phangnga 36 C1
Phanh Bho Ho Chi Minh ... 36 B3
Phatthalung ... 36 C2
Phetchabun ... 36 A2
Philadelphia .. 73 E8
Philippines ■ .. 38 B2
Philippopolis = Plovdiv 22 C5
Phillip I. 63 C4
Phillott 63 A4
Phitsanulok ... 36 A2
Phnom Dangrek 36 B2
Phnom Penh .. 36 B2
Phoenix 83 D6
Phra Nakhon Si Ayutthaya ... 36 B2
Phuket 36 C1
Phumiphon, Khuan 36 A1
Piacenza 20 B2
Pialba 63 A5
Pian Cr. → ... 63 B4
Pianosa 20 C3
Piatra Neamţ .. 17 E8
Piauí □ 92 D5
Piave → 20 B4
Pibor Post ... 53 G5
Pica 91 H5
Picardie 13 B5
Picardy = Picardie 13 B5
Pichilemu 94 C2
Pico Truncado . 95 F3
Picton, *Australia* 63 B5
Picton, *N.Z.* ... 65 D6
Pictou 69 D4
Picún Leufú ... 95 D3
Piedmont = Piemonte □ . 20 B1
Piedras, R. de las → 91 F5
Piedras Negras 84 B4
Piemonte □ ... 20 B1
Piet Retief ... 59 D6
Pietermaritzburg 59 D6
Pietersburg .. 59 C5
Pietrosul, *Romania* 17 E7

Pietrosul,
Romania 17 E7
Pigüe 94 D4
Piketberg 58 E3
Piła 16 B3
Pilar, *Brazil* .. 93 D6
Pilar, *Paraguay* 94 B5
Pilas Group ... 38 C2
Pilcomayo → . 94 B5
Pilibhit 42 E11
Pilica → 16 C5
Pílos 23 F3
Pilsen = Plzeň . 16 D1
Pimba 62 B2
Pimenta Bueno 91 F6
Pimentel 91 B5
Pinang 36 C2
Pinar del Río .. 86 B3
Pınarhisar 22 D6
Pińczów 16 C5
Pindiga 55 G7
Pindos Óros ... 23 E3
Pindus Mts. =
Pindos Óros . 23 E3
Pine Bluff 79 B7
Pine Point ... 70 B8
Pinega → ... 28 C5
Pinerolo 20 B1
Pinetown 59 D6
Ping → ... 36 A2
Pingdong 35 D7
Pingliang 35 C5
Pingwu 35 C5
Pingxiang 35 D5
Pinhel 18 B2
Pini 37 D1
Piniós → ... 23 E4
Pinnaroo 62 C3
Pínnes, Ákra .. 23 D5
Pinrang 39 E1
Pinsk 17 B8
Pintados 91 H5
Pinyug 24 A6
Piombino 20 C3
Pioner, Os. ... 30 B7
Piorini, L. ... 90 D6
Piotrków
Trybunalski . 16 C4
Pip 45 E5
Piquiri → 94 A6
Piracicaba 94 A7
Piracuruca 92 C5
Piræus =
Piraiévs 23 F4
Piraiévs 23 F4
Pirané 94 B5
Pirapora 93 F5
Pírgos 23 F3
Pirin Planina .. 22 D4
Piripiri 92 C5
Pirmasens 14 D4
Pirot 22 C4
Piru 39 E3
Pisa 20 C3
Pisagua 91 G4
Pisciotta 21 D5
Pisco 91 F3
Písek 16 D2
Pishan 34 C2
Pising 39 F2
Pistóia 20 C3
Pisuerga → .. 18 B3
Pitarpunga, L. . 62 B3
Pitcairn I. 65 M17
Piteå 8 E12
Piteşti 22 B5
Pithapuram ... 40 J4
Pittsburg 79 A6
Pittsburgh 73 D5

Pittsworth 63 A5
Piura 90 E2
Placentia 69 D5
Placentia B. ... 69 D5
Plainview 78 B3
Pláka, Ákra ... 23 G6
Plakhino 28 C9
Plasencia 18 B2
Plata, Río de la 94 C5
Platani → ... 21 F4
Plato 90 B4
Platte → 77 F7
Plauen 15 C7
Pleiku 36 B3
Plenty, B. of ... 64 B7
Pleven 22 C5
Plevlja 22 C2
Płock 16 B4
Plöckenstein .. 15 D7
Ploieşti 22 B6
Plovdiv 22 C5
Plumtree 59 C5
Plymouth 11 F4
Plzeň 16 D1
Po → 20 B4
Po Hai = Bo Hai 35 C6
Pobedy Pik ... 29 E8
Pocatello 81 D7
Poços de Caldas 93 G4
Podgorica 22 C2
Podilska
Vysochyna .. 17 D8
Podkamennaya
Tunguska → . 30 C7
Podolsk 24 B4
Podor 55 E1
Pofadder 58 D3
Poh 39 E2
Pohnpei 64 J10
Poinsett, C. ... 96 A12
Pointe-à-Pitre .. 87 C7
Pointe Noire .. 56 E2
Poitiers 12 C4
Poitou 12 C3
Pokaran 42 F7
Pokataroo 63 A4
Pokrovsk =
Engels 24 C6
Pola = Pula ... 20 B4
Polan 45 E5
Poland ■ 16 B5
Polatsk 24 B2
Polcura 94 D2
Polesye = Pripet
Marshes 17 B9
Poli 56 C2
Police 16 B2
Polillo Is. 38 B2
Políyiros 23 D4
Pollachi 43 P10
Polnovat 28 C7
Polonne 17 C8
Polonnoye =
Polonne 17 C8
Poltava 24 D3
Polunochnoye .. 28 C7
Polynesia 65 L15
Polynésie
française =
French
Polynesia ■ . 65 M16
Pombal, *Brazil* . 92 D6
Pombal,
Portugal ... 18 C1
Pomézia 20 D4
Pomorski,
Pojezierze ... 16 B3
Ponape =
Pohnpei 64 J10

Ponca City 78 A5
Ponce 87 C6
Pond Inlet 68 A3
Pondicherry .. 43 P11
Ponferrada 18 A2
Ponnani 43 P9
Ponnyadaung . 41 F10
Ponoy → 28 C5
Ponta Grossa .. 94 B6
Ponta Pora ... 94 A5
Pontarlier 13 C7
Pontchartrain L. 79 D8
Ponte Nova ... 93 G5
Pontevedra ... 18 A1
Pontiac 72 C4
Pontianak 37 E3
Pontine Is. =
Ponziane,
Ísole 21 D4
Pontine Mts. =
Kuzey
Anadolu
Dağları 46 B3
Pontivy 12 B2
Pontoise 12 B5
Ponziane, Ísole 21 D4
Poochera 62 B1
Poole 11 F6
Poona = Pune . 43 K8
Pooncarie 62 B3
Poopelloe L. .. 63 B3
Poopó, L. de . 91 G5
Popayán 90 C3
Popigay 30 B9
Popilta, L. 62 B3
Popio L. 62 B3
Poplar Bluff ... 79 A8
Popocatépetl,
Volcán 84 D5
Popokabaka .. 56 F3
Poprád 16 D5
Porbandar 43 J6
Porcupine → . 71 B5
Pordenone ... 20 B4
Pori 8 F12
Porkkala 9 G12
Porlamar 90 A6
Poroshiri-Dake . 32 F12
Porpoise B. ... 96 A13
Porretta, Passo
di 20 B3
Port Adelaide . 62 B2
Port Alberni ... 71 D7
Port Alfred ... 59 E5
Port Angeles .. 80 A2
Port Antonio .. 86 C4
Port Arthur,
Australia 62 D4
Port Arthur,
U.S.A. 79 E7
Port-au-Prince . 87 C5
Port Augusta .. 62 B2
Port Augusta
West 62 B2
Port Bergé
Vaovao 59 H9
Port Bou 19 A7
Port Broughton 62 B2
Port-Cartier ... 69 C4
Port Chalmers . 65 F4
Port Darwin .. 95 G5
Port Davey ... 62 D4
Port Dickson .. 37 D2
Port Elizabeth . 59 E5
Port Fairy 62 C3
Port-Gentil ... 56 E1
Port Harcourt . 55 H6
Port Harrison =
Inukjuak 68 C3

Port
Hawkesbury . 69 D4
Port Hedland .. 60 E2
Port Huron ... 72 C4
Port Kembla .. 63 B5
Port Kenny ... 62 B1
Port Lairge =
Waterford ... 11 E3
Port Laoise ... 11 E3
Port Lincoln .. 62 B2
Port Loko 55 G2
Port MacDonnell 62 C3
Port Macquarie 63 B5
Port Moresby . 61 B8
Port Nelson ... 71 C10
Port Nolloth .. 58 D3
Port Nouveau-
Québec =
Kangiqsualujjuaq
.... 68 C4
Port of Spain .. 87 D7
Port Pegasus . 65 G2
Port Phillip B. . 63 C3
Port Pirie 62 B2
Port Safaga =
Bûr Safâga . 52 C5
Port Said = Bûr
Sa'îd 52 B5
Port Shepstone 59 E6
Port Stanley =
Stanley 95 G5
Port Sudan =
Bûr Sûdân . 53 E6
Port Talbot 11 F5
Port-Vendres .. 13 E5
Port Wakefield . 62 B2
Port Weld 36 D2
Porta Orientalis 22 B4
Portachuelo .. 91 G6
Portadown 11 D3
Portage La
Prairie 71 D10
Portalegre 18 C2
Portile de Fier . 22 B4
Portimão 18 D1
Portland,
N.S.W.,
Australia 63 B4
Portland, *Vic.,*
Australia 62 C3
Portland, *Maine,*
U.S.A. 73 C10
Portland, *Oreg.,*
U.S.A. 80 C2
Portland B. ... 62 C3
Portland Prom. 68 C3
Porto 18 B1
Pôrto Alegre .. 94 C6
Porto Amboim
= Gunza ... 56 G2
Pôrto de Móz . 92 C3
Porto
Empédocle .. 21 F4
Pôrto Esperança 91 G7
Pôrto Franco . 92 D4
Porto Mendes . 94 A6
Pôrto Murtinho 91 H7
Pôrto Nacional 93 E4
Porto Novo ... 55 G5
Porto Santo ... 54 B1
Pôrto Seguro . 93 F6
Pôrto Tórres . 20 D2
Pôrto União .. 94 B6
Pôrto Válter . 91 E4
Porto-Vecchio . 13 F8
Pôrto Velho .. 91 E6
Portoferráio .. 20 C3
Portoscuso ... 21 E2
Portovíejo 90 D2

Portree

Portree 10 C3
Portsmouth,
  U.K. 11 F6
Portsmouth,
  U.S.A. 73 F7
Porttipahtan
  tekojärvi 8 E13
Portugal ■ 18 C1
Porvenir 95 G2
Posadas 94 B5
Poshan =
  Boshan 35 C6
Poso 39 E2
Posse 93 E4
Possel 56 C3
Possession I. .. 96 B15
Poste Maurice
  Cortier 54 D5
Postojna 20 B5
Potchefstroom . 59 D5
Potenza 21 D5
Poteriteri, L. .. 65 G2
Potgietersrus .. 59 C5
Poti 25 E5
Potiskum 55 F7
Potomac → .. 73 E7
Potosí 91 G5
Pototan 38 B2
Potrerillos 94 B3
Potsdam 15 B7
Poughkeepsie . 73 D9
Pouso Alegre .. 93 E2
Považská
  Bystrica 16 D4
Povenets 28 C4
Poverty B. 64 C8
Póvoa de Varzim 18 B1
Powder → ... 76 B2
Powell L. 83 B7
Poyang Hu .. 35 D6
Požarevac 22 B3
Poznań 16 B3
Pozo Almonte . 91 H5
Pozoblanco ... 18 C3
Pozzuoli 21 D5
Prachuap Khiri
  Khan 36 B1
Prado 93 F6
Prague = Praha 16 C2
Praha 16 C2
Prainha,
  Amazonas,
  Brazil 91 E6
Prainha, Pará,
  Brazil 92 C3
Prairies 71 C9
Prapat 37 D1
Prata 93 F4
Prato 20 C3
Pravia 18 A2
Praya 37 F5
Preparis North
  Channel 41 K9
Preparis South
  Channel 41 K9
Přerov 16 D3
Preservation
  Inlet 65 G2
Presidencia
  Roque Saenz
  Peña 94 B4
Presidente
  Epitácio 93 G3
Presidente
  Hermes 91 F6
Presidente
  Prudente 93 G3
Prešov 16 D5

Prespa, L. =
  Prespansko
  Jezero 23 D3
Prespansko
  Jezero 23 D3
Presque Isle .. 73 A11
Preston 11 E5
Pretoria 59 D5
Préveza 23 E3
Příbram 16 D2
Prichard 74 D3
Prieska 58 D4
Prievidza 16 D4
Prilep 22 D3
Priluki = Pryluky 24 C3
Prime Seal I. .. 62 D4
Prince Albert . 71 C9
Prince Albert
  Mts. 96 B15
Prince Albert
  Pen. 70 A8
Prince Albert Sd. 70 A8
Prince Charles I. 68 B3
Prince Charles
  Mts. 96 B10
Prince Edward
  I. □ 69 D4
Prince of Wales
  I., Canada .. 70 A10
Prince of Wales
  I., U.S.A. ... 71 C6
Principe da
  Beira 91 F6
Prins Harald
  Kyst 96 B8
Prinsesse Astrid
  Kyst 96 B7
Prinsesse
  Ragnhild Kyst 96 B8
Pripet → =
  Prypyat → .. 17 C10
Pripet Marshes 17 B9
Pripyat Marshes
  = Pripet
  Marshes 17 B9
Pripyats =
  Prypyat → .. 17 C10
Priština 22 C3
Privas 13 D6
Privolzhskaya
  Vozvyshennost
  .... 24 C6
Prizren 22 C3
Proddatur 43 M11
Progreso 85 C7
Prokopyevsk .. 29 D9
Prokuplje 22 C3
Prome = Pyè .. 41 H10
Propriá 93 E6
Prostějov 16 D3
Proston 63 A5
Provence 13 E6
Providence ... 73 D10
Providencia, I.
  de 86 D3
Provins 13 B5
Provo 81 E8
Prozna → 16 B3
Pruszków 16 B5
Prut → 17 F9
Prydz B. 96 A10
Pryluky 24 C3
Prypyat → .. 17 C10
Przemyśl 17 D6
Przhevalsk 29 E8
Psará 23 E5
Pskov 24 B2

Ptich =
  Ptsich → ... 17 B9
Ptsich → 17 B9
Puán 94 D4
Pucallpa 91 E4
Pudukkottai .. 43 P11
Puebla 84 D5
Pueblo 76 F2
Pueblo Hundido 94 B2
Puelches 94 D3
Puente Alto ... 94 C2
Puente-Genil .. 18 D3
Puerto Aisén .. 95 F2
Puerto
  Ayacucho ... 90 B5
Puerto Barrios . 85 D7
Puerto Bermejo 94 B5
Puerto
  Bermúdez .. 91 F4
Puerto Bolívar . 90 D3
Puerto Cabello 90 A5
Puerto Cabezas 86 D3
Puerto Carreño 90 B5
Puerto Chicama 91 E3
Puerto Coig ... 95 G3
Puerto Cortés . 85 D7
Puerto
  Cumarebo .. 90 A5
Puerto del
  Rosario 54 C2
Puerto Deseado 95 F3
Puerto Heath .. 91 F5
Puerto La Cruz 90 A6
Puerto
  Leguízamo .. 90 D4
Puerto Lobos .. 95 E3
Puerto Madryn 95 E3
Puerto
  Maldonado .. 91 F5
Puerto Montt .. 95 E2
Puerto Morelos 85 C7
Puerto Natales 95 G2
Puerto Páez ... 90 B5
Puerto Pinasco 94 A5
Puerto
  Pirámides ... 95 E4
Puerto Plata ... 87 C5
Puerto Princesa 38 C1
Puerto Quellón 95 E2
Puerto Rico ■ . 87 C6
Puerto Sastre . 91 H7
Puerto Suárez . 91 G7
Puerto Wilches 90 B4
Puertollano ... 18 C3
Pueyrredón, L. . 95 F2
Pugachev 24 C6
Puget Sound .. 80 B2
Puigcerdá 19 A6
Pukaki L. 65 F4
Pukapuka 65 L14
Pukekohe 64 B6
Pula 20 B4
Puławy 16 C5
Pullman 80 B5
Pulog 38 A2
Pułtusk 16 B5
Puna 91 G5
Puná, I. 90 D2
Punakha 41 D7
Punata 91 G5
Punch 42 C9
Pune 43 K8
Punjab □, India 42 D9
Punjab □,
  Pakistan 42 D8
Puno 91 G4
Punta Alta 94 D4
Punta Arenas .. 95 G2
Punta de Díaz . 94 B2

Puntabie 62 B1
Puntarenas ... 86 E3
Punto Fijo 90 A4
Puquio 91 F4
Pur → 28 C8
Purace, Vol. ... 90 C3
Puralia =
  Puruliya 40 F6
Puri 40 H5
Purnia 40 E6
Purukcahu ... 37 E4
Puruliya 40 F6
Purus → 90 D5
Pusan 35 C7
Pushkino 24 C6
Putao 41 D11
Putaruru 64 C6
Puthein Myit → 41 K10
Putignano ... 21 D6
Puting, Tanjung 37 E4
Putorana, Gory 30 C7
Puttalam 43 Q11
Puttgarden ... 15 A6
Putumayo → . 90 D5
Putussibau ... 37 D4
Puy-de-Dôme . 13 D4
Pweto 57 F5
Pwllheli 11 E4
Pyapon 41 J10
Pyasina → ... 30 B6
Pyatigorsk ... 25 E5
Pyè 41 H10
Pyetrikaw 17 B9
Pyinmana 41 H11
P'yŏngyang .. 35 C7
Pyrénées 19 A6
Pyu 41 H11

Q

Qādib 49 E5
Qā'emshahr ... 44 B3
Qahremānshahr
  = Bākhtarān . 46 D6
Qaidam Pendi . 34 C4
Qalāt 45 C6
Qal'at al Akhḍar 47 E4
Qal'eh Shaharak 42 B4
Qal'eh-ye Now 45 C5
Qamar, Ghubbat
  al 49 D5
Qamruddin
  Karez 42 D6
Qandahār 42 D4
Qandahār □ .. 42 D4
Qapshaghay . 29 E8
Qâra 52 C4
Qarabutaq ... 29 E7
Qarachuk 46 C5
Qaraghandy .. 29 E8
Qārah 47 E5
Qarataū 29 E8
Qarqan 34 C3
Qarqan He → . 34 C3
Qarqaraly 29 E8
Qarshi 29 F7
Qaṣr-e Qand .. 45 E5
Qasr Farâfra .. 52 C4
Qatar ■ 44 E2
Qattâra,
  Munkhafed el 52 C4
Qattâra
  Depression =
  Qattâra,
  Munkhafed el 52 C4
Qāyen 44 C4

Qazaqstan =
Kazakstan ■ . 29 E7
Qazvin 46 C7
Qena 52 C5
Qeshm 44 E4
Qila Safed . . 42 E2
Qila Saifullāh . . 42 D6
Qilian Shan . . . 34 C4
Qingdao 35 C7
Qinghai □ 34 C4
Qinghai Hu . . . 34 C5
Qinhuangdao . 35 C6
Qinzhou 35 D5
Qiqihar 35 B7
Qitai 34 B3
Qom 44 C2
Qomsheh 44 C2
Qondūz 45 B7
Qondūz □ 45 B7
Qostanay 29 D7
Qu Xian,
Sichuan,
China 35 C5
Qu Xian,
Zhejiang,
China 35 D6
Quambatook . . 62 C3
Quambone 63 B4
Quan Long . . . 36 C3
Quandialla 63 B4
Quang Ngai . . . 36 A3
Quanzhou 35 D6
Quaqtaq 68 B4
Quaraí 94 C5
Quartu
Sant'Elena . . 21 E2
Quatsino 71 C2
Quba 25 E6
Qūchān 44 B4
Queanbeyan . . 63 C4
Québec 69 D3
Québec □ 69 D3
Queen
Alexandra Ra. 96 C15
Queen Charlotte
Is. 70 C6
Queen Charlotte
Str. 71 C7
Queen Elizabeth
Is. 66 B9
Queen Mary
Land 96 B11
Queen Maud G. 70 B9
Queen Maud
Land 96 B7
Queen Maud
Mts. 96 C16
Queenscliff . . 63 C3
Queensland □ . 61 E7
Queenstown,
Australia . . 62 D4
Queenstown,
N.Z. 65 F3
Queenstown,
S. Africa 59 E5
Queimadas . . 93 E6
Quela 56 F3
Quelimane 59 B7
Quelpart =
Cheju Do . . 35 C7
Quequén 94 D5
Querétaro 84 C4
Quetta 42 D5
Quezon City . . 38 B2
Qui Nhon 36 B3
Quibaxe 56 F2
Quibdo 90 B3
Quiberon 12 C2
Quilán, C. 95 E2

Quilengues . . . 58 A2
Quillabamba . . 91 F4
Quillagua 91 H5
Quillota 94 C2
Quilon 43 Q10
Quilpie 63 A3
Quimilí 94 B4
Quimper 12 B1
Quimperlé . . . 12 C2
Quincy 77 F9
Quines 94 C3
Quinga 59 H8
Quintanar de la
Orden 18 C4
Quintero 94 C2
Quinyambie . . . 62 B3
Quipungo 58 A2
Quirindi 63 B5
Quissanga . . . 59 G8
Quitilipi 94 B4
Quito 90 D3
Quixadá 92 C6
Qŭnghirot 29 E6
Quondong 62 B3
Quorn 62 B2
Qŭqon 29 E8
Qûs 52 C5
Quseir 52 C5
Qyzylorda 29 E7

R

Raahe 8 F12
Raba → 39 F1
Rába → 16 E3
Rabat, Malta . . 21 G5
Rabat, Morocco 54 B3
Rabaul 64 K10
Rābigh 47 G4
Race, C. 69 D5
Rach Gia 36 B3
Racibórz 16 C4
Racine 72 C2
Rădăuți 17 E7
Radekhiv 17 C7
Radekhov =
Radekhiv . . . 17 C7
Radhwa, Jabal . 47 F4
Radom 16 C5
Radomsko 16 C4
Radomyshl . . . 17 C9
Radstock, C. . . 62 B1
Rae 70 B8
Rae Bareli 40 D3
Rae Isthmus . . 68 B2
Raetihi 64 C6
Rafaela 94 C4
Rafai 56 D4
Rafḥā 47 E5
Rafsanján 44 D4
Ragachow 17 B10
Ragama 43 R11
Raglan 64 B6
Ragusa 21 F5
Raha 39 E2
Rahad al Bardī . 53 F3
Rahaeng = Tak . 36 A1
Rahimyar Khan . 42 E7
Raichur 43 L10
Raigarh 40 G4
Railton 62 D4
Rainier, Mt. . . . 80 B3
Raipur 40 G3
Ra'is 47 G4
Raj Nandgaon . 40 G3
Raja, Ujung . . . 37 D1

Raja Ampat,
Kepulauan . . 39 E3
Rajahmundry . . 40 J3
Rajang → 37 D4
Rajapalaiyam . . 43 Q10
Rajasthan □ . . 42 F8
Rajasthan Canal 42 F8
Rajgarh 43 G10
Rajkot 43 H7
Rajpipla 43 J8
Rajshahi 41 E7
Rakaia 65 E5
Rakaia → 65 E5
Rakan, Ra's . . 44 E2
Rakaposhi . . . 42 A9
Rakata, Pulau . . 37 F3
Rakhiv 17 D7
Raleigh 75 B8
Ram Hd. 63 C4
Ramanathapuram
. . . . 43 Q11
Ramechhap . . . 40 D6
Ramelau 39 F3
Ramgarh, Bihar,
India 40 F5
Ramgarh, Raj.,
India 42 F7
Rāmhormoz . . 47 E7
Ramnad =
Ramanathapuram
. . . . 43 Q11
Rampur 42 E11
Rampur Hat . . 40 E6
Ramree I. =
Ramree Kyun 41 H9
Ramree Kyun . 41 H9
Ramtek 43 J11
Ranaghat 41 F7
Ranau 36 C5
Rancagua 94 C2
Ranchi 40 F5
Randers 9 G10
Rangaunu B. . . 64 A5
Rangia 41 D8
Rangiora 65 E5
Rangitaiki → . . 64 B7
Rangitata → . . 65 E4
Rangon → 41 J11
Rangoon 41 J11
Rangpur 41 E7
Ranibennur . . . 43 M9
Raniganj 40 F6
Raniwara 43 G8
Rankin Inlet . . 70 B10
Rankins Springs 63 B4
Ranong 36 C1
Ransiki 39 E4
Rantau 37 E5
Rantauprapat . . 37 D1
Rantemario . . . 39 E1
Rapa 65 M16
Rapallo 20 B2
Rāpch 45 E4
Rapid City . . . 76 C3
Rarotonga 65 M14
Ra's al Khaymah 44 E4
Ra's al-Unuf . . 52 B2
Ras Bânâs . . . 52 D6
Ras Dashen . . 57 B7
Râs Timirist . . 55 E1
Rasa, Punta . . 95 E4
Rashad 53 F5
Rashîd 52 B5
Rasht 46 C7
Rat Islands . . . 71 C1
Ratangarh 42 E9
Rathenow 15 B7
Ratibor =
Racibórz 16 C4

Ratlam 43 H9
Ratnagiri 43 L8
Raukumara Ra. 64 C7
Raurkela 40 F5
Rava-Ruska . . . 17 C6
Rava Russkaya
= Rava-Ruska 17 C6
Rāvar 44 D4
Ravenna 20 B4
Ravensburg . . 14 E5
Ravi → 42 D7
Rawalpindi . . . 42 C8
Rawāndūz 46 C6
Rawene 64 A5
Rawlins 81 E10
Rawson 95 E3
Ray, C. 69 D5
Rayadurg . . . 43 M10
Rayagada 40 H4
Raymond 71 D8
Raz, Pte. du . . 12 B1
Razdel'naya =
Rozdilna . . . 17 E10
Razelm, Lacul . 22 B7
Razgrad 22 C6
Ré, I. de 12 C3
Reading, U.K. . 11 F6
Reading, U.S.A. 73 D8
Realicó 94 D4
Rebi 39 F4
Rebiana 52 D3
Rebun-Tō . . . 32 E12
Rechytsa 17 B10
Recife 92 D7
Reconquista . . 94 B5
Recreo 94 B3
Red →, La.,
U.S.A. 79 D8
Red →, N. Dak.,
U.S.A. 76 A6
Red Cliffs 62 B3
Red Deer 71 C8
Red Sea 48 C2
Red Tower Pass
= Turnu Roşu,
P. 17 F7
Redcar 11 D6
Redcliffe 63 A5
Redding 80 E2
Redditch 11 E6
Redon 12 C2
Redondela . . . 18 A1
Redwood City . 82 B1
Ree, L. 11 E3
Reefton 65 E4
Regensburg . . 15 D7
Réggio di
Calábria 21 E5
Réggio
nell'Emília . . 20 B3
Reghin 17 E7
Regina 71 C9
Rehoboth 58 C3
Rehovot 46 E3
Rei-Bouba . . . 53 G1
Reichenbach . . 15 C7
Reidsville 75 A8
Reigate 11 F6
Reims 13 B6
Reina Adelaida,
Arch. 95 G2
Reindeer L. . . . 71 C9
Reinga, C. 64 A5
Reinosa 18 A3
Reliance 70 B9
Remarkable, Mt. 62 B2
Remeshk 45 E4
Rendsburg . . . 14 A5
Rengat 37 E2

Reni	17	F9	
Renk	53	F5	
Renmark	62	B3	
Rennes	12	B3	
Reno	80	F4	
Reno →	20	B4	
Rentería	19	A5	
Republican →	76	F6	
Repulse Bay	68	B2	
Requena, *Peru*	90	E4	
Requena, *Spain*	19	C5	
Resadiye = Datça	23	F6	
Resht = Rasht	46	C7	
Resistencia	94	B5	
Reşiţa	17	F5	
Resolution I., *Canada*	68	B4	
Resolution I., *N.Z.*	65	F2	
Réthímnon	23	G5	
Réunion ■	51	J9	
Reus	19	B6	
Reutlingen	14	D5	
Reval = Tallinn	24	B1	
Revelstoke	71	C8	
Rewa	40	E3	
Rewari	42	E10	
Rey Malabo	56	D1	
Reykjavík	8	B2	
Reynosa	84	B5	
Rhein →	14	C4	
Rhein-Main-Donau-Kanal →	15	D6	
Rheine	14	B4	
Rheinland-Pfalz □	14	C4	
Rhin = Rhein →	14	C4	
Rhine = Rhein →	14	C4	
Rhineland-Palatinate □ = Rheinland-Pfalz □	14	C4	
Rhode Island □	73	D10	
Rhodes = Ródhos	23	F7	
Rhodope Mts. = Rhodopi Planina	22	D5	
Rhodopi Planina	22	D5	
Rhön = Hohe Rhön	14	C5	
Rhondda	11	F5	
Rhône →	13	E6	
Rhum	10	C3	
Riachão	92	D4	
Riasi	42	C9	
Riau □	37	E2	
Riau, Kepulauan	37	D2	
Riau Arch. = Riau, Kepulauan	37	D2	
Ribadeo	18	A2	
Ribeirão Prêto	93	G4	
Riberalta	91	F5	
Ribniţa	17	E9	
Riccarton	65	E5	
Richland	80	B4	
Richmond, *Australia*	63	B5	
Richmond, *N.Z.*	65	D5	
Richmond, *Calif., U.S.A.*	80	G2	
Richmond, *Va., U.S.A.*	73	F7	
Richmond Ra.	63	A5	
Ridder	29	D9	
Ried	15	D7	
Riesa	15	C7	
Rieti	20	C4	
Rig Rig	53	F1	
Riga	24	B1	
Riga, G. of	24	B1	
Rīgas Jūras Līcis = Riga, G. of	24	B1	
Rīgestān □	42	D4	
Rigolet	68	C5	
Riiser-Larsen-halvøya	96	A8	
Rijeka	20	B5	
Rimah, Wadi ar →	47	F5	
Rímini	20	B4	
Rîmnicu Sărat	17	F8	
Rîmnicu Vîlcea	17	F7	
Rimouski	69	D4	
Rinca	39	F1	
Rinconada	94	A3	
Rinjani	37	F5	
Rio Branco, *Brazil*	91	E5	
Río Branco, *Uruguay*	94	C6	
Río Cuarto	94	C4	
Rio de Janeiro	93	G5	
Rio de Janeiro □	93	G5	
Rio do Sul	94	B7	
Río Gallegos	95	G3	
Río Grande, *Argentina*	95	G3	
Rio Grande, *Brazil*	94	C6	
Rio Grande →	78	E3	
Rio Grande do Norte □	92	D6	
Rio Grande do Sul □	94	B6	
Rio Largo	93	D6	
Río Mulatos	91	G5	
Río Muni = Mbini □	56	D2	
Rio Negro	94	B7	
Rio Verde	93	F3	
Ríobamba	90	D3	
Ríohacha	90	A4	
Ríosucio, *Caldas, Colombia*	90	B3	
Ríosucio, *Choco, Colombia*	90	B3	
Rishiri-Tō	32	E12	
Riva del Garda	20	B3	
Rivadavia	94	B2	
Rivera	94	C5	
Riversdale	58	E4	
Riverside	82	D4	
Riverton, *Australia*	62	B2	
Riverton, *Canada*	71	C10	
Riverton, *N.Z.*	65	G2	
Riviera di Levante	20	B2	
Riviera di Ponente	20	B2	
Rivière-du-Loup	69	D4	
Rivne	17	C8	
Rívoli	20	B1	
Rivoli B.	62	C3	
Riyadh = Ar Riyāḍ	47	F6	
Rize	46	B5	
Rizzuto, C.	21	E6	
Roanne	13	C6	
Roanoke	73	F6	
Robbins I.	62	D4	
Robertson	58	E3	
Robertson I.	96	A4	
Robertsport	55	G2	
Robertstown	62	B2	
Robinvale	62	B3	
Roboré	91	G7	
Robson, Mt.	71	C8	
Roca, C. da	18	C1	
Rocas, I.	92	C7	
Rocha	94	C6	
Rochefort	12	D3	
Rochester, *Minn., U.S.A.*	77	C8	
Rochester, *N.Y., U.S.A.*	73	C7	
Rock Hill	75	B7	
Rock Island	77	E9	
Rock Springs	81	E9	
Rockefeller Plateau	96	C18	
Rockford	77	D10	
Rockhampton	61	E9	
Rocky Mount	75	B9	
Rocky Mts.	70	C7	
Rod	42	E3	
Rodez	13	D5	
Ródhos	23	F7	
Rodney, C.	64	B6	
Roes Welcome Sd.	68	B2	
Rogachev = Ragachow	17	B10	
Rogaguá, L.	91	F5	
Rogatyn	17	D7	
Rogoaguado, L.	91	F5	
Rohri	42	F6	
Rohtak	42	E10	
Roi Et	36	A2	
Rojas	94	C4	
Rojo, C.	84	C5	
Rokan →	37	D2	
Rolândia	94	A6	
Roma, *Australia*	63	A4	
Roma, *Italy*	20	D4	
Roman	17	E8	
Romang	39	F3	
Romania ■	22	B5	
Romanovka = Basarabeasca	17	E9	
Romans-sur-Isère	13	D6	
Romblon	38	B2	
Rome = Roma	20	D4	
Rome	74	B5	
Romorantin-Lanthenay	12	C4	
Roncador, Serra do	93	E3	
Ronda	18	D3	
Rondônia □	91	F6	
Rondonópolis	93	F3	
Ronge, L. la	71	C9	
Ronne Ice Shelf	96	B3	
Roodepoort	59	D5	
Roorkee	42	E10	
Roosevelt →	91	E6	
Roosevelt I.	96	B16	
Roquetas de Mar	19	D4	
Roraima □	90	C6	
Roraima, Mt.	90	B6	
Rosa	57	F6	
Rosa, Monte	13	D7	
Rosario, *Argentina*	94	C4	
Rosário, *Brazil*	92	C5	
Rosario, *Mexico*	84	C3	
Rosario, *Paraguay*	94	A5	
Rosario de la Frontera	94	B3	
Rosário do Sul	94	C6	
Rosas, G. de	19	A7	
Roscommon	11	E2	
Roseau, *Domin.*	87	C7	
Roseau, *U.S.A.*	77	A7	
Rosebery	62	D4	
Roseburg	80	D2	
Rosenheim	15	E7	
Rosetown	71	C9	
Rosetta = Rashîd	52	B5	
Roseville	80	F3	
Rosewood	63	A5	
Rosignano Maríttimo	20	C3	
Rosignol	90	B7	
Roşiori-de-Vede	22	B5	
Roslavl	24	C3	
Roslyn	63	B4	
Ross, *Australia*	62	D4	
Ross, *N.Z.*	65	E4	
Ross I.	96	B15	
Ross Ice Shelf	96	C16	
Ross Sea	96	B15	
Rossano Cálabro	21	E6	
Rossland	71	D8	
Rosslare	11	E3	
Rosso	55	E1	
Rossosh	24	C4	
Rosthern	71	C9	
Rostock	15	A7	
Rostov, *Don, Russia*	25	D4	
Rostov, *Yarosl., Russia*	24	B4	
Roswell	78	C1	
Rotherham	11	E6	
Roto	63	B4	
Rotondo Mte.	13	E8	
Rotorua	64	C7	
Rotorua, L.	64	C7	
Rotterdam	14	C3	
Rottweil	14	D5	
Rotuma	64	L12	
Roubaix	13	A5	
Rouen	12	B4	
Round Mt.	63	B5	
Roussillon	13	E5	
Rouyn	69	D3	
Rovaniemi	8	E13	
Rovereto	20	B3	
Rovigo	20	B3	
Rovinj	20	B4	
Rovno = Rivne	17	C8	
Rovuma →	57	G8	
Rowena	63	A4	
Roxas	38	B2	
Roxburgh	65	F3	
Royal Leamington Spa	11	E6	
Royan	12	D3	
Rozdilna	17	E10	
Rozhyshche	17	C7	
Rtishchevo	24	C5	
Ruahine Ra.	64	C7	
Ruapehu	64	C6	
Ruapuke I.	65	G3	
Rub' al Khali	48	D4	
Rubio	90	B4	
Rubtsovsk	29	D9	
Rudall	62	B2	
Rudolf, Ostrov	28	A6	
Rufa'a	53	F5	
Rufiji →	57	F7	

Rufino 94 C4
Rufisque 55 F1
Rugby 11 E6
Rügen 15 A7
Ruhr → 14 C4
Rukwa L. 57 F6
Rum = Rhum . 10 C3
Rumāḥ 47 F6
Rumania =
　Romania ■ .. 22 B5
Rumbêk 53 G4
Rumia 16 A4
Rumoi 32 F12
Runanga 65 E4
Runaway, C. . 64 B7
Rungwa 57 F6
Ruoqiang 34 C3
Rupa 41 D9
Rupat 37 D2
Rupert → 69 C3
Rupert House =
　Waskaganish 69 C3
Rurrenabaque . 91 F5
Rusape 59 B6
Ruschuk = Ruse 22 C5
Ruse 22 C5
Rushworth 63 C4
Russas 92 C6
Russellkonda .. 40 H5
Russkaya
　Polyana 29 D8
Rustavi 25 E6
Rustenburg ... 59 D5
Ruteng 39 F2
Rutshuru 57 E5
Ruwenzori ... 57 D5
Ružomberok .. 16 D4
Rwanda ■ ... 57 E6
Ryazan 24 C4
Ryazhsk 24 C5
Rybache =
　Rybachye .. 29 E9
Rybachye 29 E9
Rybinsk 24 B4
Rybinskoye
　Vdkhr. 24 B4
Rybnitsa =
　Rîbniţa =
　Rîbniţa 17 E9
Rylstone 63 B4
Rypin 16 B4
Ryūkyū Is. =
　Ryūkyū-rettō 35 D7
Ryūkyū-rettō .. 35 D7
Rzeszów 16 C5
Rzhev 24 B3

S

Sa Dec 36 B3
Sa'ādatābād .. 44 D3
Saale → 15 C6
Saalfeld 15 C6
Saar → 13 B7
Saarbrücken .. 14 D4
Saaremaa 24 B1
Šabac 22 B2
Sabadell 19 B7
Sabah □ 36 C5
Ṣabāḥ, Wadi → 47 G7
Sabalān, Kūhhā-
　ye 46 C6
Sábanalarga .. 90 A4
Sabang 36 C1
Sabará 93 F5
Saberania 39 E5
Sabhah 52 C1

Sabinas 84 B4
Sabinas Hidalgo 84 B4
Sablayan 38 B2
Sable, C.,
　Canada 69 D4
Sable, C., U.S.A. 86 A3
Sable I. 69 D5
Sabrina Coast . 96 A12
Sabulubek 37 E1
Sabzevār 44 B4
Sabzvārān 44 D4
Săcele 17 F7
Sachsen □ 15 C7
Sachsen-
　Anhalt □ 15 C7
Sacramento .. 80 F3
Sacramento → 80 F3
Sacramento
　Mts. 83 D10
Sadani 57 F7
Sadd el Aali ... 52 D5
Sado 33 G11
Sadon 41 E11
Safi 54 B3
Safid Kūh 42 B3
Saga, Indonesia 39 E4
Saga, Japan .. 32 C2
Sagala 55 F3
Sagar 43 M9
Sagil 34 A4
Saginaw 72 C4
Şagīr, Zāb aş → 46 D5
Saglouc =
　Salluit 68 B3
Sagua la Grande 86 B3
Saguenay → . 69 D3
Sagunto 19 C5
Sahagún 18 A3
Sahand, Kūh-e 46 C6
Sahara 54 D6
Saharan Atlas =
　Saharien,
　Atlas 54 B5
Saharanpur .. 42 E10
Saharien, Atlas 54 B5
Sahiwal 42 D8
Sa'id Bundas .. 53 G3
Saïda 54 B5
Sa'īdābād 44 D3
Sa'īdīyeh 46 C7
Saidpur 41 E7
Saidu 42 B8
Saigon = Phanh
　Bho Ho Chi
　Minh → 36 B3
Saijō 32 C3
Saikhoa Ghat . 41 D10
Saiki 32 C2
Sailolof 39 E4
St. Andrews .. 10 C5
St. Arnaud ... 62 C3
St-Augustin-
　Saguenay .. 69 C5
St. Augustine . 75 E7
St. Austell ... 11 F4
St. Boniface ... 71 D10
St.-Brieuc ... 12 B2
St. Catharines . 69 D3
St.-Chamond .. 13 D6
St. Christopher-
　Nevis ■ = St.
　Kitts & Nevis 87 C7
St. Cloud 77 C7
St. Croix 87 C7
St.-Denis 12 B5
St.-Dizier → . 13 B6
St. Elias, Mt. . 70 B5
St.-Étienne .. 13 D6
St.-Flour 13 D5

St. Francis, C. . 58 E4
St. Gallen =
　Sankt Gallen 13 C8
St.-Gaudens .. 12 E4
St. George 63 A4
St.-Georges,
　Fr. Guiana .. 92 B3
St. George's,
　Grenada 87 D7
St. George's
　Channel 11 F3
St. Georges Hd. 63 C5
St. Gotthard P.
　= San
　Gottardo, P.
　del 13 C8
St. Helena ■ . 51 H3
St. Helena B. .. 58 E3
St. Helens 62 D4
St. Helier 11 G5
St-Hyacinthe .. 69 D3
St. John 69 D4
St. John's,
　Antigua 87 C7
St. John's,
　Canada 69 D5
St. Johns → .. 75 D7
St. Joseph ... 77 F7
St. Joseph, L. . 69 C1
St. Kilda, N.Z. . 65 F4
St. Kilda, U.K. . 10 C2
St. Kitts & Nevis 87 C7
St.-Laurent → 92 A3
St. Lawrence → 69 D4
St. Lawrence,
　Gulf of ... 69 D4
St. Lawrence I. 71 B2
St. Leonard .. 69 D4
St.-Lô 12 B3
St-Louis 55 E1
St. Louis 77 F9
St. Lucia ■ .. 87 D7
St. Lucia, L. .. 59 D6
St. Maarten .. 87 C7
St.-Malo 12 B2
St-Marc 87 C5
St. Mary Pk. .. 62 B2
St. Marys 62 D4
St.-Mathieu, Pte. 12 B1
St. Matthews, I.
　= Zadetkyi
　Kyun 36 C1
St.-Nazaire 12 C2
St.-Omer 12 A5
St. Paul 77 C8
St. Peter Port .. 11 G5
St. Petersburg =
　Sankt-
　Peterburg ... 24 B3
St. Petersburg . 75 F6
St.-Pierre et
　Miquelon □ . 69 D5
St.-Quentin ... 13 B5
St. Thomas I. .. 87 C7
St.-Tropez ... 13 E7
St. Vincent, G. . 62 C2
St. Vincent &
　the
　Grenadines ■ 87 D7
Saintes 12 D3
Saintonge 12 D3
Saipan 64 H9
Sairang 41 F9
Sairecábur,
　Cerro 94 A3
Sajama 91 G5
Sajó 16 D5
Sakai 32 B4
Sakākah 47 E5

Sakakawea, L. . 76 B4
Sakarya =
　Adapazarı ... 46 B2
Sakarya → ... 25 E3
Sakata 33 G11
Sakhalin 31 D12
Sakon Nakhon . 36 A2
Sala 9 G11
Sala Consilina . 21 D5
Saladillo 94 D5
Salado →,
　Buenos Aires,
　Argentina ... 94 D5
Salado →,
　La Pampa,
　Argentina ... 94 D3
Salado →,
　Santa Fe,
　Argentina ... 94 C4
Salaga 55 G4
Salālah 49 D5
Salamanca,
　Chile 94 C2
Salamanca,
　Spain 18 B3
Salamís 23 F4
Salar de
　Atacama 94 A3
Salar de Uyuni 91 H5
Salaverry 91 E3
Salawati 39 E4
Salayar 39 F2
Saldanha 58 E3
Sale, Australia . 63 C4
Salé, Morocco . 54 B3
Salekhard ... 28 C7
Salem, India .. 43 P11
Salem, U.S.A. . 80 C2
Salerno 21 D5
Salgótarján ... 16 D4
Salihli 23 E7
Salihorsk 17 B8
Salima 59 A6
Salina, Italy .. 21 E5
Salina, U.S.A. . 76 F6
Salina Cruz .. 85 D5
Salinas, Brazil . 93 F5
Salinas, Ecuador 90 D2
Salinas, U.S.A. 82 B2
Salinas Grandes 94 C3
Salinópolis ... 92 C4
Salisbury,
　Australia 62 B2
Salisbury, U.K. 11 F6
Salisbury, U.S.A. 75 B7
Salluit 68 B3
Salmās 46 C6
Salmon → ... 80 C5
Salmon River
　Mts. 81 C6
Salon-de-
　Provence ... 13 E6
Salonica =
　Thessaloníki . 23 D4
Salonta 16 E5
Salsk 25 D5
Salso → 21 F4
Salt Creek ... 62 C2
Salt Lake City . 81 E8
Salta 94 A3
Saltillo 84 B4
Salto 94 C5
Salto → 20 C4
Salton Sea ... 82 D5
Saltpond 55 G4
Salûm 52 B4
Salûm, Khâlig el 52 B4
Salur 40 H4
Salvador 93 E6

Salwa	44	E2
Salween →	41	J11
Salyan	25	F6
Salzach →	15	D7
Salzburg	15	E7
Salzgitter	14	B6
Salzwedel	15	B6
Sama	28	C7
Sama de Langreo	18	A3
Samales Group	38	C2
Samangán □	45	B7
Samar	38	B3
Samara	24	C7
Samarinda	37	E5
Samarkand = Samarqand	29	F7
Samarqand	29	F7
Sāmarrā	46	D5
Sambalpur	40	G5
Sambar, Tanjung	37	E4
Sambas	37	D3
Sambava	59	G10
Sambhal	42	E11
Sambhar	42	F9
Sambiase	21	E6
Sambir	17	D6
Same	57	E7
Samokov	22	C4
Sámos	23	F6
Samothráki, Évros, Greece	23	D5
Samothráki, Kérkira, Greece	23	E2
Sampacho	94	C4
Sampit	37	E4
Sampit, Teluk	37	E4
Samra	47	F5
Samsun	46	B4
Samui, Ko	36	C2
Samut Prakan	36	B2
Samut Songkhram →	36	B2
San	55	F4
San →	16	C5
San Agustin, C.	38	C3
San Andrés, I. de	86	D3
San Andrés Tuxtla	85	D5
San Angelo	78	D3
San Antonio, Chile	94	C2
San Antonio, Spain	19	C6
San Antonio, U.S.A.	78	E4
San Antonio, C., Argentina	94	D5
San Antonio, C., Cuba	85	C8
San Antonio Oeste	95	E4
San Benedetto del Tronto	20	C4
San Bernardino	82	C4
San Bernardino Str.	38	B2
San Bernardo	94	C2
San Bernardo, I. de	90	B3
San Borja	91	F5
San Carlos, Argentina	94	C3
San Carlos, Chile	94	D2
San Carlos, Phil.	38	B2
San Carlos, Amazonas, Venezuela	90	C5
San Carlos, Cojedes, Venezuela	90	B5
San Carlos de Bariloche	95	E2
San Carlos del Zulia	90	B4
San Cristóbal, Argentina	94	C4
San Cristóbal, Mexico	85	D6
San Cristóbal, Venezuela	90	B4
San Diego	82	D4
San Diego, C.	95	G3
San Felipe, Chile	94	C2
San Felipe, Venezuela	90	A5
San Felíu de Guíxols	19	B7
San Fernando, Chile	94	C2
San Fernando, La Union, Phil.	38	A2
San Fernando, Pampanga, Phil.	38	A2
San Fernando, Spain	18	D2
San Fernando, Trin. & Tob.	87	D7
San Fernando, U.S.A.	82	C3
San Fernando de Apure	90	B5
San Fernando de Atabapo	90	C5
San Francisco, Argentina	94	C4
San Francisco, U.S.A.	82	B1
San Francisco de Macorîs	87	C5
San Francisco del Monte de Oro	94	C3
San Gil	90	B4
San Gottardo, P. del	13	C8
San Ignacio, Bolivia	91	G6
San Ignacio, Paraguay	94	B5
San Ildefonso, C.	38	A2
San Javier, Argentina	94	C5
San Javier, Bolivia	91	G6
San Joaquin →	82	A2
San Jorge, B. de	84	A2
San Jorge, G.	95	F3
San Jorge, G. de	19	B6
San José, Bolivia	91	G6
San José, Costa Rica	86	E3
San José, Guatemala	85	E6
San Jose, Phil.	38	A2
San Jose, U.S.A.	82	B2
San Jose de Buenovista	38	B2
San José de Jáchal	94	C3
San José de Mayo	94	C5
San José de Ocune	90	C4
San José del Guaviare	90	C4
San Juan, Argentina	94	C3
San Juan, Phil.	38	C3
San Juan, Puerto Rico	87	C6
San Juan →	86	D3
San Juan, C.	56	D1
San Juan de los Morros	90	B5
San Juan Mts.	83	B9
San Julián	95	F3
San Justo	94	C4
San Leandro	82	B1
San Lorenzo	90	C3
San Lorenzo, I.	91	F3
San Lorenzo, Mt.	95	F2
San Lucas	91	H5
San Lucas, C.	84	C2
San Luis	94	C3
San Luis Obispo	82	C2
San Luis Potosí	84	C4
San Marino	20	C4
San Marino ■	20	C4
San Martín, L.	95	F2
San Matías	91	G7
San Matías, G.	95	E4
San Miguel	85	E7
San Miguel →	91	F6
San Miguel de Tucumán	94	B3
San Narciso	38	A2
San Nicolás de los Arroyas	94	C4
San-Pédro	55	H3
San Pedro de las Colonias	84	B4
San Pedro de Lloc	91	E3
San Pedro de Macorís	87	C6
San Pedro del Paraná	94	B5
San Pedro Sula	85	D7
San Pieto	21	E2
San Rafael, Argentina	94	C3
San Rafael, U.S.A.	80	G2
San Ramón de la Nueva Orán	94	A4
San Remo	20	C1
San Roque, Argentina	94	B5
San Roque, Spain	18	D3
San Rosendo	94	D2
San Salvador, Bahamas	86	B5
San Salvador, El Salv.	85	E7
San Salvador de Jujuy	94	A4
San Sebastián, Argentina	95	G3
San Sebastián, Spain	19	A5
San Severo	20	D5
San Valentin, Mte.	95	F2
San Vicente de la Barquera	18	A3
Sana'	49	D3
Sana →	20	B6
Sanaga →	56	D1
Sanana	39	E3
Sanandaj	46	D6
Sanco Pt.	38	C3
Sancti-Spíritus	86	B4
Sancy, Puy de	13	D5
Sandakan	36	C5
Sandanski	22	D4
Sanday	10	B5
Sandgate	63	A5
Sandía	91	F5
Sandıklı	46	C2
Sandoa	56	F4
Sandomierz	16	C5
Sandoway	41	H10
Sandusky	72	D4
Sandwip Chan.	41	F8
Sandy C.	62	D3
Sandy L.	71	C10
Sanford, Fla., U.S.A.	75	E7
Sanford, N.C., U.S.A.	75	B8
Sanford, Mt.	71	B5
Sanga →	56	E3
Sangamner	43	K9
Sangar	31	C10
Sangasangadalam	37	E5
Sangeang	39	F1
Sangerhausen	15	C6
Sanggau	37	D4
Sangihe, P.	39	D3
Sangkapura	37	F4
Sangli	43	L9
Sangmélina	56	D2
Sangre de Cristo Mts.	78	A1
Sankt Gallen	13	C8
Sankt Moritz	13	C8
Sankt-Peterburg	24	B3
Sankt Pölten	15	D8
Sankuru →	56	E4
Sanliurfa	46	C4
Sanlúcar de Barrameda	18	D2
Sannicandro Gargánico	20	D5
Sanok	17	D6
Sanshui	35	D6
Santa Ana, Bolivia	91	F5
Santa Ana, Ecuador	90	D2
Santa Ana, El Salv.	85	E7
Santa Ana, U.S.A.	82	D4
Santa Barbara	82	C3
Santa Catalina I.	82	D3
Santa Catarina □	94	B7
Santa Clara	86	B4
Santa Clotilde	90	D4
Santa Coloma de Gramanet	19	B7
Santa Cruz, Argentina	95	G3
Santa Cruz, Bolivia	91	G6
Santa Cruz, Phil.	38	B2
Santa Cruz, U.S.A.	82	B1
Santa Cruz →	95	G3
Santa Cruz de Tenerife	54	C1

Santa Cruz do
  Sul 94 B6
Santa Cruz I. .. 64 L11
Santa Elena ... 90 D2
Santa Eugenia,
  Pta. 84 B1
Santa Fe,
  Argentina ... 94 C4
Santa Fe, *U.S.A.* 83 C10
Santa Filomena 93 D4
Santa Inés, I. .. 95 G2
Santa Isabel =
  Rey Malabo . 56 D1
Santa Isabel,
  Argentina ... 94 D3
Santa Isabel,
  Brazil 93 E3
Santa Lucia
  Range 82 C2
Santa Maria,
  Brazil 94 B6
Santa Maria,
  U.S.A. 82 C2
Santa María → 84 A3
Santa Maria da
  Vitória 93 E5
Santa Maria di
  Leuca, C. ... 23 E2
Santa Marta ... 90 A4
Santa Marta,
  Sierra Nevada
  de 90 A4
Santa Maura =
  Levkás 23 E3
Santa Rosa,
  La Pampa,
  Argentina ... 94 D4
Santa Rosa,
  San Luis,
  Argentina ... 94 C3
Santa Rosa,
  Bolivia 91 F5
Santa Rosa,
  Brazil 94 B6
Santa Rosa,
  U.S.A. 80 F2
Santa Rosa de
  Copán 85 E7
Santa Vitória do
  Palmar 94 C6
Santai 35 C5
Santana, Coxilha
  de 94 C5
Santana do
  Livramento .. 94 C5
Santander ... 18 A4
Sant'Antíoco .. 21 E2
Santarém, *Brazil* 92 C3
Santarém,
  Portugal 18 C1
Santiago, *Brazil* 94 B6
Santiago, *Chile* 94 C2
Santiago → .. 90 D3
Santiago de
  Compostela . 18 A1
Santiago de
  Cuba 86 C4
Santiago de los
  Cabelleros .. 87 C5
Santiago del
  Estero 94 B4
Santo Amaro . 93 E6
Santo Ângelo . 94 B6
Santo Antonio . 93 F2
Santo Corazón 91 G7
Santo Domingo 87 C6
Santo Tomás . 91 F4
Santo Tomé ... 94 B5

Santo Tomé de
  Guayana =
  Ciudad
  Guayana 90 B6
Santoña 18 A4
Santoríni =
  Thíra 23 F5
Santos 94 A7
Santos Dumont 94 A8
Sanza Pombo . 56 F3
São Anastácio . 94 A6
São Bernado de
  Campo 93 G4
São Borja 94 B5
São Carlos 94 A7
São Cristóvão . 93 E6
São Domingos 93 E4
São Francisco . 93 F5
São
  Francisco → 93 E6
São Francisco
  do Sul 94 B7
São Gabriel ... 94 C6
São João da
  Madeira 18 B1
São João del
  Rei 93 G5
São João do
  Araguaia 92 D4
São João do
  Piauí 92 D5
São José do Rio
  Prêto 94 A7
São Leopoldo . 94 B6
São Lourenço . 93 G4
São
  Lourenço → 93 F2
São Luís 92 C5
São Marcos → 93 F4
São Marcos, B.
  de 92 C5
São Mateus .. 93 F6
São Paulo 94 A7
São Paulo □ .. 94 A7
São Roque, C.
  de 92 D6
São Sebastião, I.
  de 94 A7
São Tomé &
  Principe ■ .. 51 F4
São Vicente, C.
  de 18 D1
Saône → 13 D6
Saonek 39 E4
Saparua 39 E3
Sapele 55 G6
Saposoa 91 E3
Sapporo 32 F12
Sapulpa 79 B5
Saqqez 46 C6
Sar-e Pol 45 B6
Sar-e Pol □ .. 45 B6
Sar Planina ... 22 C3
Sarāb 46 C6
Sarada → 40 D3
Saragossa =
  Zaragoza .. 19 B5
Saraguro 90 D3
Sarajevo 20 C7
Saran, G. 37 E4
Sarandí del Yi . 94 C5
Sarangani B. .. 38 C3
Sarangani Is. .. 38 C3
Sarangarh 40 G4
Saransk 24 C6
Sarapul 29 D6
Sarasota 75 F6
Saratov 24 C6
Saravane 36 A3

Sarawak □ 37 D4
Saray 22 D6
Sarayköy 23 F7
Sarbāz 45 E5
Sarbīsheh 45 C4
Sarda → =
  Sarada → .. 40 D3
Sardalas 54 C7
Sardarshahr .. 42 E9
Sardegna □ ... 21 D2
Sardinia =
  Sardegna □ . 21 D2
Sardis 23 E7
Sargodha 42 C8
Sarh 53 G2
Sārī 44 B3
Sangöl 23 E7
Sarikamiş 46 B5
Sarikei 37 D4
Sark 11 G5
Şarköy 23 D6
Sarlat-la-Canéda 12 D4
Sarmi 39 E5
Sarmiento 95 F3
Sarnia 69 D2
Sarny 24 C2
Sarolangun ... 37 E2
Saronikós
  Kólpos 23 F4
Sarre = Saar → 13 B7
Sarreguemines 13 B7
Sarro 55 F3
Sarthe → 12 C3
Sartynya 28 C7
Sarvestän 44 D3
Sary-Tash 29 F8
Saryshagan ... 29 E8
Sasabeneh 49 F3
Sasaram 40 E5
Sasebo 32 C1
Saser 42 B10
Saskatchewan □ 71 C9
Saskatchewan →
  .... 71 C9
Saskatoon 71 C9
Saskylakh 30 B9
Sasovo 24 C5
Sassandra 55 H3
Sassandra → .. 55 H3
Sássari 21 D2
Sassnitz 15 A7
Sassuolo 20 B3
Sasyk, Ozero .. 17 F9
Sata-Misaki .. 33 D2
Satadougou ... 55 F2
Satara 43 L8
Satmala Hills . 43 J9
Satna 40 E3
Sátoraljaújhely 16 D5
Satpura Ra. ... 43 J10
Satu Mare 17 E6
Satui 37 E5
Saturnina → .. 91 F7
Saudi Arabia ■ 47 F6
Sauerland 14 C4
Sault Ste. Marie,
  Canada 69 D2
Sault Ste. Marie,
  U.S.A. 72 A3
Saumlaki 39 F4
Saumur 12 C3
Saunders C. ... 65 F4
Sauri 55 F6
Saurimo 56 F4
Sava → 22 B3
Savage I. =
  Niue 65 L14
Savalou 55 G5
Savanna la Mar 86 C4

Savannah 75 C7
Savannah → . 75 C7
Savannakhet .. 36 A2
Savanur 43 M9
Savé 55 G5
Save → 59 C6
Säveh 46 D7
Savelugu 55 G4
Savoie □ 13 D7
Savona 20 B2
Sawahlunto .. 37 E2
Sawai 39 E3
Sawai
  Madhopur .. 42 F10
Sawara 32 B7
Sawatch Mts. . 83 A9
Sawmills 59 B5
Sawu 39 F2
Sawu Sea ... 39 F2
Saxony, Lower
  =
  Niedersachsen □
  .... 14 B5
Say 55 F5
Sayán 91 F3
Sayan,
  Zapadnyy ... 30 D7
Saydā 46 D3
Sayghān 45 C6
Sayhut 49 D5
Saynshand 35 B6
Sazanit 23 D2
Sázava → 16 D2
Sazin 42 B8
Scandicci 20 C3
Scandinavia .. 8 F10
Scarborough .. 11 D6
Scebeli,
  Wabi → ... 49 G3
Schaffhausen . 13 C8
Schefferville .. 69 C4
Schelde → ... 14 C3
Schenectady .. 73 C9
Schiermonnikoog
  .... 14 B4
Schio 20 B3
Schleswig 14 A5
Schleswig-
  Holstein □ .. 14 A5
Schouten I. ... 62 D4
Schouten Is. =
  Supiori 39 E5
Schwäbische
  Alb 14 D5
Schwaner,
  Pegunungan 37 E4
Schwarzwald .. 14 D5
Schwedt 15 B8
Schweinfurt .. 14 C6
Schwenningen
  = Villingen-
  Schwenningen 14 D5
Schwerin 15 B6
Schwyz 13 C8
Sciacca 21 F4
Scilla 21 E5
Scilly, Isles of . 11 G3
Scone 63 B5
Scotland □ ... 10 C4
Scott Glacier .. 96 A12
Scott Inlet ... 68 A3
Scottsbluff ... 76 E3
Scottsdale ... 62 D4
Scranton 73 D8
Scunthorpe .. 11 E6
Scusciuban .. 49 E5
Scutari =
  Üsküdar 25 E2
Seaspray 63 C4

Seattle

Seattle	80	B2
Sebastopol =		
Sevastopol ..	25	E3
Sebha =		
Sabhah	52	C1
Şebinkarahisar	46	B4
Sebta = Ceuta .	18	E3
Sebuku	37	E5
Sebuku, Teluk .	37	D5
Secretary I. ...	65	F2
Secunderabad .	43	L11
Sedalia	77	F8
Sedan, Australia	62	B2
Sedan, France .	13	B6
Seddon	65	D6
Seddonville ...	65	D5
Sedhiou	55	F1
Sedova, Pik ...	28	B6
Seeheim	58	D3
Seferihisar ...	23	E6
Segamat	37	D2
Segesta	21	F4
Seget	39	E4
Ségou	55	F3
Segovia =		
Coco ➤	86	D3
Segovia	18	B3
Segre ➤	19	B6
Séguéla	55	G3
Segura ➤	19	C5
Sehitwa	58	C4
Sehore	43	H10
Seine ➤	12	B4
Seistan	45	D5
Seistan,		
Daryācheh-ye	45	D5
Sekayu	37	E2
Sekondi-		
Takoradi	55	H4
Selaru	39	F4
Selçuk	23	F6
Sele ➤	21	D5
Selenga =		
Selenge		
Mörön ➤ .	35	A5
Selenge		
Mörön ➤ .	35	A5
Seletan, Tg. ...	37	E4
Sélibabi	55	E2
Selîma, El		
Wâhât el	52	D4
Selkirk	71	C10
Selkirk Mts. ..	71	C8
Selma	75	B8
Selpele	39	E4
Selu	39	F4
Selva	94	B4
Selvas	91	E5
Semani ➤ ..	23	D2
Semarang	37	F4
Semey	29	D9
Semiozernoye .	29	D7
Semipalatinsk =		
Semey	29	D9
Semirara Is. ..	38	B2
Semisopochnoi		
I.	71	C1
Semitau	37	D4
Semiyarka	29	D8
Semiyarskoye =		
Semiyarka ..	29	D8
Semmering P. .	15	E8
Semnān	44	C3
Semnān □	44	C3
Semporna	38	D1
Semuda	37	E4
Sena Madureira	91	E5
Senador		
Pompeu	92	D6

Senaja	36	C5
Senanga	58	B4
Sendai,		
Kagoshima,		
Japan	32	D2
Sendai, Miyagi,		
Japan	33	G12
Senegal ■	55	F2
Senegal ➤ ...	55	E1
Senge Khambab		
= Indus ➤ .	43	G5
Sengkang	39	E2
Senhor-do-		
Bonfim	93	E5
Senigállia	20	C4
Senj	20	B5
Senja	8	E11
Senlis	13	B5
Senmonorom .	36	B3
Sennâr	53	F5
Senneterre	69	D3
Sens	13	B5
Senta	22	B3
Sentani	39	E6
Seo de Urgel ..	19	A6
Seoul = Sŏul ..	35	C7
Sepīdān	44	D3
Sept-Îles	69	C4
Sequoia		
National Park	82	B3
Seram	39	E3
Seram Laut,		
Kepulauan .	39	E4
Seram Sea	39	E3
Serasan	37	D3
Serbia □	22	C3
Serdobsk	24	C5
Seremban ...	37	D2
Serenje	59	A6
Sereth =		
Siret ➤	17	F9
Sergino	28	C7
Sergipe □	93	E6
Sergiyev Posad	24	B4
Seria	36	D4
Serian	37	D4
Seribu,		
Kepulauan ..	37	F3
Sérifos	23	F5
Sermata	39	F3
Serny Zavod ..	29	F6
Serov	28	D7
Serowe	59	C5
Serpukhov	24	C4
Sérrai	22	D4
Serrezuela	94	C3
Serrinha	93	E6
Sertânia	92	D6
Serua	39	F4
Serui	39	E5
Serule	59	C5
Sesepe	39	E3
Sesfontein	58	B2
Sesheke	58	B4
Sète	13	E5
Sete Lagôas ...	93	F5
Sétif	54	A6
Seto	32	B5
Settat	54	B3
Setté-Cama ...	56	E1
Setúbal	18	C1
Setúbal, B. de .	18	C1
Seulimeum ...	36	C1
Sevan, Ozero =		
Sevana Lich .	25	E6
Sevana Lich ...	25	E6
Sevastopol	25	E3
Severn ➤,		
Canada	69	C2

Severn ➤, U.K.	11	F5
Severnaya		
Zemlya	30	B7
Severnyye Uvaly	24	B6
Severo-		
Yeniseyskiy .	30	C7
Severodvinsk ..	28	C4
Sevilla	18	D2
Seville = Sevilla	18	D2
Sevlievo	22	C5
Seward	71	B5
Seward Pen. ..	71	B3
Sewer	39	F4
Seychelles ■ ..	26	K9
Seymour	63	C4
Sfax	52	B1
Sfîntu Gheorghe	17	F7
Shaanxi □	35	C5
Shaba □	57	F4
Shabunda	57	E5
Shache	34	C2
Shackleton Ice		
Shelf	96	A11
Shackleton Inlet	96	C15
Shadrinsk	29	D7
Shāhābād	44	B4
Shahdād	44	D4
Shahdād,		
Namakzar-e .	44	D4
Shahdadkot ...	42	F5
Shahgarh	42	F6
Shaḩḩāt	52	B3
Shahjahanpur .	42	F11
Shahr Kord ...	44	C2
Shahrig	42	D5
Shaikhabad ...	42	B6
Shajapur	43	H10
Shakhty	25	D5
Shakhunya	24	B6
Shaki	55	G5
Shala, L.	53	G6
Shalqar	29	E6
Sham, J. ash ..	44	F4
Shamīl	44	E4
Shammar, Jabal	47	F5
Shamo = Gobi	35	B6
Shamo, L.	53	G6
Shamva	59	B6
Shan □	41	G12
Shandong □ ..	35	C6
Shanghai	35	C7
Shangqiu	35	C6
Shangrao	35	D6
Shangshui	35	C6
Shannon	64	D6
Shannon ➤ ..	11	E2
Shansi =		
Shanxi □	35	C6
Shantou	35	D6
Shantung =		
Shandong □	35	C6
Shanxi □	35	C6
Shaoguan	35	D6
Shaowu	35	D6
Shaoxing	35	C7
Shaoyang	35	D6
Shaqra',		
Si. Arabia ...	47	F6
Shaqrā', Yemen	49	E4
Sharjah = Ash		
Shāriqah ...	44	E3
Sharon	72	D5
Sharya	24	B6
Shashi,		
Botswana ..	59	C5
Shashi, China .	35	C6
Shasta, Mt. ...	80	E2
Shatt al'Arab ➤	47	E7
Shaunavon ...	71	D9

Shawan	34	B3
Shawinigan ...	69	D3
Shcherbakov =		
Rybinsk ...	24	B4
Shchuchinsk ..	29	D8
Shebele =		
Scebeli,		
Wabi ➤ ...	49	G3
Sheboygan ...	72	C2
Sheffield	11	E6
Shekhupura ...	42	D8
Shelburne	69	D4
Shelby	75	B7
Shelbyville	74	B4
Shellharbour .	63	B5
Shenandoah ➤	73	E7
Shendam	55	G6
Shendî	53	E5
Shensi =		
Shaanxi □ ..	35	C5
Shenyang	35	B7
Sheopur Kalan	42	G10
Shepetivka ...	17	C8
Shepetivka =		
Shepetivka ..	17	C8
Shepparton ..	63	C4
Sherbro I.	55	G2
Sherbrooke ...	69	D3
Sheridan	81	C10
Sherman	79	C5
Sherridon	71	C9
Shetland Is. ...	10	A6
Shibām	49	D4
Shigaib	53	E3
Shihchiachuangi		
=		
Shijiazhuang	35	C6
Shijiazhuang ..	35	C6
Shikarpur	42	F6
Shikoku	32	C3
Shikoku □	33	H10
Shikoku-Sanchi	32	C3
Shilabo	49	F3
Shiliguri	40	D7
Shillong	41	E8
Shilong	35	D6
Shimada	32	B6
Shimizu	32	B6
Shimoga	43	N9
Shimonoseki ..	32	C2
Shinano ➤ ...	32	A6
Shīndand	42	C3
Shingū	32	C4
Shinyanga	57	E6
Shipehenski		
Prokhod	22	C5
Shipki La	42	D11
Shīr Kūh	44	D3
Shirane-San ..	32	A6
Shīrāz	44	D3
Shire ➤	59	B7
Shiretoko-Misaki	32	F13
Shiriya-Zaki ...	32	F12
Shīrvān	44	B4
Shirwa, L. =		
Chilwa, L. ...	59	B7
Shivpuri	43	G10
Shizuoka	32	B6
Shklov =		
Shklow	17	A10
Shklow	17	A10
Shkoder =		
Shkodra	22	C2
Shkodra	22	C2
Shkumbini ➤ .	23	D2
Shmidta, O. ...	30	A7
Sholapur =		
Solapur	43	L9
Shoshone Mts.	80	F5

Shoshong	59	C5
Shreveport	79	C7
Shrewsbury ...	11	E5
Shrirampur ...	40	F7
Shu	29	E8
Shuangyashan	35	B8
Shule	34	C2
Shumagin Is. . .	71	C4
Shungnak	71	B4
Shūr ⟶	44	D3
Shurugwi ...	59	B5
Shūsf	45	D5
Shūshtar	47	D7
Shwebo	41	F10
Shwegu	41	E11
Shweli ⟶	41	F11
Shymkent ...	29	E7
Shyok	42	B11
Shyok ⟶	42	B9
Si Kiang = Xi		
Jiang ⟶ ...	35	D6
Si-ngan = Xi'an	35	C5
Si Racha	36	B2
Siah	47	G6
Siahan Range .	42	F4
Siaksriindrapura	37	D2
Sialkot	42	C9
Siantan, P.	37	D3
Siāreh	45	D5
Siargao	38	C3
Siasi	38	C2
Siau	39	D3
Šiauliai	24	B1
Šibenik	20	C5
Siberut	37	E1
Sibi	42	E5
Sibil	39	E6
Sibiti	56	E2
Sibiu	17	F7
Sibolga	37	D1
Sibsagar	41	D10
Sibu	37	D4
Sibuco	38	C2
Sibuguey B. ...	38	C2
Sibut	56	C3
Sibutu	38	D1
Sibutu Passage	38	D1
Sibuyan	38	B2
Sibuyan Sea . .	38	B2
Sichuan □ ...	34	C5
Sicilia	21	F5
Sicily = Sicilia .	21	F5
Sicuani	91	F4
Sidāipet	43	K11
Sidéradougou .	55	F4
Sîdi Barrâni ...	52	B4
Sidi-bel-Abbès .	54	A4
Sidley, Mt.	96	B18
Sidon = Saydā	46	D3
Sidra, G. of =		
Surt, Khalīj .	52	B2
Siedlce	17	B6
Sieg ⟶	14	C4
Siegen	14	C5
Siem Reap ...	36	B2
Siena	20	C3
Sieradz	16	C4
Sierra Colorada	95	E3
Sierra Gorda .	94	A3
Sierra Leone ■	55	G2
Sífnos	23	F5
Sighetu-		
Marmatiei ...	17	E6
Sighişoara	17	E7
Sigli	36	C1
Sigsig	90	D3
Sigüenza	19	B4
Siguiri	55	F3

Sihanoukville =		
Kompong		
Som	36	B2
Siirt	46	C5
Sikar	42	F9
Sikasso	55	F3
Sikhote Alin,		
Khrebet	31	E11
Sikhote Alin Ra.		
= Sikhote		
Alin, Khrebet	31	E11
Síkinos	23	F5
Sikkim □	40	D7
Sil ⟶	18	A2
Silchar	41	E9
Silesia = Śląsk	16	C3
Silgarhi Doti . .	40	C3
Silghat	41	D9
Silifke	46	C3
Siliguri =		
Shiliguri	40	D7
Siling Co	34	C3
Silistra	22	B6
Silivri	22	D7
Sillajhuay,		
Cordillera ..	91	G5
Silva Porto =		
Kuito	58	A3
Simanggang ..	37	D4
Simav	23	E7
Simbirsk	24	C6
Simeria	17	F6
Simeulue	37	D1
Simferopol	25	E3
Sími	23	F6
Simikot	40	C3
Simla	42	D10
Simplon P.	13	C8
Simpson Desert	60	F6
Simunjan	37	D4
Sinabang	37	D1
Sinadogo	49	F4
Sinai = Es Sînâ'	52	C5
Sinai, Mt. =		
Mûsa, G.	47	E3
Sinaloa de		
Levya	84	B3
Sînâwan	54	B7
Sincelejo	90	B3
Sincorá, Serra		
do	93	E5
Sind □	42	F6
Sind Sagar		
Doab	42	D7
Sindangan	38	C2
Sines	18	D1
Sines, C. de ...	18	D1
Singa	53	F5
Singapore ■ ..	37	D2
Singaraja	37	F5
Singida	57	E6
Singitikós		
Kólpos	23	D5
Singkaling		
Hkamti	41	E10
Singkawang ...	37	D3
Singleton	63	B5
Singora =		
Songkhla ...	36	C2
Sinjai	39	F2
Sinjār	46	C5
Sinkat	53	E6
Sinkiang Uighur		
= Xinjiang		
Uygur		
Zizhiqu □ ...	34	B3
Sinni ⟶	21	D6
Sinnuris	52	C5
Sinop	46	B3

Sintang	37	D4
Sintra	18	C1
Sinūiju	35	B7
Siocon	38	C2
Siófok	16	E4
Sioma	58	B4
Sion	13	C7
Sioux City	77	D6
Sioux Falls ...	76	D6
Sioux Lookout .	71	C10
Siping	35	B7
Sipura	37	E1
Siquijor	38	C2
Siracusa	21	F5
Sirajganj	41	E7
Sirdaryo =		
Syrdarya ⟶	29	E7
Siret ⟶	17	F9
Sirohi	43	G8
Sironj	43	G10
Síros	23	F5
Sirsa	42	E9
Sisak	20	B6
Sisaket	36	A2
Sisophon	36	B2
Sīstān va		
Balūchestān □	45	E5
Sitapur	40	D3
Sitges	19	B6
Sitía	23	G6
Sitka	71	C6
Sittang Myit ⟶	41	J11
Sittwe	41	G9
Siuri	40	F6
Sivand	44	D3
Sivas	46	C4
Siverek	46	C4
Sivrihisar	46	C2
Sîwa	52	C4
Siwalik Range .	40	D4
Siwan	40	D5
Sjælland	9	G10
Sjumen =		
Šumen	22	C6
Skadarsko		
Jezero	22	C2
Skagerrak	9	G9
Skagway	71	C6
Skala-Podilska .	17	D8
Skala		
Podolskaya =		
Skala-Podilska	17	D8
Skalat	17	D7
Skardu	42	B9
Skarzysko-		
Kamienna ..	16	C5
Skeena ⟶ ...	70	C6
Skegness	11	E7
Skeldon	90	B7
Skellefte älv ⟶	8	F12
Skellefteå	8	F12
Skíathos	23	E4
Skien	9	G9
Skierniewice ...	16	C5
Skikda	54	A6
Skipton	62	C3
Skíros	23	E5
Skole	17	D6
Skópelos	23	E4
Skopje	22	C3
Skvyra	17	D9
Skye	10	C3
Skyros = Skíros	23	E5
Slamet	37	F3
Śląsk	16	C3
Slatina	22	B5
Slave ⟶	70	B8
Slavgorod	29	D8
Slavonski Brod	20	B7

Slavuta	17	C8
Slavyansk =		
Slovyansk ...	25	D4
Slawharad ...	17	B10
Sleaford B.	62	B2
Sleeper Is.	68	C2
Sligeach = Sligo	11	D2
Sligo	11	D2
Sliven	22	C6
Slobodskoy ...	24	B7
Slobozia	22	B6
Slonim	17	B7
Slough	11	F6
Slovak Rep. ■ .	16	D5
Slovakia =		
Slovak Rep. ■	16	D5
Slovakian Ore		
Mts. =		
Slovenské		
Rudohorie ..	16	D4
Slovenia ■	20	B5
Slovenija =		
Slovenia ■ ..	20	B5
Slovenská		
Republika =		
Slovak Rep. ■	16	D5
Slovenské		
Rudohorie ..	16	D4
Slovyansk ...	25	D4
Sluch ⟶	17	C8
Słupsk	16	A3
Slutsk	17	B8
Smara	54	B3
Smarhon	17	A8
Smederevo ...	22	B3
Smith Arm ..	70	B7
Smithton	62	D4
Smithtown	63	B5
Smoky Bay ...	62	B1
Smolensk	24	C3
Smolikas, Óros	23	D3
Smolyan	22	D5
Smorgon =		
Smarhon ...	17	A8
Smyrna = İzmir	23	E6
Snake ⟶	80	B4
Snake I.	63	C4
Snake River		
Plain	81	D7
Sneek	14	B3
Sněžka	16	C2
Snøhetta	8	F9
Snowdon	11	E4
Snowdrift	70	B8
Snowtown	62	B2
Snowy ⟶ ...	63	C4
Snowy Mts. ...	63	C4
Snyatyn	17	D7
Snyder	78	C3
Soalala	59	H9
Sobat, Nahr ⟶	53	G5
Sobral	92	C5
Soc Trang ...	36	C3
Soch'e =		
Shache	34	C2
Sochi	25	E5
Société, Is. de la	65	L15
Society Is. =		
Société, Is. de		
la	65	L15
Socompa,		
Portezuelo de	94	A3
Socorro	90	B4
Socotra	49	E5
Soda Plains ..	42	B11
Söderhamn ..	8	F11
Sodiri	53	F4
Sodo	53	G6

Sofia

Sofia = Sofiya . 22 C4
Sofia → 59 H9
Sofiya 22 C4
Sogamoso 90 B4
Sognefjorden . 8 F9
Sohâg 52 C5
Soissons 13 B5
Sokal 17 C7
Söke 23 F6
Sokhumi 25 E5
Sokodé 55 G5
Sokol 24 B5
Sokółka 17 B6
Sokolo 55 F3
Sokolów
  Podlaski 17 B6
Sokoto 55 F6
Solano 38 A2
Solapur 43 L9
Soledad 90 B6
Soligalich 24 B5
Soligorsk =
  Salihorsk ... 17 B8
Solikamsk 28 D6
Solimões =
  Amazonas → 92 C3
Solingen 14 C4
Sóller 19 C7
Sologne 12 C4
Solok 37 E2
Solomon Is. ■ . 64 K10
Solomon Sea . 61 B9
Solon 35 B7
Solor 39 F2
Solothurn 13 C7
Šolta 20 C6
Solţānābād ... 44 B4
Solunska Glava 22 D3
Solwezi 59 A5
Soma 23 E6
Somali Rep. ■ . 49 F4
Somalia =
  Somali Rep. ■ 49 F4
Sombor 22 B2
Sombrerete ... 84 C4
Sombrero 87 C7
Somerset East . 59 E5
Somerset I. ... 70 A10
Someş → 17 E6
Sommariva ... 63 A4
Somme → ... 12 A4
Somosierra,
  Puerto de ... 18 B4
Somport, Puerto
  de 12 E3
Søndre
  Strømfjord .. 68 B5
Sóndrio 20 A2
Sonepur 40 G4
Song Cau 36 B3
Songea 57 G7
Songhua
  Jiang → ... 35 B8
Songkhla 36 C2
Songpan 34 C5
Sonipat 42 E10
Sonmiani 42 G5
Sono → 93 D4
Sonora → 84 B2
Sonsonate 85 E7
Soochow =
  Suzhou 35 C7
Sopi 39 D3
Sopot 16 A4
Sopron 16 E3
Sør-Rondane .. 96 B8
Sorata 91 G5

Sorel 69 D3
Sorgono 21 D2
Soria 19 B4
Sorkh, Kuh-e .. 44 C4
Soroca 17 D9
Sorocaba 94 A7
Soroki = Soroca 17 D9
Sorong 39 E4
Soroti 57 D6
Sørøya 8 D12
Sorrento 63 C3
Sorsogon 38 B2
Sosnowiec 16 C4
Souanké 56 D2
Soúdhas, Kólpos 23 G5
Sŏul 35 C7
Sound, The =
  Øresund 9 G10
Sources, Mt. aux 59 D5
Soure 92 C4
Souris 71 D9
Souris → 71 D10
Sousa 92 D6
Sousel 92 C3
Sousse 52 A1
South Africa ■ 58 E4
South
  Australia □ .. 62 B2
South Bend ... 72 D2
South
  Carolina □ .. 75 C7
South China Sea 36 C4
South Dakota □ 76 C4
South East C. . 62 D4
South Horr ... 57 D7
South I. 65 E3
South
  Invercargill . . 65 G3
South Korea ■ 35 C7
South Magnetic
  Pole 96 A13
South
  Nahanni → . 70 B7
South Natuna Is.
  = Natuna
  Selatan,
  Kepulauan . . 37 D3
South Orkney Is. 96 A4
South Pagai, I.
  = Pagai
  Selatan, P. . . 37 E2
South Platte → 76 E4
South Pole .... 96 C
South
  Ronaldsay . . 10 B5
South
  Saskatchewan →
  .... 71 C9
South Shetland
  Is. 96 A4
South Shields . 11 D6
South Taranaki
  Bight 64 C6
South Uist ... 10 C3
South West C. . 62 D4
Southampton . 11 F6
Southampton I. 68 B2
Southbridge .. 65 E5
Southend-on-
  Sea 11 F7
Southern Alps . 65 E4
Southern Indian
  L. 71 C10
Southern Ocean 96 A9
Southern
  Uplands 10 D5
Southport 63 A5

Southwest C. .. 65 G2
Soutpansberg . 59 C5
Sovetsk,
  Kaliningd.,
  Russia 24 B1
Sovetsk, Kirov,
  Russia 24 B6
Sōya-Misaki ... 32 E12
Soyo 56 F2
Sozh → 17 C10
Spain ■ 18 C4
Spalding 62 B2
Spanish Town . 86 C4
Sparta = Spárti 23 F4
Spartanburg .. 75 B7
Spárti 23 F4
Spartivento, C.,
  Calabria, Italy 21 F6
Spartivento, C.,
  Sard., Italy .. 21 E2
Spátha, Ákra . 23 G4
Spence Bay ... 70 B10
Spencer, C. ... 62 C2
Spencer G. 62 B2
Spenser Mts. .. 65 E5
Spey → 10 C5
Speyer 14 D5
Split 20 C6
Spokane 80 B5
Spoleto 20 C4
Sporyy Navolok,
  Mys 28 B7
Spratly I. 36 C4
Spree → 15 B7
Springbok 58 D3
Springfield, N.Z. 65 E4
Springfield, Ill.,
  U.S.A. 77 F10
Springfield,
  Mass., U.S.A. 73 C9
Springfield, Mo.,
  U.S.A. 79 A7
Springfield,
  Ohio, U.S.A. . 72 E4
Springfield,
  Oreg., U.S.A. 80 C2
Springfontein . 59 E5
Springhill 69 D4
Springhurst .. 63 C4
Springs 59 D5
Squamish 71 D7
Srbija =
  Serbia □ 22 C3
Sre Umbell ... 36 B2
Srebrnica 20 B7
Sredinny Ra. =
  Sredinnyy
  Khrebet 31 D14
Sredinnyy
  Khrebet 31 D14
Srednevilyuysk 31 C10
Śrem 16 B3
Sremska
  Mitrovica ... 22 B2
Sri Lanka ■ ... 43 R12
Srikakulam ... 40 H4
Srinagar 42 B9
Stade 14 B5
Stadlandet .. 8 F9
Stafford 11 E5
Stakhanov 25 D4
Stalingrad =
  Volgograd .. 25 D5
Staliniri =
  Tskhinvali ... 25 E5
Stalino =
  Donetsk 25 D4

Stalinogorsk =
  Novomoskovsk
  .... 24 C4
Stalowa Wola . 17 C6
Stamford 73 D9
Stanislav =
  Ivano-
  Frankivsk ... 17 D7
Stanke Dimitrov 22 C4
Stanley,
  Australia 62 D4
Stanley, Canada 71 C9
Stanley, Falk. Is. 95 G5
Stanovoy
  Khrebet 31 D10
Stanovoy Ra. =
  Stanovoy
  Khrebet 31 D10
Stanthorpe 63 A5
Stara Planina .. 22 C4
Stara Zagora .. 22 C5
Starachowice .. 16 C5
Staraya Russa . 24 B3
Starbuck I. 65 K15
Stargard
  Szczeciński .. 16 B2
Staritsa 24 B3
Starogard
  Gdański 16 B4
Starokonstantinov
  =
  Starokonstyantyniv
  .... 17 D8
Starokonstyantyniv
  .... 17 D8
Staryy
  Chartoriysk . . 17 C7
Staryy Oskol .. 24 C4
State College .. 73 D7
Staten, I. =
  Estados, I. de
  Los 95 G4
Statesville 75 B7
Stavanger 9 G9
Staveley 65 E4
Stavropol 25 D5
Stawell 62 C3
Steenkool =
  Bintuni 39 E4
Stefanie L. =
  Chew Bahir . 53 H6
Stefansson Bay 96 A9
Steiermark □ .. 15 E8
Steinkjer 8 F10
Stellenbosch .. 58 E3
Stendal 15 B6
Steornabhaigh
  = Stornoway 10 B3
Stepanakert =
  Xankändi ... 25 F6
Stephens Creek 62 B3
Stepnoi = Elista 25 D5
Stepnyak 29 D8
Sterlitamak ... 29 D6
Stettin =
  Szczecin 16 B2
Stettiner Haff . 15 B8
Stettler 71 C8
Steubenville .. 72 D5
Stevenage 11 F6
Stewart 70 B6
Stewart, I. 95 G2
Stewart I. 65 G2
Steyr 15 D8
Stillwater, N.Z. 65 E4
Stillwater,
  U.S.A. 78 A5

150

Štip 22 D4
Stirling 10 C5
Stockerau 15 D9
Stockholm 9 G11
Stockport 11 E5
Stockton 82 B2
Stockton-on-
Tees 11 D6
Stoke on Trent . 11 E5
Stokes Pt. 62 D3
Stolac 20 C6
Stolbovoy,
Ostrov 30 B11
Stolbtsy =
Stowbtsy . . . 17 B8
Stolin 17 C8
Stonehaven . . . 10 C5
Stonewall 71 C10
Stony Tunguska
=
Podkamennaya
Tunguska → 30 C7
Stora Lulevatten 8 E11
Storavan 8 E11
Store Bælt 9 G10
Store Creek . . . 63 B4
Storm B. 62 D4
Stormberge . . . 59 E5
Stornoway 10 B3
Storozhinets =
Storozhynets 17 D7
Storozhynets . . 17 D7
Storsjön 8 F10
Storuman 8 E11
Stowbtsy 17 B8
Strahan 62 D4
Stralsund 15 A7
Strand 58 E3
Stranraer 11 D4
Strasbourg 13 B7
Stratford,
Canada 69 D2
Stratford, N.Z. . 64 C6
Strathalbyn . . . 62 C2
Straubing 15 D7
Streaky B. 62 B1
Streaky Bay . . . 62 B1
Strelka 30 D7
Strezhevoy 29 C8
Strimón → . . . 23 D4
Strimonikós
Kólpos 23 D5
Strómboli 21 E5
Stronsay 10 B5
Stroud Road . . 63 B5
Strumica 22 D4
Stryy 17 D6
Strzelecki Cr. → 62 A2
Stuart Ra. 62 A1
Stung Treng . . 36 B3
Stutterheim . . . 59 E5
Stuttgart 14 D5
Styria =
Steiermark □ 15 E8
Suakin 53 E6
Subansiri → . . 41 D9
Subi 37 D3
Subotica 16 E4
Suceava 17 E8
Suchou =
Suzhou 35 C7
Süchow =
Xuzhou 35 C6
Sucre 91 G5
Sudair 47 F6
Sudan ■ 53 E4
Sudbury 69 D2

Sûdd 53 G4
Sudeten Mts. =
Sudety 16 C3
Sudety 16 C3
Sudirman,
Pegunungan 39 E5
Sueca 19 C5
Suez = El
Suweis 52 C5
Suez, G. of =
Suweis, Khalîg
el 52 C5
Sugluk = Salluit 68 B3
Suhãr 44 E4
Suhbaatar 35 A5
Suhl 14 C6
Suihua 35 B7
Sukadana,
Kalimantan,
Indonesia . . . 37 E3
Sukadana,
Sumatera,
Indonesia . . . 37 F3
Sukaraja 37 E4
Sukarnapura =
Jayapura . . . 39 E6
Sukhona → . . 28 C5
Sukhumi =
Sokhumi 25 E5
Sukkur 42 F6
Sukkur Barrage 42 F6
Sula, Kepulauan 39 E3
Sulaiman Range 42 D6
Sulawesi □ . . . 39 E2
Sulawesi Sea =
Celebes Sea . 39 D2
Sulima 55 G2
Sulina 17 F9
Sulitjelma 8 E11
Sullana 90 D2
Sultanpur 40 D4
Sulu Arch. 38 C2
Sulu Sea 38 C2
Suluq 52 B3
Sulzberger Ice
Shelf 96 B17
Sumalata 39 D2
Sumatera □ . . . 37 D2
Sumatra =
Sumatera □ . 37 D2
Sumba 39 F1
Sumba, Selat . 39 F1
Sumbawa 37 F5
Sumbawa Besar 37 F5
Sumbe 56 G2
Šumen 22 C6
Sumgait =
Sumqayıt . . . 25 E6
Summerside . . 69 D4
Sumperk 16 D3
Sumqayıt 25 E6
Sumter 75 C7
Sumy 24 C3
Sunbury 63 C3
Sunda, Selat . 37 F3
Sunda Str. =
Sunda, Selat 37 F3
Sundarbans,
The 41 G7
Sundargarh . . . 40 F5
Sunderland . . . 11 D6
Sundsvall 8 F11
Sungaigerong . 37 E2
Sungailiat 37 E3
Sungaipakning 37 D2
Sungaipenuh . . 37 E2
Sungaitiram . . . 37 E5

Sungari =
Songhua
Jiang → . . . 35 B8
Sungguminasa 39 F1
Sunghua Chiang
= Songhua
Jiang → . . . 35 B8
Sungurlu 46 B3
Sunnyvale 82 B1
Suntar 30 C9
Supaul 40 D6
Superior 77 B8
Superior, L. . . . 69 D2
Suphan Dağı . . 46 C5
Supiori 39 E5
Sûr, Lebanon . . 46 D3
Sûr, Oman 45 F4
Sura → 24 B6
Surabaja =
Surabaya . . . 37 F4
Surabaya 37 F4
Surakarta 37 F4
Surat, Australia 63 A4
Surat, India . . . 43 J8
Surat Thani . . . 36 C1
Suratgarh 42 E8
Surgut 29 C8
Suriapet 43 L11
Surigao 38 C3
Surinam ■ . . . 92 B2
Suriname ■ =
Surinam ■ . . 92 B2
Suriname → . . 92 A2
Surt 52 B2
Surt, Khalîj . . . 52 B2
Suruga-Wan . . 32 B6
Süsangerd 47 E7
Susanville 80 E3
Susquehanna →
. . . . 73 E7
Susques 94 A3
Susunu 39 E4
Susurluk 23 E7
Sutherland Falls 65 F2
Sutlej → 42 E7
Suva 64 L12
Suva Planina . . 22 C4
Suvorov Is. =
Suwarrow Is. 65 L14
Suwałki 17 A6
Suwarrow Is. . . 65 L14
Suweis, Khalîg
el 52 C5
Suzdal 24 B5
Suzhou 35 C7
Suzu-Misaki . . . 32 A5
Svealand □ . . . 9 G10
Sverdlovsk =
Yekaterinburg 29 D7
Svetlogorsk =
Svyetlahorsk 17 B9
Svetozarevo . . 22 B3
Svishtov 22 C5
Svislach 17 B7
Svyetlahorsk . . 17 B9
Swabian Alps =
Schwäbische
Alb 14 D5
Swakopmund . 58 C2
Swan Hill 62 C3
Swansea,
Australia 63 B5
Swansea, U.K. . 11 F5
Swatow =
Shantou 35 D6
Swaziland ■ . . 59 D6
Sweden ■ 9 G10

Sweetwater . . . 78 C3
Swellendam . . 58 E4
Świdnica 16 C3
Świdnik 17 C6
Świebodzin . . . 16 B2
Świecie 16 B4
Swift Current . . 71 C9
Swindon 11 F6
Swinemünde =
Świnoujście . 16 B2
Świnoujście . . . 16 B2
Switzerland ■ . 13 C8
Sydney,
Australia 63 B5
Sydney, Canada 69 D4
Sydra, G. of =
Surt, Khalîj . . 52 B2
Syktyvkar 28 C6
Sylhet 41 E8
Sylt 14 A5
Sym 29 C9
Syracuse 73 C7
Syrdarya → . . 29 E7
Syria ■ 46 D4
Syrian Desert =
Ash Shām,
Bādiyat 47 D4
Syzran 24 C6
Szczecin 16 B2
Szczecinek . . . 16 B3
Szczytno 16 B5
Szechwan =
Sichuan □ . . 34 C5
Szeged 16 E5
Székesfehérvár 16 E4
Szekszárd 16 E4
Szentes 16 E5
Szolnok 16 E5
Szombathely . . 16 E3

T

Tabacal 94 A4
Tabaco 38 B2
Ţābah 47 F5
Tabarka 54 A6
Ţabas,
Khorāsān, Iran 44 C4
Ţabas,
Khorāsān, Iran 45 C5
Tabatinga, Serra
da 93 E5
Tablas 38 B2
Table B. =
Tafelbaai 58 E3
Table Mt. 58 E3
Tábor 16 D2
Tabora 57 F6
Tabou 55 H3
Tabrīz 46 C6
Tabuk 47 E4
Tacheng 34 B3
Tacloban 38 B2
Tacna 91 G4
Tacoma 80 B2
Tacuarembó . . 94 C5
Tademaït,
Plateau du . . 54 C5
Tadjoura 49 E3
Tadmor 65 D5
Tadoussac . . . 69 D4
Tadzhikistan =
Tajikistan ■ . . 29 F8
Taegu 35 C7

Taejŏn 35 C7	Talayan 38 C2	Tanganyika, L. . 57 F5	Tarfaya 54 C2
Tafalla 19 A5	Talbragar → .. 63 B4	Tanger =	Tarifa 18 D3
Tafelbaai 58 E3	Talca 94 D2	Tangier 54 A3	Tarija 94 A4
Tafermaar 39 F4	Talcahuano ... 94 D2	Tanggula Shan 34 C4	Tariku → 39 E5
Taft 38 B3	Talcher 40 G5	Tangier 54 A3	Tarim Basin =
Taga Dzong ... 41 D7	Taldy Kurgan =	Tangshan 35 C6	Tarim Pendi . 34 C3
Taganrog 25 D4	Taldyqorghan 29 E8	Tanimbar,	Tarim He → .. 34 C3
Tagbilaran 38 C2	Taldyqorghan . 29 E8	Kepulauan .. 39 F4	Tarim Pendi ... 34 C3
Tagish 70 B6	Ţalesh, Kūhhā-	Tanimbar Is. =	Taritatu → ... 39 E5
Tagish L. 70 B6	ye 46 C7	Tanimbar,	Tarkhankut, Mys 25 D3
Tagliamento → 20 B4	Tali Post 53 G5	Kepulauan .. 39 F4	Tarko Sale ... 28 C8
Taguatinga ... 93 E5	Talibon 38 B2	Tanjay 38 C2	Tarkwa 55 G4
Tagum 38 C3	Taliwang 37 F5	Tanjore =	Tarlac 38 A2
Tagus =	Tall 'Afar 46 C5	Thanjavur ... 43 P11	Tarma 91 F3
Tejo → 18 C1	Talladega 74 C4	Tanjung 37 E5	Tarn → 12 D4
Tahakopa 65 G3	Tallahassee ... 75 D5	Tanjungbalai .. 37 D1	Tarnobrzeg ... 16 C5
Tahan, Gunong 36 D2	Tallangatta 63 C4	Tanjungbatu .. 37 D5	Tarnów 16 C5
Tahat 54 D6	Tallarook 63 C4	Tanjungkarang	Tarnowskie
Tāherī 44 E3	Tallinn 24 B1	Telukbetung . 37 F3	Góry 16 C4
Tahiti 65 L16	Talodi 53 F5	Tanjungpandan 37 E3	Taroom 63 A4
Tahoe, L. 80 F3	Taltal 94 B2	Tanjungpinang 37 D2	Taroudannt ... 54 B3
Tahoua 55 F6	Talwood 63 A4	Tanjungredeb . 37 D5	Tarragona 19 B6
Tahta 52 C5	Talyawalka	Tanjungselor .. 37 D5	Tarrasa 19 B7
Tahulandang .. 39 D3	Cr. → 62 B3	Tanout 55 F6	Tarso Emissi .. 52 D2
Tahuna 39 D3	Tamale 55 G4	Tanta 52 B5	Tarsus 46 C3
Taï 55 G3	Tamanrasset .. 54 D6	Tantung =	Tartagal 94 A4
Tai Hu 35 C7	Tamarang 63 B5	Dandong ... 35 B7	Tartu 24 B2
Taibei 35 D7	Tamaské 55 F6	Tanunda 62 B2	Tarţūs 46 D3
T'aichung =	Tambacounda . 55 F2	Tanzania ■ ... 57 F6	Tarutung 37 D1
Taizhong ... 35 D7	Tambelan,	Taolanaro 59 K9	Tasäwah 52 C1
Taidong 35 D7	Kepulauan .. 37 D3	Taoudenni 54 D4	Taschereau ... 69 D3
Taieri → 65 G4	Tambo de Mora 91 F3	Taourirt 54 B4	Tash-Kömür ... 29 E8
Taihape 64 C6	Tambora 37 F5	Tapa Shan =	Tash-Kumyr =
Tailem Bend .. 62 C2	Tambov 24 C5	Daba Shan .. 35 C5	Tash-Kömür . 29 E8
Taimyr	Tamburâ 53 G4	Tapah 37 D2	Tashauz =
Peninsula =	Tâmchekket .. 55 E2	Tapajós → ... 92 C3	Dashhowuz . 29 E6
Taymyr,	Tamega → .. 18 B1	Tapaktuan ... 37 D1	Tashi Chho
Poluostrov .. 30 B7	Tamenglong .. 41 E9	Tapanui 65 F3	Dzong =
Tain 10 C4	Tamgak, Mts. . 55 E6	Tapauá → 91 E6	Thimphu 41 D7
Tainan 35 D7	Tamil Nadu □ . 43 P10	Tapeta 55 G3	Tashkent =
Taínaron, Ákra 23 F4	Tammerfors =	Tapi → 43 J8	Toshkent ... 29 E7
T'aipei = Taibei 35 D7	Tampere 8 F12	Tapirapecó,	Tashtagol 29 D9
Taiping 36 D2	Tamo Abu,	Serra 90 C6	Tasman B. 65 D5
Taitao, Pen. de 95 F1	Pegunungan 37 D5	Tapuaenuku, Mt. 65 D5	Tasman Mts. .. 65 D5
Taiwan ■ 35 D7	Tampa 75 F6	Tapul Group .. 38 C2	Tasman Pen. .. 62 D4
Taïyetos Óros . 23 F4	Tampere 8 F12	Taquari → ... 91 G7	Tasmania □ ... 62 D4
Taiyuan 35 C6	Tampico 84 C5	Tara, Australia . 63 A5	Tatabánya 16 E4
Taizhong 35 D7	Tamrida =	Tara, Russia .. 29 D8	Tatar Republic □
Ta'izz 49 E3	Qādib 49 E5	Tara →,	= Tatarstan □ 29 D6
Tajikistan ■ ... 29 F8	Tamsagbulag . 35 B6	Montenegro, Yug.	Tatarbunary ... 17 F9
Tajo = Tejo → 18 C1	Tamu 41 E10	.... 22 C2	Tatarsk 29 D8
Tājūrā 52 B1	Tamworth 63 B5	Tara →, Russia 29 D8	Tatarstan □ ... 29 D6
Tak 36 A1	Tana →, Kenya 57 E8	Tarabagatay,	Tateyama 32 B6
Takada 32 A6	Tana →,	Khrebet 29 E9	Tathra 63 C4
Takaka 65 D5	Norway 8 D13	Tarābulus,	Tatra = Tatry . 16 D4
Takamatsu 32 B4	Tana, L. 53 F6	Lebanon 46 D3	Tatry 16 D4
Takaoka 32 A5	Tanahbala 37 E1	Tarābulus, Libya 52 B1	Tatta 43 G5
Takapuna 64 B6	Tanahgrogot .. 37 E5	Tarakan 37 D5	Tatuï 94 A7
Takasaki 32 A6	Tanahjampea . 39 F2	Taralga 63 B4	Tat'ung =
Takatsuki 32 B4	Tanahmasa ... 37 E1	Taranaki □ 64 C6	Datong 35 B6
Takaungu 57 E7	Tanahmerah .. 39 F6	Tarancón 19 B4	Tatvan 46 C5
Takayama 32 A5	Tanami Desert . 60 D5	Taranga Hill .. 43 H8	Taubaté 94 A7
Takefu 32 B5	Tanana 71 B4	Táranto 21 D6	Tauern 15 E7
Takengon 36 D1	Tanana → 71 B4	Táranto, G. di . 21 D6	Taumarunui ... 64 C6
Takeo 36 B2	Tananarive =	Tarapacá 90 D5	Taumaturgo .. 91 E4
Täkestän 46 D7	Antananarivo 59 H9	Tararua Ra. ... 64 D6	Taungdwingyi . 41 G10
Takhār □ 45 B7	Táncharo → .. 20 B2	Tarashcha ... 17 D10	Taunggyi 41 G11
Takum 55 G6	Tandag 38 C3	Tarauacá 91 E4	Taungup 41 H10
Talara 90 D2	Tandil 94 D5	Tarauacá → .. 91 E5	Taungup Pass . 41 H10
Talas 29 E8	Tandil, Sa. del . 94 D5	Tarawera 64 C7	Taungup
Talaud,	Tando Adam .. 42 G6	Tarawera L. ... 64 C7	Taunggya ... 41 H9
Kepulauan .. 38 D3	Tandou L. 62 B3	Tarazona 19 B5	Taunton 11 F5
Talaud Is. =	Tane-ga-Shima 33 D2	Tarbela Dam .. 42 B8	Taunus 14 C5
Talaud,	Taneatua 64 C7	Tarbes 12 E4	Taupo 64 C7
Kepulauan .. 38 D3	Tanen Tong Dan 41 J12	Tarcoola 62 B1	Taupo, L. 64 C6
Talavera de la	Tanezrouft 54 D5	Tarcoon 63 B4	Tauranga 64 B7
Reina 18 C3	Tanga 57 F7	Taree 63 B5	Tauranga Harb. 64 B7

Taurianova ... 21 E6
Taurus Mts. =
   Toros Dağları 46 C3
Tavda 29 D7
Tavda → 29 D7
Taveta 57 E7
Tavira 18 D2
Tavoy 36 B1
Tawau 36 D5
Tawitawi 38 C2
Tay → 10 C5
Tayabamba .. 91 E3
Taylakova 29 D8
Taylakovy =
   Taylakova ... 29 D8
Taymā 47 F4
Taymyr, Oz. ... 30 B8
Taymyr,
   Poluostrov .. 30 B7
Taytay 38 B1
Taz → 28 C8
Taza 54 B4
Tazovskiy ... 28 C8
Tbilisi 25 E5
Tchad = Chad ■ 53 E2
Tchad, L. 53 F1
Tch'eng-tou =
   Chengdu 35 C5
Tchibanga ... 56 E2
Tch'ong-k'ing =
   Chongqing .. 35 D5
Tczew 16 A4
Te Anau, L. ... 65 F2
Te Aroha 64 B6
Te Awamutu . 64 C6
Te Kuiti 64 C6
Te Puke 64 B7
Te Waewae B. . 65 G2
Tebakang 37 D4
Tébessa 54 A6
Tebicuary → . 94 B5
Tebingtinggi . 37 D1
Tecuci 22 B6
Tedzhen =
   Tejen 29 F7
Tefé 90 D6
Tegal 37 F3
Tegina 55 F6
Tegucigalpa .. 85 E7
Tehachapi ... 82 C3
Tehrān 44 C2
Tehuantepec .. 85 D5
Tehuantepec, G.
   de 85 D5
Tehuantepec,
   Istmo de 85 D6
Tejen 29 F7
Tejo → 18 C1
Tekapo, L. ... 65 E4
Tekeli 29 E8
Tekirdağ 22 D6
Tekkali 40 H5
Tel Aviv-Yafo . 46 D3
Tela 85 D7
Telanaipura =
   Jambi 37 E2
Telavi 25 E6
Telekhany =
   Tsyelyakhany 17 B7
Teles Pires → . 91 E7
Telford 11 E5
Télimélé 55 F2
Tellicherry .. 43 P9
Telsen 95 E3
Teluk Anson . 37 D2
Teluk Betung =
   Tanjungkarang
   Telukbetung . 37 F3

Teluk Intan =
   Teluk Anson . 37 D2
Telukbutun ... 37 D3
Telukdalem ... 37 D1
Tema 55 G4
Temax 85 C7
Temerloh 37 D2
Temir 29 E6
Temirtau,
   Kazakstan ... 29 D8
Temirtau, Russia 29 D9
Temma 62 D3
Temora 63 B4
Tempe 83 D7
Temple 78 D5
Temuco 94 D2
Temuka 65 F4
Tenali 40 J3
Tenasserim ... 36 B1
Tenda, Col di .. 20 B1
Tendaho 49 E3
Tenerife 54 C1
Teng Xian ... 35 C6
Tengah □ 39 E2
Tengah
   Kepulauan .. 37 F5
Tengchong ... 34 D4
Tenggara □ ... 39 E2
Tenggarong ... 37 E5
Tengiz, Ozero . 29 D7
Tenkasi 43 Q10
Tenke 57 G5
Tenkodogo ... 55 F4
Tennessee □ .. 74 B4
Tennessee → . 72 F1
Tenom 36 C5
Tenryū-
   Gawa → ... 32 B5
Tenterfield 63 A5
Teófilo Otoni . 93 F5
Tepa 39 F3
Tepic 84 C4
Teplice 16 C1
Ter → 19 A7
Téra 55 F5
Téramo 20 C4
Terang 62 C3
Terebovlya ... 17 D7
Terek → 25 E6
Teresina 92 D5
Terewah, L. .. 63 A4
Terhazza 54 D3
Teridgerie
   Cr. → 63 B4
Termez =
   Termiz 29 F7
Términi Imerese 21 F4
Termiz 29 F7
Térmoli 20 C5
Ternate 39 D3
Terni 20 C4
Ternopil 17 D7
Ternopol =
   Ternopil 17 D7
Terowie, N.S.W.,
   Australia ... 63 B4
Terowie,
   S. Austral.,
   Australia 62 B2
Terracina 20 D4
Terralba 21 E2
Terranova =
   Ólbia 20 D2
Terrassa =
   Tarrasa 19 B7
Terre Haute ... 72 E2
Terschelling ... 14 B3

Teruel 19 B5
Teryaweyna L. . 62 B3
Teshio-Gawa → 32 F12
Tesiyn Gol → . 34 A4
Teslin 70 B6
Tessalit 54 D5
Tessaoua 55 F6
Tetas, Pta. ... 94 A2
Tete 59 B6
Teterev → ... 17 C10
Teteven 22 C5
Tetiyev 17 D9
Tétouan 54 A3
Tetovo 22 C3
Teuco → 94 B4
Teun 39 F3
Teutoburger
   Wald 14 B5
Tévere → ... 20 D4
Tewantin 63 A5
Texarkana ... 79 C6
Texas 63 A5
Texas □ 78 D4
Texas City ... 79 E6
Texel 14 B3
Teyvareh 45 C6
Tezpur 41 D9
Thabana
   Ntlenyana ... 59 D5
Thabazimbi ... 59 C5
Thailand ■ ... 36 A2
Thailand, G. of 36 B2
Thakhek 36 A2
Thal 42 C7
Thala La 41 C11
Thallon 63 A4
Thames 64 B6
Thames → ... 11 F7
Thane 43 K8
Thanh Pho Ho
   Chi Minh =
   Phanh Bho Ho
   Chi Minh ... 36 B3
Thanjavur ... 43 P11
Thar Desert .. 42 F7
Tharad 43 G7
Thargomindah 63 A3
Tharrawaddy .. 41 J10
Thásos 23 D5
Thaton 41 J11
Thaungdut ... 41 E10
Thayetmyo ... 41 H10
Thazi 41 G11
The Alberga → 62 A2
The Coorong . 62 C2
The Frome → . 62 A2
The Grampians 62 C3
The Great Divide
   = Great
   Dividing Ra. . 61 E8
The Hague = 's-
   Gravenhage . 14 B3
The
   Hamilton → 62 A2
The
   Macumba → 62 A2
The Neales → . 62 A2
The Pas 71 C9
The Rock 63 C4
The Salt L. ... 62 B3
The
   Stevenson → 62 A2
The
   Warburton → 62 A2
Thebes = Thívai 23 E4
Theebine 63 A5
Thermaïkós
   Kólpos 23 D4
Thermopylae P. 23 E4

Thessaloníki .. 23 D4
Thessaloniki,
   Gulf of =
   Thermaïkós
   Kólpos 23 D4
Thetford 11 E7
Thetford Mines 69 D3
Thevenard 62 B1
Thiel Mts. ... 96 C2
Thiers 13 D5
Thies 55 F1
Thika 57 E7
Thimphu 41 D7
Thionville 13 B7
Thíra 23 F5
Thistle I. 62 C2
Thívai 23 E4
Thomas, L. ... 62 A2
Thomasville,
   Ga., U.S.A. .. 75 D6
Thomasville,
   N.C., U.S.A. . 75 B7
Thompson 71 C10
Thomson's Falls
   = Nyahururu 57 D7
Thrace 22 D6
Three Hummock
   I. 62 D3
Three Points, C. 55 H4
Thun 13 C7
Thunder Bay .. 69 D2
Thung Song . 36 C1
Thunkar 41 D8
Thüringer Wald 14 C6
Thurles 11 E3
Thurloo Downs 63 A3
Thurso 10 B5
Thurston I. .. 96 B2
Thylungra ... 63 A3
Tia 63 B5
Tian Shan ... 34 B3
Tianjin 35 C6
Tianshui 35 C5
Tiaret 54 A5
Tiassalé 55 G4
Tibati 53 G1
Tiber =
   Tévere → .. 20 D4
Tiberias, L. =
   Yam Kinneret 46 D3
Tibesti 52 D2
Tibet =
   Xizang □ ... 34 C3
Tibooburra ... 62 A3
Tiburón 84 B2
Tîchît 55 E3
Ticino → 20 B2
Tiddim 41 F9
Tidjikja 55 E2
Tidore 39 D3
Tiel 55 F1
Tierra de
   Campos 18 A3
Tierra del
   Fuego, I. Gr.
   de 95 G3
Tiétar → 18 C2
Tieyon 62 A1
Tiflis = Tbilisi . 25 E5
Tifu 39 E3
Tighina 17 E9
Tignish 69 D4
Tigre → 90 D4
Tigris = Dijlah,
   Nahr → 47 E6
Tigyaing 41 F11
Tîh, Gebel el . 52 C5
Tijuana 84 A1
Tikamgarh ... 43 G11

Tikhoretsk 25 D5
Tikrīt 46 D5
Tiksi 30 B10
Tilamuta 39 D2
Tilburg 14 C3
Tillabéri 55 F5
Tílos 23 F6
Tilpa 63 B3
Tilsit = Sovetsk 24 B1
Timaru 65 F4
Timbedgha ... 55 E3
Timboon 62 C3
Timbuktu =
Tombouctou 55 E4
Timimoun 54 C5
Timişoara 16 F5
Timmins 69 D2
Timok → 22 B4
Timon 92 D5
Timor 39 F2
Tinaca Pt. 38 C3
Tindouf 54 C3
Tingo Maria .. 91 E3
Tinjoub 54 C3
Tinnevelly =
Tirunelveli ... 43 Q10
Tinogasta 94 B3
Tínos 23 F5
Tintinara 62 C3
Tioman, Pulau . 37 D2
Tipongpani ... 41 D10
Tipperary 11 E2
Tīrān 44 C2
Tirana 22 D2
Tiranë = Tirana 22 D2
Tiraspol 17 E9
Tire 23 E6
Tirebolu 46 B4
Tiree 10 C3
Tîrgovişte 22 B5
Tîrgu-Jiu 17 F6
Tîrgu Mureş .. 17 E7
Tirich Mir 42 A7
Tîrnăveni 17 E7
Tírnavos 23 E4
Tirodi 43 J11
Tirol □ 14 E6
Tirso → 21 E2
Tiruchchirappalli 43 P11
Tirunelveli 43 Q10
Tirupati 43 N11
Tiruppur 43 P10
Tiruvannamalai 43 N11
Tisa → 22 B3
Tisdale 71 C9
Tisza = Tisa → 22 B3
Tit-Ary 30 B10
Titicaca, L. 91 G5
Titograd =
Podgorica ... 22 C2
Titov Veles 22 D3
Titova-Mitrovica 22 C3
Titovo Užice .. 22 C2
Titule 57 D5
Tivaouane 55 F1
Tívoli 20 D4
Tiwi 45 F4
Tizi-Ouzou 54 A5
Tiznit 54 C3
Tjirebon =
Cirebon 37 F3
Tlaxcala 84 D5
Tlaxiaco 84 D5
Tlemcen 54 B4
Tmassah 52 C2
Toamasina ... 59 H9
Toay 94 D4
Toba 32 B5
Toba Kakar ... 42 D6

Tobago 87 D7
Tobelo 39 D3
Tobermory 10 C3
Toboali 37 E3
Tobol → 29 D7
Toboli 39 E2
Tobolsk 29 D7
Tobruk =
Tubruq 52 B3
Tobyl =
Tobol → ... 29 D7
Tocantinópolis 92 D4
Tocantins □ .. 93 E4
Tocantins → .. 92 C4
Tochigi 32 A6
Tocopilla 94 A2
Tocumwal 63 C4
Tocuyo → 90 A5
Todeli 39 E2
Todenyang ... 57 D7
Todos os
Santos, B. de 93 E6
Togba 55 E2
Togian,
Kepulauan .. 39 E2
Togliatti 24 C6
Togo ■ 55 G5
Toinya 53 G4
Tojikiston =
Tajikistan ■ . 29 F8
Tojo 39 E2
Tokala 39 E2
Tōkamachi ... 32 A6
Tokanui 65 G3
Tokar 53 E6
Tokarahi 65 F4
Tokat 46 B4
Tokelau Is. 65 K13
Tokmak 29 E8
Tokushima 32 B4
Tokuyama 32 B2
Tōkyō 32 B6
Tolaga Bay ... 64 C8
Tolbukhin =
Dobrich 22 C6
Toledo, Spain . 18 C3
Toledo, U.S.A. . 72 D4
Toledo, Montes
de 18 C3
Tolga 54 B6
Toliara 59 J8
Tolima 90 C3
Tolitoli 39 D2
Tolo 56 E3
Tolo, Teluk ... 39 E2
Toluca 84 D5
Tomakomai ... 32 F12
Tomar 18 C1
Tomaszów
Mazowiecki . 16 C4
Tombé 53 G5
Tombouctou .. 55 E4
Tombua 58 B2
Tomelloso 18 C4
Tomingley 63 B4
Tomini 39 D2
Tomini, Teluk . 39 E2
Tomorit 23 D3
Tomsk 29 D9
Tonantins 90 D5
Tonawanda ... 73 C6
Tondano 39 D2
Tonekābon ... 44 B2
Tong Xian 35 C6
Tonga ■ 65 L13
Tonga Trench . 65 L13
Tongareva 65 K15
Tongchuan ... 35 C5
Tonghua 35 B7

Tongjiang 35 B8
Tongking, G. of
= Tonkin, G.
of 35 D5
Tongoy 94 C2
Tongren 35 D5
Tongsa Dzong . 41 D8
Tongue → ... 76 B2
Tonk 42 F9
Tonkin, G. of .. 35 D5
Tonlé Sap 36 B2
Toompine 63 A3
Toora 63 C4
Toowoomba .. 63 A5
Topeka 77 F7
Topki 29 D9
Topol'čany ... 16 D4
Topolobampo . 84 B3
Torata 91 G4
Torbalı 23 E6
Torbat-e
Heydārīyeh .. 44 C4
Torbat-e Jām .. 45 C5
Torbay 11 F5
Tordesillas ... 18 B3
Torgau 15 C7
Torino 20 B1
Torit 53 H5
Tormes → ... 18 B2
Torne älv → .. 8 E12
Torneå = Tornio 8 E12
Torneträsk 8 E11
Tornio 8 E12
Tornquist 94 D4
Toro, Cerro del 94 B3
Toroníios
Kólpos 23 D4
Toronto,
Australia 63 B5
Toronto, Canada 69 D3
Toropets 24 B3
Tororo 57 D6
Toros Dağları .. 46 C3
Tôrre de
Moncorvo ... 18 B2
Torre del Greco 21 D5
Torrejón de
Ardoz 18 B4
Torrelavega .. 18 A3
Torremolinos .. 18 D3
Torrens, L. ... 62 B2
Torrente 19 C5
Torreón 84 B4
Torres 84 B2
Torres Vedras . 18 C1
Torrevieja ... 19 D5
Tortosa 19 B6
Tortosa, C. de . 19 B6
Ţorūd 44 C3
Toruń 16 B4
Tosa-Wan 32 C3
Toscana □ ... 20 C3
Toshkent 29 E7
Tostado 94 B4
Tosya 46 B3
Toteng 58 C4
Totma 24 A5
Totten Glacier . 96 A12
Tottenham ... 63 B4
Tottori 32 B4
Touba 55 G3
Toubkal, Djebel 54 B3
Tougan 55 F4
Touggourt 54 B6
Tougué 55 F2
Toul 13 B6
Toulepleu 55 G3
Toulon 13 E6
Toulouse 12 E4

Toummo 52 D1
Toungoo 41 H11
Touraine ... 12 C4
Tourane = Da
Nang 36 A3
Tourcoing ... 13 A5
Touriñán, C. .. 18 A1
Tournai 14 C2
Tournon 13 D6
Tours 12 C4
Towamba 63 C4
Towang 41 D8
Townsville .. 61 D8
Toyama 32 A5
Toyohashi ... 32 B5
Toyonaka ... 32 B4
Toyooka ... 32 B4
Toyota 32 B5
Tozeur 54 B6
Trá Li = Tralee 11 E2
Trabzon 46 B4
Trafalgar, C. .. 18 D2
Trail 71 D8
Tralee 11 E2
Trancas 94 B3
Trang 36 C1
Trangan 39 F4
Trangie 63 B4
Trani 20 D6
Transantarctic
Mts. 96 C3
Transilvania .. 17 F7
Transilvanian
Alps =
Carpaţii
Meridionali .. 17 F7
Transylvania =
Transilvania . 17 F7
Trápani 21 E4
Traralgon 63 C4
Trasimeno, L. . 20 C4
Traun 15 D8
Travemünde . 15 B6
Travers, Mt. .. 65 E5
Travnik 20 B6
Trébbia → ... 20 B2
Třebíč 16 D2
Trebinje 20 C7
Treinta y Tres . 94 C6
Trelew 95 E3
Tremp 19 A6
Trenčín 16 D4
Trenque
Lauquen 94 D4
Trent → 11 E6
Trento 20 A3
Trenton 73 D8
Trepassey 69 D5
Tres Arroyos . 94 D4
Três Corações . 93 G4
Três Lagoas .. 93 G3
Tres Montes, C. 95 F1
Tres Puentes . 94 B2
Tres Puntas, C. 95 F3
Três Rios 93 G5
Treviso 20 B4
Triabunna 62 D4
Trichinopoly =
Tiruchchirappalli
.... 43 P11
Trichur 43 P10
Trida 63 B4
Trier 14 D4
Trieste 20 B4
Triglav 20 A4
Tríkkala 23 E3
Trikora, Puncak 39 E5
Trinidad, Bolivia 91 F6
Trinidad,
Colombia ... 90 B4

154

Name	Page	Grid
Trinidad, *Cuba*	86	B3
Trinidad, *Uruguay*	94	C5
Trinidad, *U.S.A.*	78	A1
Trinidad, I.	95	D4
Trinidad & Tobago ■	87	D7
Trinity →	79	E6
Trinity B.	69	D5
Trinity Range	80	E4
Trinkitat	53	E6
Tripoli = Tarābulus, *Lebanon*	46	D3
Tripoli = Tarābulus, *Libya*	52	B1
Trípolis	23	F4
Tripura □	41	F8
Trivandrum	43	Q10
Trnava	16	D3
Troglav	20	C6
Trois-Rivières	69	D3
Troitsko Pechorsk	28	C6
Trollhättan	9	G10
Tronador	95	E2
Trondheim	8	F10
Trondheimsfjorden	8	F10
Trout L.	70	B7
Trouvile-sur-Mer	12	B4
Troy, *Turkey*	23	E6
Troy, *Ala., U.S.A.*	74	D5
Troy, *N.Y., U.S.A.*	73	C9
Troyes	13	B6
Trucial States = United Arab Emirates ■	44	F3
Trujillo, *Honduras*	85	D7
Trujillo, *Peru*	91	E3
Trujillo, *Spain*	18	C3
Trujillo, = *Venezuela*	90	B4
Truk	64	J10
Trundle	63	B4
Truro, *Canada*	69	D4
Truro, *U.K.*	11	F4
Truskavets	17	D6
Trutnov	16	C2
Tsaratanana	59	I9
Tsarevo = Michurin	22	C6
Tsau	58	C4
Tselinograd = Aqmola	29	D8
Tsetserleg	34	B5
Tshabong	58	D4
Tshane	58	C4
Tshela	56	E2
Tshikapa	56	F4
Tshofa	57	F5
Tshwane	58	C4
Tsihombe	59	K9
Tsimlyansk Res. = Tsimlyanskoye Vdkhr.	25	D5
Tsimlyanskoye Vdkhr.	25	D5
Tsinan = Jinan	35	C6
Tsinghai = Qinghai □	34	C4
Tsingtao = Qingdao	35	C7
Tskhinvali	25	E5
Tsna →	24	C5
Tsu	32	B5
Tsuchiura	32	A7
Tsugaru-Kaikyō	32	F12
Tsumeb	58	B3
Tsumis	58	C3
Tsuruga	32	B5
Tsurugi-San	32	C4
Tsuruoka	33	G11
Tsyelyakhany	17	B7
Tual	39	F4
Tuamotu Arch. = Tuamotu Is.	65	L16
Tuamotu Is.	65	L16
Tuamotu Ridge	65	M17
Tuao	38	A2
Tuapse	25	E4
Tuatapere	65	G2
Tubarão	94	B7
Tubau	37	D4
Tübingen	14	D5
Tubruq	52	B3
Tubuai Is.	65	M16
Tucacas	90	A5
Tucson	83	D7
Tucumcari	78	B2
Tucupita	90	B6
Tucuruí	92	C4
Tucuruí, Reprêsa de	92	C4
Tudela	19	A5
Tudmur	46	D4
Tuen	63	A4
Tuguegarao	38	A2
Tukangbesi, Kepulauan	39	F2
Tükrah	52	B3
Tuktoyaktuk	70	B6
Tukuyu	57	F6
Tula, *Mexico*	84	C5
Tula, *Russia*	24	C4
Tulak	45	C5
Tulancingo	84	C5
Tulare	82	B3
Tulcán	90	C3
Tulcea	17	F9
Tulchyn	17	D9
Tuli, *Indonesia*	39	E2
Tuli, *Zimbabwe*	59	C5
Tullamore, *Australia*	63	B4
Tullamore, *Ireland*	11	E3
Tulle	12	D4
Tullibigeal	63	B4
Tulmaythah	52	B3
Tulsa	79	A6
Tulua	90	C3
Tulungagung	37	F4
Tum	39	E4
Tumaco	90	C3
Tumatumari	90	B7
Tumba, L.	56	E3
Tumbarumba	63	C4
Túmbes	90	D2
Tumby Bay	62	B2
Tumeremo	90	B6
Tumkur	43	N10
Tump	42	F3
Tumpat	36	C2
Tumu	55	F4
Tumucumaque, Serra	92	B3
Tumut	63	C4
Tuncurry	63	B5
Tunduru	57	G7
Tundzha →	22	D6
Tunga Pass	41	C10
Tungabhadra →	43	M11
Tungaru	53	F5
Tunguska, Nizhnyaya →	30	C6
Tunis	54	A7
Tunisia ■	54	B6
Tunja	90	B4
Tunuyán →	94	C3
Tunxi	35	D6
Tupelo	74	B3
Tupinambaranas	90	D7
Tupiza	94	A3
Tupungato, Cerro	94	C3
Túquerres	90	C3
Tura	30	C8
Turabah	47	E5
Turān	44	C4
Turayf	47	E4
Turda	17	E6
Turek	16	B4
Turfan = Turpan	34	B3
Turfan Depression = Turpan Hami	34	B3
Tŭrgovishte	22	C6
Turgutlu	23	E6
Turhal	46	B4
Turia →	19	C5
Turiaçu	92	C4
Turiaçu →	92	C4
Turin = Torino	20	B1
Turkana, L.	57	D7
Turkestan = Türkistan	29	E7
Turkey ■	46	C3
Türkistan	29	E7
Turkmenistan ■	29	F6
Turks & Caicos Is. ■	87	B5
Turku	9	F12
Turnagain, C.	64	D7
Turneffe Is.	85	D7
Turnhout	14	C3
Tŭrnovo = Veliko Tŭrnovo	22	C5
Turnu Măgurele	22	C5
Turnu Roşu, P.	17	F7
Turpan	34	B3
Turpan Hami	34	B3
Turukhansk	30	C6
Tuscaloosa	74	C4
Tuscany = Toscana □	20	C3
Tuticorin	43	Q11
Tutóia	92	C5
Tutong	36	D4
Tutrakan	22	B6
Tuttlingen	14	E5
Tutuala	39	F3
Tutuila	65	L13
Tuvalu ■	64	K12
Tuxpan	84	C5
Tuxtla Gutiérrez	85	D6
Tuy	18	A1
Tuz Gölü	46	C3
Ţūz Khurmātū	46	D6
Tuzla	20	B7
Tver	24	B4
Tweed Heads	63	A5
Twillingate	69	D5
Twin Falls	81	D6
Twofold B.	63	C4
Tyachiv	17	D6
Tychy	16	C4
Tyler	79	C6
Tyre = Şūr	46	D3
Tyrol = Tirol □	14	E6
Tyrrell →	62	C3
Tyrrell, L.	62	C3
Tyrrhenian Sea	21	E4
Tyumen	29	D7
Tzaneen	59	C6
Tzukong = Zigong	35	D5

U

Name	Page	Grid
U.S.A. = United States of America ■	67	F10
Uarsciek	49	G4
Uato-Udo	39	F3
Uatumã →	90	D7
Uaupés	90	D5
Uaupés →	90	C5
Ubá	93	G5
Ubaitaba	93	E6
Ubangi = Oubangi →	56	E3
Ubauro	42	E6
Ube	32	C2
Ubeda	18	C4
Uberaba	93	F4
Uberlândia	93	F4
Ubon Ratchathani	36	A2
Ubort →	17	B9
Ubundu	57	E5
Ucayali →	90	D4
Uchiura-Wan	32	F12
Udaipur	43	G8
Udaipur Garhi	40	D6
Udgir	43	K10
Udhampur	42	C9
Udi	55	G6
Údine	20	A4
Udmurtia □	28	D6
Udon Thani	36	A2
Udupi	43	N9
Ueda	32	A6
Uele →	57	D4
Uelzen	14	B6
Ufa	29	D6
Ugab →	58	C2
Ugalla →	57	F6
Uganda ■	57	D6
Ugljane	20	B5
Ugolyak	30	C10
Uibhist a Deas = South Uist	10	C3
Uibhist a Tuath = North Uist	10	C3
Uíge	56	F2
Uinta Mts.	81	E8
Uitenhage	59	E5
Ujjain	43	H9
Ujung Pandang	39	F1
Ukerewe I.	57	E6
Ukhrul	41	E10
Ukhta	28	C6
Ukraine ■	25	D3
Ulaanbaatar	35	B5
Ulaangom	34	A4
Ulan Bator = Ulaanbaatar	35	B5
Ulan Ude	30	D8
Ulcinj	22	D2
Ulhasnagar	43	K8
Ulladulla	63	C5
Ullapool	10	C4
Ulm	14	D5
Ulmarra	63	A5

Ulonguè

Ulonguè 59 A6
Ulster □ 11 D3
Ulubat Gölü ... 23 D7
Uludağ 23 D7
Ulungur He ⇢ 34 B3
Uluru = Ayers
  Rock 60 F5
Ulutau 29 E7
Ulverstone .. 62 D4
Ulyanovsk =
  Simbirsk 24 C6
Ulyasutay 34 B4
Umala 91 G5
Uman 17 D10
Umaria 40 F3
Umarkot 43 G6
Umba 28 C4
Umbrella Mts. . 65 F3
Ume älv ⇢ .. 8 F12
Umeå 8 F12
Umera 39 E3
Umm al
  Qaywayn .. 44 E3
Umm Bel 53 F4
Umm Lajj 47 F4
Umm Ruwaba . 53 F5
Umnak I. 71 C3
Umniati ⇢ ... 59 B5
Umtata 59 E5
Umuarama ... 93 G3
Una ⇢ 20 B6
Unalaska 71 C3
Uncía 91 G5
Underbool 62 C3
Ungarie 63 B4
Ungarra 62 B2
Ungava B. 68 C4
Ungava Pen. .. 68 B3
Ungeny =
  Ungheni 17 E8
Ungheni 17 E8
União da Vitória 94 B6
Unimak I. 71 C3
United Arab
  Emirates ■ .. 44 F3
United
  Kingdom ■ . 11 E5
United States of
  America ■ .. 67 F10
Unnao 40 D3
Unst 10 A6
Ünye 46 B4
Upata 90 B6
Upemba, L. ... 57 F5
Upington 58 D4
Upper Hutt 65 D6
Upper Volta =
  Burkina
  Faso ■ 55 F4
Uppsala 9 G11
Ur 47 E6
Uracara 90 D7
Ural =
  Zhayyq ⇢ .. 29 E6
Ural 63 B4
Ural Mts. =
  Uralskie Gory 28 D6
Uralla 63 B5
Uralsk = Oral . 24 C7
Uralskie Gory . 28 D6
Uranquinty ... 63 C4
Urawa 32 B6
Uray 29 C7
Urbana 72 D1
Urbino 20 C4
Urbión, Picos de 19 A4
Urcos 91 F4
Urda 25 D6
Urdzhar 29 E9

Ures 84 B2
Urfa = Sanliurfa 46 C4
Urganch 29 E7
Urgench =
  Urganch 29 E7
Uribia 90 A4
Urla 23 E6
Urmia =
  Orūmīyeh ... 46 C6
Urmia, L. =
  Orūmīyeh,
  Daryācheh-ye 46 C6
Uroševac 22 C3
Uruana 93 F4
Urubamba ... 91 F4
Urubamba ⇢ . 91 F4
Uruçuí 92 D5
Uruguai ⇢ ... 94 B6
Uruguaiana .. 94 B5
Uruguay ■ ... 94 C5
Uruguay ⇢ .. 94 C5
Urumchi =
  Ürümqi 34 B3
Ürümqi 34 B3
Usakos 58 C3
Usedom 15 B8
Ush-Tobe 29 E8
Ushant =
  Ouessant, I. d' 12 B1
Ushuaia 95 G3
Üsküdar 25 E2
Usman 24 C4
Usoke 57 F6
Uspallata, P. de 94 C3
Uspenskiy 29 E8
Ussuriysk 31 E11
Ust-Aldan =
  Batamay 31 C10
Ust Ilimpeya =
  Yukti 30 C8
Ust Ishim 29 D8
Ust-
  Kamenogorsk
  = Öskemen . 29 E9
Ust-Karenga . 31 D9
Ust Kuyga 31 B11
Ust Muya 30 D9
Ust Olenek ... 30 B10
Ust Port 28 C9
Ust Tsilma 28 C6
Ust Urt =
  Ustyurt,
  Plateau 29 E6
Ust Vorkuta ... 28 C7
Ústí nad Labem 16 C2
Ustica 21 E4
Ustinov =
  Izhevsk 28 D6
Ustye 30 D7
Ustyurt, Plateau 29 E6
Usu 34 B3
Usuki 32 C2
Usumacinta ⇢ 85 D6
Usumbura =
  Bujumbura . 57 E5
Uta 39 E5
Utah □ 81 F8
Utete 57 F7
Utiariti 91 F7
Utica 73 C8
Utrecht 14 B3
Utrera 18 D3
Utsunomiya .. 32 A6
Uttar Pradesh □ 42 F11
Uttaradit 36 A2
Uummannarsuaq
  = Farvel, Kap 68 C6
Uusikaupunki . 8 F12
Uvat 29 D7

Uvinza 57 F6
Uvira 57 E5
Uvs Nuur 34 A4
Uwajima 32 C3
Uyuni 91 H5
Uzbekistan ■ .. 29 E7
Uzerche 12 D4
Uzh ⇢ 17 C10
Uzhgorod =
  Uzhhorod .. 17 D6
Uzhhorod 17 D6
Uzunköprü 22 D6

V

Vaal ⇢ 58 D4
Vaasa 8 F12
Vác 16 E4
Vach ⇢ =
  Vakh ⇢ 29 C8
Vadodara 43 H8
Vadsø 8 D13
Vaduz 13 C8
Váh ⇢ 16 E4
Vahsel B. 96 B5
Vaigach 28 B6
Vakh ⇢ 29 C8
Val d'Or 69 D3
Valahia 22 B5
Valandovo 22 D4
Valcheta 95 E3
Valdayskaya
  Vozvyshennost
  .... 24 B3
Valdepeñas ... 18 C4
Valdés, Pen. .. 95 E4
Valdez 71 B5
Valdivia 95 D2
Valdosta 75 D6
Valença 93 E6
Valença do Piauí 92 D5
Valence 13 D6
Valencia, Spain 19 C5
Valencia,
  Venezuela .. 90 A5
Valencia □ 19 C5
Valencia, G. de 19 C6
Valencia de
  Alcántara ... 18 C2
Valencia I. 11 F1
Valenciennes . 13 A5
Valentim, Sa. do 92 D5
Valera 90 B4
Valjevo 22 B2
Vall de Uxó ... 19 C5
Valladolid,
  Mexico 85 C7
Valladolid, Spain 18 B3
Valle de la
  Pascua 90 B5
Valledupar ... 90 A4
Vallejo 80 F2
Vallenar 94 B2
Valletta 21 G5
Valls 19 B6
Valognes 12 B3
Valona = Vlóra 23 D2
Valozhyn 17 A8
Valparaíso 94 C2
Vals, Tanjung . 39 F5
Valsad 43 J8
Valverde del
  Camino 18 D2
Van 46 C5
Van, L. = Van
  Gölü 46 C5

Van Gölü 46 C5
Van Rees,
  Pegunungan 39 E5
Vanadzor 25 E5
Vancouver,
  Canada 71 D7
Vancouver,
  U.S.A. 80 C2
Vancouver I. .. 71 D7
Vänern 9 G10
Vanga 57 E7
Vangaindrano . 59 J9
Vännäs 8 F11
Vannes 12 C2
Vanrhynsdorp . 58 E3
Vanua Levu ... 64 L12
Vanuatu ■ 64 L11
Vapnyarka ... 17 D9
Varanasi 40 E4
Varangerfjorden 8 D13
Varaždin 20 A6
Varberg 9 G10
Vardak □ 45 C7
Vardar =
  Axiós ⇢ ... 23 D4
Varese 20 B2
Varna 22 C6
Varzaneh 44 C3
Vasa Barris ⇢ 93 E6
Vascongadas =
  País Vasco □ 19 A4
Vasht = Khāsh 42 E2
Vasilevichi ... 17 B9
Vasilkov =
  Vasylkiv ... 17 C10
Vaslui 17 E8
Västerås 9 G11
Västervik 9 G11
Vasto 20 C5
Vasylkiv 17 C10
Vatican City ■ . 20 D4
Vatomandry .. 59 H9
Vatra-Dornei .. 17 E7
Vättern 9 G10
Vaupés =
  Uaupés ⇢ .. 90 C5
Vawkavysk 17 B7
Vaygach, Ostrov 28 C6
Vedea ⇢ 22 C5
Vega 8 E10
Vegreville 71 C8
Vejer de la
  Frontera 18 D3
Velebit Planina 20 B5
Vélez 90 B4
Vélez Málaga .. 18 D3
Vélez Rubio ... 19 D4
Velhas ⇢ 93 F5
Velika Kapela .. 20 B5
Velikaya ⇢ ... 24 B2
Veliki Ustyug .. 24 A6
Velikiye Luki .. 24 B3
Veliko Tŭrnovo 22 C5
Velikonda Range 43 M11
Velletri 20 D4
Vellore 43 N11
Venado Tuerto 94 C4
Vendée □ 12 C3
Vendôme 12 C4
Venézia 20 B4
Venézia, G. di . 20 B4
Venezuela ■ .. 90 B5
Venezuela, G. de 90 A4
Vengurla 43 M8
Venice =
  Venézia 20 B4
Venkatapuram . 40 H3
Ventotene 21 D4
Ventoux, Mt. .. 13 D6

Ventspils 24 B1
Ventuarí → .. 90 C5
Ventura 82 C3
Venus B. 63 C4
Vera, *Argentina* 94 B4
Vera, *Spain* ... 19 D5
Veracruz 84 D5
Veraval 43 J7
Verbánia 20 B2
Vercelli 20 B2
Verde → 95 E3
Verden 14 B5
Verdun 13 B6
Vereeniging ... 59 D5
Verga, C. 55 F2
Verín 18 B2
Verkhnevilyuysk 30 C10
Verkhniy
 Baskunchak . 25 D6
Verkhoyansk .. 31 C11
Verkhoyansk Ra.
 =
Verkhoyanskiy
 Khrebet 31 C10
Verkhoyanskiy
 Khrebet 31 C10
Vermilion 71 C8
Vermont □ 73 C9
Vernon, *Canada* 71 C8
Vernon, *U.S.A.* 78 B4
Véroia 23 D4
Verona 20 B3
Versailles 12 B5
Vert, C. 55 F1
Verviers 14 C3
Veselovskoye
 Vdkhr. 25 D5
Vesoul 13 C7
Vesterålen 8 E11
Vestfjorden ... 8 E10
Vesuvio 21 D5
Vesuvius, Mt. =
 Vesuvio ... 21 D5
Veszprém ... 16 E3
Vetlugu → ... 28 D5
Vettore, Mte. .. 20 C4
Veys 47 E7
Vezhen 22 C5
Viacha 91 G5
Viamão 94 C6
Viana 92 C5
Viana do
 Alentejo 18 C2
Viana do Castelo 18 B1
Vianópolis 93 F4
Viaréggio 20 C3
Vibo Valéntia .. 21 E6
Vicenza 20 B3
Vich 19 B7
Vichy 13 C5
Vicksburg 79 C8
Viçosa 93 D6
Victor 43 J7
Victor Harbor .. 62 C2
Victoria, *Canada* 71 D7
Victoria, *Chile* . 94 D2
Victoria, *Guinea* 55 F2
Victoria,
 Malaysia .. 36 C5
Victoria □ 62 C3
Victoria, L.,
 Africa 57 E6
Victoria, L.,
 Australia ... 62 B3
Victoria Beach . 71 C10
Victoria de
 Durango 84 C4
Victoria Falls . 59 B5
Victoria I. 70 A8
Victoria Ld. ... 96 B14

Victoria
 Taungdeik .. 41 G9
Victoria West .. 58 E4
Victorica 94 D3
Vicuña 94 C2
Vidin 22 C4
Vidisha 43 H10
Viedma 95 E4
Viedma, L. 95 F2
Vienna = Wien 15 D9
Vienne 13 D6
Vienne → 12 C4
Vientiane 36 A2
Vierzon 12 C5
Vietnam ■ 36 A3
Vigan 38 A2
Vigévano 20 B2
Vigia 92 C4
Vigo 18 A1
Vijayawada ... 40 J3
Vikeke 39 F3
Vikna 8 F10
Vila da Maganja 59 B7
Vila de João
 Belo = Xai-Xai 59 D6
Vila do Bispo .. 18 D1
Vila Franca de
 Xira 18 C1
Vila Machado . 59 B6
Vila Nova de
 Gaia 18 B1
Vila Real 18 B2
Vila Real de
 Santo António 18 D2
Vila Velha 93 G5
Vilaine → 12 C2
Vilanculos 59 C7
Vileyka 17 A8
Vilhelmina 8 F11
Vilhena 91 F6
Vilkitskogo,
 Proliv 30 B8
Vilkovo =
 Vylkove 17 F9
Villa Ahumada 84 A3
Villa Ángela .. 94 B4
Villa Bella 91 F5
Villa Bens =
 Tarfaya 54 C2
Villa Colón ... 94 C4
Villa de María . 94 B4
Villa Dolores . 94 C3
Villa Hayes ... 94 B5
Villa María ... 94 C4
Villa Mazán .. 94 B3
Villa Montes .. 94 A4
Villa Ocampo . 94 B5
Villacarrillo ... 18 C4
Villach 15 E7
Villagarcía de
 Arosa 18 A1
Villaguay 94 C5
Villahermosa .. 85 D6
Villajoyosa 19 C5
Villalba 18 A2
Villanueva de la
 Serena 18 C3
Villanueva y
 Geltrú 19 B6
Villarreal 19 C5
Villarrica, *Chile* 95 D2
Villarrica,
 Paraguay ... 94 B5
Villarrobledo . 19 C4
Villavicencio .. 90 C4
Villaviciosa .. 18 A3
Villazón 94 A3
Villena 19 C5
Villeneuve-
 d'Ascq 13 A5

Villeneuve-sur-
 Lot 12 D4
Villingen-
 Schwenningen 14 D5
Vilnius 24 C2
Vilyuy → 31 C10
Vilyuysk 31 C10
Viña del Mar .. 94 C2
Vinaroz 19 B6
Vindhya Ra. ... 43 H10
Vineland 73 E8
Vinkovci 20 B7
Vinnitsa =
 Vinnytsya ... 17 D9
Vinnytsya 17 D9
Virac 38 B2
Viramgam 43 H8
Viranşehir 46 C4
Virden 71 D9
Vire 12 B3
Vírgenes, C. ... 95 G3
Virgin Is.
 (British) ■ .. 87 C7
Virginia □ 73 F6
Virginia Beach . 73 F8
Virovitica 20 B6
Virudunagar .. 43 Q10
Vis 20 C6
Visalia 82 B3
Visayan Sea ... 38 B2
Visby 9 G11
Višegrad 20 C7
Viseu, *Brazil* ... 92 C4
Viseu, *Portugal* 18 B2
Vishakhapatnam 40 J4
Viso, Mte. 20 B1
Vistula =
 Wisła → ... 16 A4
Vitebsk =
 Vitsyebsk ... 24 B3
Viterbo 20 C4
Viti Levu 64 L12
Vitigudino 18 B2
Vitória, *Brazil* . 93 G5
Vitoria, *Spain* . 19 A4
Vitória da
 Conquista ... 93 E5
Vitsyebsk 24 B3
Vittória 21 F5
Vittório Véneto 20 B4
Vivero 18 A2
Vize 22 D6
Vizianagaram . 40 H4
Vjosa → 23 D2
Vladikavkaz .. 25 E5
Vladimir 24 B5
Vladimir
 Volynskiy =
 Volodymyr-
 Volynskyy ... 17 C7
Vladivostok ... 31 E11
Vlissingen 14 C2
Vlóra 23 D2
Vltava → 16 C2
Vogelsberg ... 14 C5
Voghera 20 B2
Vohibinany ... 59 H9
Vohimarina .. 59 G10
Vohimena,
 Tanjon' i ... 59 K9
Vohipeno 59 J9
Voi 57 E7
Voiron 13 D6
Vojvodina □ .. 22 B2
Volga → 25 D6
Volga Hts. =
 Privolzhskaya
 Vozvyshennost
 .... 24 C6

Volgodonsk ... 25 D5
Volgograd ... 25 D5
Volgogradskoye
 Vdkhr. 24 C6
Volkhov → ... 24 A3
Volkovysk =
 Vawkavysk .. 17 B7
Volochanka .. 30 B7
Volodymyr-
 Volynskyy ... 17 C7
Vologda 24 B4
Vólos 23 E4
Volovets 17 D6
Volozhin =
 Valozhyn ... 17 A8
Volsk 24 C6
Volta → 55 G5
Volta, L. 55 G5
Volta Redonda 93 G5
Volterra 20 C3
Volturno → .. 20 D4
Volzhskiy ... 25 D5
Voríai
 Sporádhes .. 23 E4
Vorkuta 28 C7
Voronezh ... 24 C4
Voroshilovgrad
 = Luhansk ... 25 D4
Voroshilovsk =
 Alchevsk 25 D4
Vosges 13 B7
Vostok I. 65 L15
Vouga → ... 18 B1
Voznesensk .. 25 D3
Voznesenye .. 28 C4
Vrangelya,
 Ostrov 31 B15
Vranje 22 C3
Vratsa 22 C4
Vrbas → 20 B6
Vrede 59 D5
Vredenburg .. 58 E3
Vršac 22 B3
Vryburg 58 D4
Vryheid 59 D6
Vukovar 20 B7
Vulcan 17 F6
Vulcaneşti ... 17 F9
Vulcano 21 E5
Vulkaneshty =
 Vulcaneşti .. 17 F9
Vung Tau ... 36 B3
Vyatka = Kirov 28 D5
Vyatskiy
 Polyany ... 24 B7
Vyazma 24 B3
Vyborg 28 C3
Vychegda → . 28 C5
Vychodné
 Beskydy 17 D5
Vylkove 17 F9
Vynohradiv ... 17 D6
Vyshniy
 Volochek .. 24 B3
Vyškov 16 D3

W

Wa 55 F4
Waal → 14 C3
Wabash → .. 72 F1
Waco 78 D5
Wad Banda .. 53 F4
Wad Hamid .. 53 E5
Wâd Medanî .. 53 F5
Waddington, Mt. 71 C7

Waddy Pt. 63 A5
Wadi Halfa 52 D5
Wafrah 47 E6
Wager B. 68 B2
Wager Bay 68 B1
Wagga Wagga .. 63 C4
Waghete 39 E5
Wah 42 C8
Wahai 39 E3
Waiau → 65 E5
Waibeem 39 E4
Waigeo 39 E4
Waihi 64 B6
Waihou → ... 64 B6
Waikari 65 E5
Waikato → ... 64 B6
Waikerie 62 B2
Waikokopu 64 C7
Waikouaiti ... 65 F4
Waimakariri → 65 E5
Waimate 65 F4
Wainganga → 43 K11
Waingapu 39 F2
Wainwright, Canada ... 71 C8
Wainwright, U.S.A. 71 A3
Waiouru 64 C6
Waipara 65 E5
Waipawa 64 C7
Waipiro 64 C8
Waipu 64 A6
Waipukurau ... 64 D7
Wairakei 64 C7
Wairarapa, L. .. 65 D6
Wairoa 64 C7
Waitaki → 65 F4
Waitara 64 C6
Waiuku 64 B6
Wajima 32 A5
Wajir 57 D8
Wakasa-Wan .. 32 B4
Wakatipu, L. .. 65 F3
Wakayama ... 32 B4
Wakefield 65 D5
Wakema 41 J10
Wakkanai 32 E12
Wakool 63 C3
Wakool → ... 62 C3
Wakre 39 E4
Wałbrzych ... 16 C3
Walcha 63 B5
Wałcz 16 B3
Wales □ 11 E4
Walgett 63 A4
Walgreen Coast 96 B1
Walhalla 63 C4
Walla Walla .. 80 B4
Wallachia = Valahia 22 B5
Wallal 63 A4
Wallaroo 62 B2
Wallerawang . 63 B5
Wallis & Futuna, Is. 64 L13
Wallowa Mts. .. 80 C5
Wallsend 63 B5
Wallumbilla .. 63 A4
Walvisbaai ... 58 C2
Wamba 57 D5
Wamena 39 E5
Wamulan 39 E3
Wana 42 C6
Wanaaring ... 63 A3
Wanaka 65 F3
Wanaka L. ... 65 F3
Wanapiri 39 E5

Wanbi 62 B3
Wandoan 63 A4
Wangal 39 F4
Wanganella .. 63 C3
Wanganui 64 C6
Wangaratta .. 63 C4
Wangary 62 B2
Wangerooge .. 14 B4
Wangiwangi .. 39 F2
Wanxian 35 C5
Warangal 43 L11
Waratah 62 D4
Waratah B. ... 63 C4
Warburton 63 C4
Ward 65 D6
Ward → 63 A4
Wardha 43 J11
Wardha → ... 43 K11
Warialda 63 A5
Wariap 39 E4
Warkopi 39 E4
Warmbad, Namibia 58 D3
Warmbad, S. Africa 59 C5
Warner Mts. .. 80 E3
Warracknabeal 62 C3
Warragul 63 C4
Warrego → .. 63 B4
Warren 63 B4
Warrenton ... 58 D4
Warrenville ... 63 A4
Warri 55 G6
Warrina 62 A2
Warrington, U.K. 11 E5
Warrington, U.S.A. 74 D4
Warrnambool . 62 C3
Warsa 39 E5
Warsaw = Warszawa ... 16 B5
Warszawa ... 16 B5
Warta → 16 B2
Warthe = Warta → 16 B2
Waru 39 E4
Warwick 63 A5
Wasatch Ra. .. 81 E8
Wash, The 11 E7
Washim 43 J10
Washington .. 73 E7
Washington □ . 80 B3
Washington, Mt. 73 B10
Wasian 39 E4
Wasior 39 E4
Waskaganish .. 69 C3
Wasserkuppe .. 14 C5
Watangpone .. 39 E2
Waterbury ... 73 D9
Waterford ... 11 E3
Waterloo, S. Leone 55 G2
Waterloo, U.S.A. 77 D8
Waterton-Glacier International Peace Park .. 81 A6
Watertown ... 73 C8
Waterville ... 73 B11
Watford 11 F6
Watling I. = San Salvador 86 B5
Watrous 71 C9
Watsa 57 D5
Watson Lake .. 70 B7
Watsonville ... 82 B2
Wattiwarriganna Cr. → 62 A2
Watuata = Batuata 39 F2

Watubela, Kepulauan .. 39 E4
Watubela Is. = Watubela, Kepulauan .. 39 E4
Wauchope 63 B5
Waukegan 72 C2
Waukesha 72 C1
Wausau 77 C10
Wauwatosa ... 72 C2
Waverley 64 C6
Wâw 53 G4
Wâw al Kabïr .. 52 C2
Waxahachie .. 79 C5
Wayabula Rau . 39 D3
Wayatinah 62 D4
Waycross 75 D6
Wazirabad 42 C9
We 36 C1
Weda 39 D3
Weda, Teluk . 39 D3
Weddell I. ... 95 G4
Weddell Sea .. 96 A4
Wedderburn .. 62 C3
Wee Waa 63 B4
Weemelah 63 A4
Weiden 15 D7
Weifang 35 C6
Weimar 15 C6
Weipa 61 C7
Weir → 63 A4
Wejherowo ... 16 A4
Welbourn Hill . 62 A1
Welkom 59 D5
Wellesley Is. .. 60 D6
Wellington, Australia 63 B4
Wellington, N.Z. 65 D6
Wellington, I. .. 95 F1
Wellington, L. . 63 C4
Wels 15 D8
Welshpool ... 11 E5
Wemindji 69 C3
Wenatchee ... 80 B3
Wenchi 55 G4
Wenchow = Wenzhou 35 D7
Wendesi 39 E4
Wensu 34 B3
Wentworth ... 62 B3
Wenut 39 E4
Wenzhou 35 D7
Werda 58 D4
Werder 49 F4
Weri 39 E4
Werra → 14 C5
Werribee 63 C3
Werrimull 62 B3
Werris Creek . 63 B5
Wersar 39 E4
Weser → 14 B5
Wesiri 39 F3
West Bengal □ 40 F7
West Beskids = Západné Beskydy ... 16 D4
West Falkland . 95 G4
West Fjord = Vestfjorden . 8 E10
West Ice Shelf . 96 A11
West Nicholson 59 C5
West Palm Beach 75 F7
West Pt. 62 C2
West Virginia □ 72 E5
West Wyalong 63 B4
Westall Pt. ... 62 B1
Westbury 62 D4

Westerland ... 14 A5
Western Australia □ .. 60 F3
Western Dvina = Daugava → . 24 B1
Western Ghats 43 N9
Western Sahara ■ ... 54 D2
Western Samoa ■ ... 65 L13
Westerwald .. 14 C4
Westland Bight 65 E4
Weston 36 C5
Weston-super-Mare 11 F5
Westport, Ireland 11 E2
Westport, N.Z. 65 D4
Westray 10 B5
Wetar 39 F3
Wetaskiwin .. 71 C8
Wetzlar 14 C5
Wexford 11 E3
Weyburn 71 D9
Weymouth ... 11 F5
Whakatane ... 64 B7
Whale → 68 C4
Whale Cove .. 70 B10
Whales, B. of . 96 B16
Whangamomona 64 C6
Whangarei ... 64 A6
Whangarei Harb. 64 A6
Wheeling 72 D5
White →, Ark., U.S.A. 79 C8
White →, Ind., U.S.A. 72 E2
White Cliffs .. 62 B3
White I. 64 B7
White Nile = Nîl el Abyad → . 53 E5
White Russia = Belarus ■ ... 24 C2
White Sea = Beloye More . 28 C4
Whitecliffs ... 65 E4
Whitehaven .. 11 D5
Whitehorse .. 70 B6
Whitemark ... 62 D4
Whitfield 63 C4
Whitianga ... 64 B6
Whitney, Mt. . 82 B3
Whittlesea ... 63 C4
Wholdaia L. .. 70 B9
Whyalla 62 B2
Whyjonta 62 A3
Wichita 78 A5
Wichita Falls . 78 C5
Wick 10 B5
Wickham, C. . 62 C3
Wicklow Mts. . 11 E3
Wieluń 16 C4
Wien 15 D9
Wiener Neustadt 15 E9
Wiesbaden .. 14 C4
Wilcannia 62 B3
Wildspitze ... 14 E6
Wilhelm, Mt. . 61 B8
Wilhelm II Coast 96 A11
Wilhelmshaven 14 B5
Wilkes-Barre . 73 D8
Wilkie 71 C9
Willandra Billabong Creek → ... 63 B4
Willemstad ... 87 D6

William Creek	62	A2	Woodville	64	D6	Xi'an	35	C5
Williams Lake	71	C7	Woolamai, C.	63	C4	Xiang Jiang →	35	D6
Williamsburg	73	F7	Woolgoolga	63	B5	Xiangfan	35	C6
Williamsport	73	D7	Woombye	63	A5	Xiangtan	35	D6
Williamstown	63	C3	Woomera	62	B2	Xiao Hinggan		
Williston	76	A3	Worcester,			Ling	35	B7
Willowmore	58	E4	S. Africa	58	E3	Xiapu	35	D6
Willunga	62	C2	Worcester, U.K.	11	E5	Xichang	34	D5
Wilmington,			Worcester,			Xigazê	34	D3
Australia	62	B2	U.S.A.	73	C10	Xinavane	59	D6
Wilmington,			Workington	11	D5	Xing'an	35	D6
Del., U.S.A.	73	E8	Worms	14	D5	Xingren	35	D5
Wilmington,			Worthing	11	F6	Xingtai	35	C6
N.C., U.S.A.	75	B9	Wosi	39	E3	Xingu →	92	C3
Wilpena Cr. →	62	B2	Wou-han =			Xining	34	C5
Wilson	75	B9	Wuhan	35	C6	Xinjiang Uygur		
Wilson →	62	A3	Wour	52	D2	Zizhiqu □	34	B3
Wilsons			Wousi = Wuxi	35	C7	Xinjin	35	C7
Promontory	63	C4	Wowoni	39	E2	Xinxiang	35	C6
Wimmera →	62	C3	Woy Woy	63	B5	Xinyang	35	C6
Winchester	11	F6	Wrangel I. =			Xinzhu	35	D7
Wind River			Vrangelya,			Xique-Xique	93	E5
Range	81	D9	Ostrov	31	B15	Xisha Qundao =		
Windau =			Wrangell	71	C6	Hsisha		
Ventspils	24	B1	Wrangell Mts.	71	B5	Chuntao	36	A4
Windhoek	58	C3	Wrath, C.	10	B4	Xixabangma		
Windsor,			Wrexham	11	E5	Feng	40	C5
Australia	63	B5	Wright	38	B3	Xizang □	34	C3
Windsor, N.S.,			Wrigley	70	B7	Xuanhua	35	B6
Canada	69	D4	Wrocław	16	C3	Xuchang	35	C6
Windsor, Ont.,			Września	16	B3	Xuzhou	35	C6
Canada	69	D2	Wu Jiang →	35	D5			
Windward Is.	87	D7	Wuhan	35	C6			
Wingen	63	B5	Wuhsi = Wuxi	35	C7	**Y**		
Wingham	63	B5	Wuhu	35	C6			
Winisk →	69	C2	Wukari	55	G6	Ya'an	34	D5
Winneba	55	G4	Wuliaru	39	F4	Yaapeet	62	C3
Winnipeg	71	D10	Wulumuchi =			Yabelo	53	H6
Winnipeg, L.	71	C10	Ürümqi	34	B3	Yablonovy Ra. =		
Winnipegosis L.	71	C9	Wum	55	G7	Yablonovyy		
Winona	77	C9	Wuntho	41	F10	Khrebet	30	D9
Winston-Salem	75	A7	Wuppertal	14	C4	Yablonovyy		
Winterthur	13	C8	Würzburg	14	D5	Khrebet	30	D9
Winton,			Wutongqiao	34	D5	Yacuiba	94	A4
Australia	61	E7	Wuwei	34	C5	Yadgir	43	L10
Winton, N.Z.	65	G3	Wuxi	35	C7	Yagoua	56	B3
Wirrulla	62	B1	Wuyi Shan	35	D6	Yahuma	56	D4
Wisconsin □	77	C10	Wuzhi Shan	35	E5	Yakima	80	B3
Wisła →	16	A4	Wuzhong	35	C5	Yakutat	71	C6
Wismar	15	B6	Wuzhou	35	D6	Yakutsk	31	C10
Witbank	59	D5	Wyandotte	72	C4	Yala	36	C2
Wittenberg	15	C7	Wyandra	63	A4	Yalinga	56	C4
Wittenberge	15	B6	Wyangala Res.	63	B4	Yalong Jiang →	34	D5
Wkra →	16	B5	Wyara, L.	63	A3	Yalova	22	D7
Włocławek	16	B4	Wycheproof	62	C3	Yalta	25	E3
Włodawa	17	C6	Wynbring	62	B1	Yalutorovsk	29	D7
Wodonga	63	C4	Wyndham,			Yam Ha Melah		
Wokam	39	F4	Australia	60	D4	= Dead Sea	47	E3
Wolfsberg	15	E8	Wyndham, N.Z.	65	G3	Yam Kinneret	46	D3
Wolfsburg	15	B6	Wynnum	63	A5	Yamagata	33	G12
Wolin	16	B2	Wynyard	62	D4	Yamaguchi	32	B2
Wollaston, Is.	95	H3	Wyoming □	81	D10	Yamal,		
Wollaston Pen.	70	B8	Wyong	63	B5	Poluostrov	28	B8
Wollongong	63	B5				Yamal Pen. =		
Wolseley	62	C3				Yamal,		
Wolverhampton	11	E5	**X**			Poluostrov	28	B8
Wondai	63	A5				Yamantau, Gora	29	D6
Wongalarroo L.	63	B3	Xai-Xai	59	D6	Yamba, N.S.W.,		
Wŏnsan	35	C7	Xainza	34	C3	Australia	63	A5
Wonthaggi	63	C4	Xangongo	58	B3	Yamba,		
Woocalla	62	B2	Xankändi	25	F6	S. Austral.,		
Woodburn	63	A5	Xánthi	22	D5	Australia	62	B3
Woodenbong	63	A5	Xapuri	91	F5	Yâmbiô	53	H4
Woodend	63	C3	Xi Jiang →	35	D6	Yambol	22	C6
Woodland	80	F3	Xiaguan	34	D5	Yamdena	39	F4
Woodroffe, Mt.	60	F5	Xiamen	35	D6	Yamethin	41	G11
Woods, L. of the	71	D10						
Woodstock	69	D4						

Yamma-Yamma,		
L.	62	A3
Yamoussoukro	55	G3
Yampil	17	D9
Yampol =		
Yampil	17	D9
Yamuna →	40	E3
Yamzho Yumco	34	D4
Yana →	31	B11
Yanac	62	C3
Yanbu 'al Baḥr	47	F4
Yancannia	62	B3
Yanco Cr. →	63	C4
Yandoon	41	J10
Yangambi	56	D4
Yangch'ü =		
Taiyuan	35	C6
Yangchun	35	D6
Yangjiang	35	D6
Yangon =		
Rangoon	41	J11
Yangquan	35	C6
Yanji	35	B7
Yanna	63	A4
Yanqi	34	B3
Yantabulla	63	A4
Yantai	35	C7
Yanykurgan	29	E7
Yao	53	F2
Yaoundé	56	D2
Yapen	39	E5
Yapen, Selat	39	E5
Yaqui →	84	B2
Yar-Sale	28	C8
Yaraka	61	E7
Yardea P.O.	62	B2
Yaremcha	17	D7
Yarensk	28	C5
Yarí →	90	D4
Yarkand =		
Shache	34	C2
Yarkhun →	42	A8
Yaroslavl	24	B4
Yarram	63	C4
Yarraman	63	A5
Yarranvale	63	A4
Yarras	63	B5
Yartsevo	30	C7
Yaselda	17	B8
Yasinya	17	D7
Yass	63	B4
Yatağan	23	F7
Yathkyed L.	70	B10
Yatsushiro	32	C2
Yauyos	91	F3
Yavari →	90	D4
Yavatmal	43	J11
Yavoriv	17	D6
Yavorov =		
Yavoriv	17	D6
Yawatahama	32	C3
Yazd	44	D3
Yazd □	44	C3
Yazdān	45	C5
Yazoo →	79	C8
Ye Xian	35	C6
Yebyu	41	K12
Yecla	19	C5
Yedintsy =		
Edinița	17	D8
Yeelanna	62	B2
Yegros	94	B5
Yei	53	H5
Yekaterinburg	29	D7
Yekaterinodar =		
Krasnodar	25	D4
Yelarbon	63	A5

Yelets 24 C4
Yelizavetgrad =
  Kirovohrad .. 25 D3
Yell 10 A6
Yellow Sea ... 35 C7
Yellowhead Pass 71 C8
Yellowknife ... 70 B8
Yellowknife ~> 70 B8
Yellowstone ~> 76 B3
Yellowstone
  National Park 81 C8
Yelsk 17 C9
Yemen ■ 49 E3
Yenangyaung . 41 G10
Yenbo = Yanbu
  'al Baḥr 47 F4
Yenda 63 B4
Yenice 23 E6
Yenisey ~> ... 28 B9
Yeniseysk 30 D7
Yeniseyskiy
  Zaliv 28 B9
Yenyuka 31 D10
Yeola 43 J9
Yeovil 11 F5
Yerbent 29 F6
Yerevan 25 E5
Yermak 29 D8
Yershov 24 C6
Yerushalayim =
  Jerusalem .. 47 E3
Yessey 30 C8
Yeu, I. d' 12 C2
Yevpatoriya ... 25 D3
Yeysk 25 D4
Yezd = Yazd .. 44 D3
Yiannitsa 23 D4
Yibin 35 D5
Yichang 35 C6
Yichun 35 B7
Yilan 35 D7
Yinchuan 35 C5
Yingkou 35 B7
Yingtan 35 D6
Yining 34 B3
Yinmabin 41 F10
Yishan 35 D5
Yíthion 23 F4
Yiyang 35 D6
Yog Pt. 38 B2
Yogyakarta ... 37 F4
Yokadouma ... 56 D2
Yokkaichi 32 B5
Yoko 53 G1
Yokohama ... 32 B6
Yokosuka 32 B6
Yola 53 G1
Yonago 32 B3
Yongji 35 C6
Yonibana 55 G2
Yonkers 73 D9
Yonne ~> 13 B5
York, U.K. 11 E6
York, U.S.A. .. 73 E7
York, C. 61 C7
Yorke Pen. 62 B2
Yorkton 71 C9
Yos Sudarso,
  Pulau 39 F5
Yosemite
  National Park 82 B3
Yoshkar Ola ... 24 B6
Youghal 11 F3
Young 63 B4
Younghusband,
  L. 62 B2
Younghusband
  Pen. 62 C2

Youngstown .. 72 D5
Yozgat 46 C3
Ysyk-Köl, Ozero 29 E8
Yu Jiang ~> .. 35 D6
Yu Shan 35 D7
Yuan Jiang ~> 35 D6
Yuanling 35 D6
Yuanyang 34 D5
Yuba City 80 F3
Yūbari 32 F12
Yucatán □ 85 C7
Yucatán, Canal
  de 85 C7
Yucatán,
  Península de 85 D7
Yucatan Str. =
  Yucatán,
  Canal de 85 C7
Yuci 35 C6
Yugoslavia ■ .. 22 B3
Yukon ~> 71 B3
Yukon
  Territory □ .. 70 B6
Yukti 30 C8
Yuma 82 D5
Yumen 34 C4
Yun Xian 35 C6
Yungas 91 G5
Yunnan □ 34 D5
Yunta 62 B2
Yurgao 29 D9
Yuribei 28 B8
Yurimaguas .. 91 E3
Yuryung Kaya . 30 B9
Yushu 34 C4
Yuzhno-
  Sakhalinsk .. 31 E12
Yvetot 12 B4

Z

Zabid 49 E3
Zābol 45 D5
Zābol □ 45 D6
Zābolī 45 E5
Zabrze 16 C4
Zacapa 85 E7
Zacatecas 84 C4
Zacoalco 84 C4
Zadar 20 B5
Zadetkyi Kyun . 36 C1
Zafra 18 C2
Żagań 16 C2
Zagazig 52 B5
Zagorsk =
  Sergiyev
  Posad 24 B4
Zagreb 20 B6
Zāgros, Kūhhā-
  ye 44 C2
Zagros Mts. =
  Zāgros,
  Kūhhā-ye .. 44 C2
Zāhedān 45 D5
Zahlah 46 D3
Zaïre ■ 56 E4
Zaïre ~> 56 F2
Zaječar 22 C4
Zakhodnaya
  Dzvina =
  Daugava ~> . 24 B1
Zākhū 46 C5
Zákinthos ... 23 F3
Zakopane 16 D4
Zalaegerszeg .. 16 E3

Zalău 17 E6
Zaleshchiki =
  Zalishchyky . 17 D7
Zalew Wiślany . 16 A4
Zalingei 53 F3
Zalishchyky ... 17 D7
Zambeze ~> .. 59 B7
Zambezi =
  Zambeze ~> 59 B7
Zambezi 58 A4
Zambia ■ 59 B5
Zamboanga ... 38 C2
Zamora, Mexico 84 D4
Zamora, Spain 18 B3
Zamość 17 C6
Zanaga 56 E2
Zanesville ... 72 E4
Zanjān 46 C7
Zanjān □ 46 C7
Zante =
  Zákinthos ... 23 F3
Zanzibar 57 F7
Zaouiet El-Kala
  = Bordj Omar
  Driss 54 C6
Zaouiet Reggane 54 C5
Zap Suyu =
  Kabīr, Zab
  al ~> 46 C5
Zapadnaya
  Dvina 28 D4
Zapadnaya
  Dvina ~> =
  Daugava ~> . 24 B1
Západné
  Beskydy ... 16 D4
Zaporizhzhya . 25 D4
Zaporozhye =
  Zaporizhzhya 25 D4
Zara 46 C4
Zaragoza 19 B5
Zarand 44 D4
Zaranj 42 D2
Zaria 55 F6
Zaruma 90 D3
Zarqā' 16 C2
Zarzis 52 B1
Zaskar Mts. ... 42 C10
Zawāreh 44 C3
Zawiercie 16 C4
Zāyandeh ~> . 44 C3
Zaysan 29 E9
Zaysan, Oz. ... 29 E9
Zbarazh 17 D7
Zdolbuniv ... 17 C8
Zdunska Wola . 16 C4
Zeebrugge ... 14 C2
Zeehan 62 D4
Zeerust 59 D5
Zeila 49 E3
Zeitz 15 C7
Zelenograd ... 24 B4
Zémio 57 C5
Zemun 22 B3
Zenica 20 B6
Žepce 20 B7
Zêzere ~> ... 18 C1
Zgorzelec 16 C2
Zhabinka 17 B7
Zhailma 29 D7
Zhambyl 29 E8
Zhangaqazaly . 29 E7
Zhanghua 35 D7
Zhangjiakou ... 35 B6
Zhangye 34 C5
Zhangzhou ... 35 D6
Zhanjiang 35 D6
Zhanyi 34 D5

Zhaotong 34 D5
Zhashkiv 17 D10
Zhayyq ~> ... 29 E6
Zhdanov =
  Mariupol 25 D4
Zhejiang □ ... 35 D7
Zhengzhou ... 35 C6
Zhenjiang 35 C6
Zhenyuan ... 35 D5
Zhezqazghan . 29 E7
Zhigansk 30 C10
Zhijiang 35 D5
Zhilinda 30 B9
Zhitomir =
  Zhytomyr ... 17 C9
Zhlobin 17 B10
Zhmerinka =
  Zhmerynka . 17 D9
Zhmerynka .. 17 D9
Zhodino =
  Zhodzina ... 17 A9
Zhodzina 17 A9
Zhongdian ... 34 D4
Zhumadian ... 35 C6
Zhuzhou 35 D6
Zhytomyr 17 C9
Zibo 35 C6
Zielona Góra . 16 C2
Zigey 53 F2
Zigong 35 D5
Zigui 35 C6
Ziguinchor ... 55 F1
Zile 46 B3
Žilina 16 D4
Zillah 52 C2
Zimbabwe ■ .. 59 B5
Zimnicea 22 C5
Zinder 55 F6
Zion National
  Park 83 B6
Zipaquirá 90 C4
Ziway, L. 53 G6
Zlatograd 22 D5
Zlatoust 29 D6
Zlin 16 D3
Zlitan 52 B1
Zmeinogorsk . 29 D9
Znojmo 16 D3
Zolochev =
  Zolochiv ... 17 D7
Zolochiv 17 D7
Zomba 59 B7
Zongo 56 D3
Zonguldak ... 46 B2
Zorritos 90 D2
Zouar 52 D2
Zouérate 54 D2
Zrenjanin 22 B3
Zuetina = Az
  Zuwaytīnah . 52 B3
Zufar 48 D5
Zug 13 C8
Zugspitze 14 E6
Zula 53 E6
Zumbo 59 B6
Zungeru 55 G6
Zunyi 35 D5
Zürich 13 C8
Zuwārah 52 B1
Zvishavane ... 59 C6
Zvolen 16 D4
Zwettl 15 D8
Zwickau 15 C7
Zwolle 14 B4
Żyrardów 16 B5
Zyryan 29 E9
Zyryanovsk =
  Zyryan ... 29 E9
Żywiec 16 D4

THE WORLD

Alaska (U.S.A.)

GREENLAND (Den.)

ICELAN

CANADA

IRELAND

UNITED STATES

NORTH

PORT.

ATLANTIC

MOROCCO

Tropic of Cancer

W. SAH.

MEXICO

CUBA

OCEAN

Hawaiian Is. (U.S.A.)

HAITI DOM. REP.

JAMAICA

MAURITANIA

BELIZE HOND.

GUATEMALA

CAPE VERDE IS.

SEN. G.

EL SAL. NICARAGUA

G.B.

GUINEA

COSTA RICA PANAMA

VENEZUELA

S.L. IVO. COAS.

PACIFIC

COLOMBIA

GUY. SUR. F. GUIANA

LIBERIA

Equator

ECUADOR

BRAZIL

PERU

FRENCH POLYNESIA

OCEAN

BOLIVIA

SOUTH

Tropic of Capricorn

PARAGUAY

ATLANTI

URUGUAY

CHILE ARGENTINA

OCEAN

S

A   n   t   a   r